Nanotechnology in Intelligent Food Packaging

Scrivener Publishing
100 Cummings Center, Suite 541J
Beverly, MA 01915-6106

Publishers at Scrivener
Martin Scrivener (martin@scrivenerpublishing.com)
Phillip Carmical (pcarmical@scrivenerpublishing.com)

Nanotechnology in Intelligent Food Packaging

Edited by

Annu

Department of Applied Sciences, Galgotias College of Engineering and Technology, Greater Noida, Uttar Pradesh, India

Tanima Bhattacharya

College of Chemistry and Chemical Engineering, Hubei University, China

and

Shakeel Ahmed

Department of Chemistry, Government Degree College Mendhar, Mendhar, India

Scrivener Publishing

WILEY

This edition first published 2022 by John Wiley & Sons, Inc., 111 River Street, Hoboken, NJ 07030, USA and Scrivener Publishing LLC, 100 Cummings Center, Suite 541J, Beverly, MA 01915, USA

For more information about Scrivener publications please visit www.scrivenerpublishing.com.

Wiley Global Headquarters
111 River Street, Hoboken, NJ 07030, USA

For details of our global editorial offices, customer services, and more information about Wiley products visit us at www.wiley.com.

Library of Congress Cataloging-in-Publication Data

ISBN 978-1-119-81895-3

Cover image: Pixabay.Com
Cover design by Russell Richardson

Set in size of 11pt and Minion Pro by Manila Typesetting Company, Makati, Philippines

Printed in the USA

10 9 8 7 6 5 4 3 2 1

Contents

Preface

Nanotechnology is well accomplished with its vast application in various fields of science and technology. Over the past few years, one of the facile applications of nanotechnology has been in food packaging. Different biopolymeric nanocomposite materials have been developed, which are especially designed for food packaging. Such materials mainly possess significant barrier properties for gas exchange, in addition to an enhanced antimicrobial efficacy against various microbes or bacteria that degrade food ingredients. Besides which, there are different methods of nanoencapsulation that also increase the shelf life of food packaging material.

This book is specially designed with an emphasis on the state of the art in nanotechnology and food packaging. It offers fascinating techniques for producing smart and active food packaging and also discusses its toxicity and the role that nanosensors play in detecting different pathogens in food packaging. The concluding chapters also explain recent developments concerning the incorporation of health supplements in food packaging and their future role in producing intelligent food packaging.

The 16 chapters of this book were contributed by experts working in their respective areas of research and are thoughtfully arranged in a systematic fashion that preserves the flow of knowledge. An attempt has been made to include all the information in a single monograph to better understand the topics and technologies involved in the application of nanotechnology in intelligent food packaging.

We hope this book will be a useful reference for academicians, scientists, researchers, chemists, biologists, polymer technologists, food packaging industrialists, biotechnologists, health experts, quality/food safety officials, and policymakers. It will also be helpful to undergraduate and graduate students in the fields of materials science, polymer science, and biotechnology to better understand the fundamental concepts involved in food packaging in relation to extending shelf life. The editor and contributors of this book are experts in their research domain and are researchers and scientists of academia and industry.

We are very thankful to the contributing authors for their enthusiastic efforts toward completing this book. We also greatly appreciate the whole team at Wiley-Scrivener for their help with the book.

Annu, Tanima Bhattacharya and Shakeel Ahmed
May 2022

1

Nanocomposite and Food Packaging

Aayeena Altaf[1], Aamir Hussain Dar[2*], Shafat Ahmad Khan[2] and Anurag Singh[3]

[1]Department of Food Technology Jamia Hamdard, (Deemed to be University), New Delhi, India
[2]Department of Food Technology Islamic University of Science and Technology Awantipora, Kashmir, India
[3]Department of Food Science and Technology, National Institute of Food Technology Entrepreneurship and Management, Kundli, Sonepat, Haryana, India

Abstract
The development of different packaging materials such as nanocomposites is a strong drive in food stuff industry for packaging resolutions that contribute to maintainable development by targeting a globular economy, which pivots around the recyclability of the packaging materials and it is a novel approach to enhance characteristics of polymers together with mechanical strength, barrier against gas and vapor as well as thermal stability, as they markedly upgrade the packaging properties because of their nanometer size dispersion. Properties such as increased water resistance, modulus and strength, and decreased gas permeability are improved. Additionally, antioxidant properties and antimicrobial can be also presented, which are very significant for food packaging applications. Although utilization of biopolymers is restricted because of their usual poor mechanical and barrier properties, but due to reinforcement of fillers, these properties can be improved. Nanocomposite is produced when fillers used are at least one nanoscale dimension (nanoparticles). This chapter will discuss many nanofillers and techniques to utilize them with different polymer matrices, focusing on the enhancement and improvement in the food packaging system.

***Keywords*:** Nanocomposites, food packaging, polymers, nanofiller

**Corresponding author*: daraamirft@gmail.com

Annu, Tanima Bhattacharya and Shakeel Ahmed (eds.) Nanotechnology in Intelligent Food Packaging, (1–24) © 2022 Scrivener Publishing LLC

1.1 Introduction

Fast deterioration of value and decrease in security of food products is because of improper or absence of packaging materials, which results in loss of important food. About 43% of produced polymers are used by packaging industry [1]. Currently, most resources that are utilized for packaging of food stuffs are virtually non-degradable, which leads to a serious worldwide environmental problem. There are different many bio-based resources that are explored to manufacture edible and biodegradable films to enhance the quality of food and extending the shelf life as well decreases the packaging waste [2]. Although, utilization of palatable and biodegradable polymers is less due to difficulties related to performance, processing, and cost such as brittleness and high/low distortion temperature. Polymers like biodegradable thermoplastic starch polymer have received considerable attention, but it showed poor performance related to water sensitivity and its less mechanical properties with extraordinary brittleness that is reason for the disordered growth of amylose crystals with time [3]. Thus, new prospects may be introduced by the use of nanotechnology to these polymers that will improve the performance quality as well as cost price efficiency of food stuffs [4]. To improve the different properties of polymers, many composites have been introduced in which reinforcing compounds are added [5]. By the utilization of nanotechnology boundaries of the biopolymer, packaging material can be personalized. In short, nanotechnology is the branch of science that pacts with the handling, preparation, characterization, and application of materials with at least one dimension in the nanometer range (10–9) [6]. For large surface-to-volume and aspect ratios, nanoparticles exhibit diverse properties compared to those of micro- and macroscale particles. The use of nanotechnology in food technology can be utilized for preservation, processing, storage, nutrient availability, and packaging and for enhancement of taste [7]. New researches highlight the use of nano composite technology in food stuff packaging application. This chapter clarifies the different biopolymer materials and reinforced nanomaterials with an opinion to make them feasible material for food packaging application. An overview of research conducted in those parts is also given for better understanding.

1.2 Nanocomposites Based on Biopolymers

1.2.1 Nanocomposites Based on Chitosan

The abundant and primary source of chitosan is crustaceans (crab and crayfish), and crustacean exoskeleton is also available in ample amount. Chitosan is second biopolymer after cellulose available naturally and is deacetylated form of chitin. Throughout deacetylation, same conditions yield a random distribution of acetyl groups, whereas complex conditions yield a block-wise sharing of acetyl groups in chitosan [70]. Chitosan exists in different polymorphic forms, whose properties differ extensively because of its unbending and definite crystalline structures, which is derived via intra- and intermolecular hydrogen bonding [71]. Structure-wise, it comprises of linear β-1,4-linked GlcN and GlcNAc units and is a heteropolysaccharide [8] as shown in Figure 1.1. Properties like biodegradability, non-toxicity,

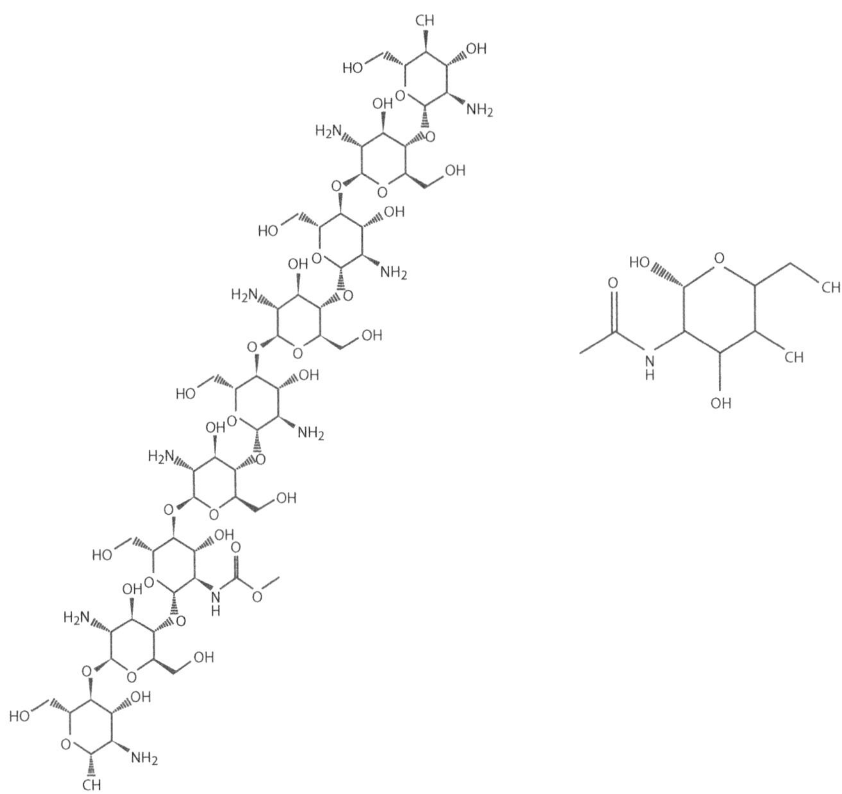

Figure 1.1 The structure of chitosan and chitin.

biocompatibility, film-forming property, and medium-changing mechanical properties made chitosan utmost explored biopolymers in field of food packaging industries. Antimicrobial properties against fungi, bacteria, and viruses and its biocide properties have also been found for many years [9]. Several types of chitosan nanocomposites have been manufacture with different kinds of nanoparticles [10]. Some researchers studied that reduced break at elongation can be improved by addition of clay into chitosan material as well as tensile strength will be maintained [11]. Synthesis of chitosan nanocomposite film with nano-ZnO as nanofiller proved that tensile strength was improved with the antimicrobial inhibitory effects [12]. The shelf life of soft white cheese produced in Egypt was improved by using chitosan, carboxymethyl, cellulose, and nano-ZnO film. Chitosan films with the addition of nano reinforcements affect not only the property of film itself but also food stuff packed inside [13]. These nanocomposites consist of good antimicrobial activity for gram-positive and gram-negative as well as for bacteria and fungi. Films that are nanocrystalline cellulose (NCC)–reinforced chitosan-based are developed by casing solution where content of NCC varies between 1% and 10%. Tensile strength and water vapor permeability of the packaging materials are enhanced by 5% of NCC. [72]. It was found that montmorillonite and nanoclay were able to increase the plasticizing effect acting as plasticizers carrier at higher concentration and recover the permeability of water vapor and the mechanical strength of chitosan films, respectively [14]. Ethyl-Na-dodecanoyl-L-arginine hydrochloride is used against micobes. Concentrations of ethyl-Na-dodecanoyl-L-arginine hydrochloride (1 g/L) and nanoclay (10 g/L) revealed that they can recover the mechanical properties and reduced the solubility of water and water vapor transmission rate. In addition to opacity of the films, it is upgraded by nanoclay that protects the food stuff from the light [15]. Antioxidant behavior, antimicrobial properties, mechanical properties, and physical properties of chitosan film are enhanced by incorporating the rosemary oil (1.5% v/v) and montmorillonite nanoclay. Moisture sensitivity and higher permeability of chitosan film are main problems particularly in food stuff packaging in which the food product with maximum moisture content is interacted with the packaging material [16]. Therefore, adding the nanofillers helps chitosan to discover its commercial application in the food sector at commercial levels.

1.2.2 Nanocomposites Based on Polylactic Acid

The most biodegradable and renewable thermoplastic polyester that is systematically explored is polylactic acid (PLA). It belongs to polyesters commonly made from α-hydroxy acids and family of aliphatic polyesters that contain polyglycolic acid or polymandelic acid [17]. Lactic acid is usually composed of two simple chiral enantiomers L- and D-lactic acid. It is well known for its biocompatibility and different physical properties. However, its main applications are its degradation of long time and increased fragmentation of crystals. Low thermal properties, permeability to gases, and multiple degradations limit its wide-scale utilization [73]. Thus, nanocomposite food packaging invention with PLA nanocomposites has emerged as an excellent material. Such materials are having good utilization in packaging of food products as well as used in medical applications and tissue cultures. Main characteristics of PLA polymer depends upon the ratio of two isomeric types of lactic acid that is (L-lactic acid and D-lactic acid) and the molecular weight. They have melting point 55°C and 175°C, respectively, with glass transition [74]. PLA is polar in nature, and its permeability coefficient of water does not change with relative humidity; it is more soluble in alcohol, anhydride compounds, and nitrogen compounds but has very poor interaction with water, aromatic hydrocarbons, ketone, esters, and sulfur compounds. Because of its good mechanical properties and it is very economic as compared to biodegradable polymers, it is applicable in food packaging sector [18]. The addition of organoclay such as C16-Mica and C16-MMT improved the thermal and mechanical aspects and showed that C16-Mica was effective of PLA nanocomposite

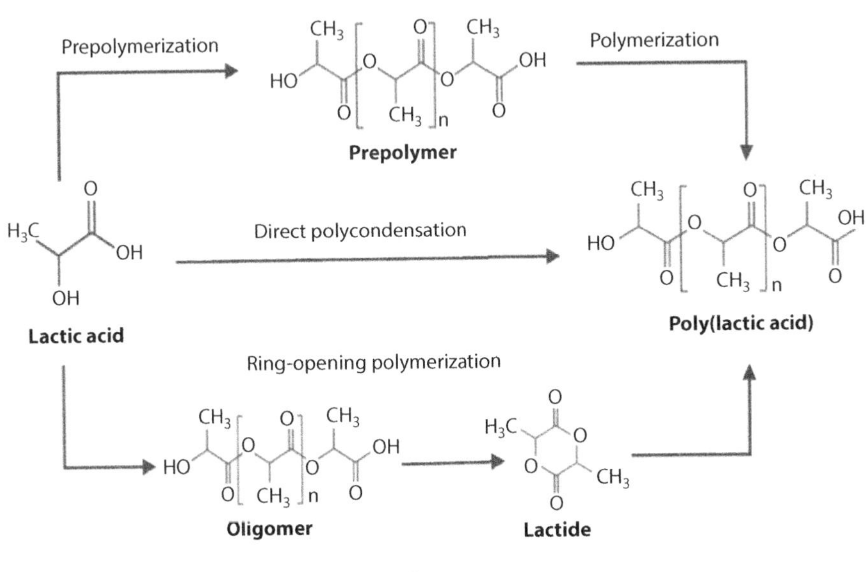

Figure 1.2 The synthesis of polylactic acid.

packaging materials [19]. TiO_2 nanoparticle and its surface modification improve transparency, photo-degradability, and dispersion in the matrix in PLA polymers, which was key for improving the optical properties of PLA-TiO_2 nanocomposite film [20, 21]. Nanosilver particles are used to improve antimicrobial properties in PLA derived from nanoclay and possessed tough activity against gram-negative *Salmonella* sp. with lowest bactericidal concentration below 1 mg per 10 ml. Optimum concentration and suitable type of nanoclay impart virtuous mechanical and barriers properties [22]. The new Cargill Dow LLC PLA plant in Blair, NB, USA, began to lessen the cost of production of PLA and escalate its demand. Danone and McDonalds' companies in Europe are previously using PLA as packaging materials for yogurt and cutleries. Also, in countries such as Japan, Europe, and the United States, PLA is utilized for packaging perishable food stuffs like fruits and vegetables.

1.2.3 Nanocomposites Based on Cellulose

Most abundant linear polysaccharide is made up of repeating monomers of anhydrous glucose linked via a β-1,4 glycosidic bond cellobiose known as cellulose [23]. Its degree of polymerization varies from 700 to 27,000 and depends upon the source of the cellulose [24]. The main sources of cellulose are found in plant materials as well as from animal source cotton, jute, bagasse, straw, potato tubers, tunicate, algae, and bacteria (*Gluconacetobacter xylinus*) [25]. Cotton seeds have maximum purity among these in the range of 90%– 99% [26].

1.2.4 Nanocomposites Based on Starch

Plants produce a biodegradable polymer, which is renewable and is stored as a source of energy for further use. The polymer mentioned above is starch and can be derived from many sources such as maize (80%–83%), wheat (7%–8%), and potatoes (3%–5%). Starch, structurally, is made up of the following constituents:

a) Granule (110–500 nm)
b) Growth rings (15–45 nm)
c) Lamellae (amorphous and crystalline) that contain amylopectin and amylose chains (0.2–1 nm)

The main constituents of starch granules are two major polysaccharides:

1. Amylose
2. Amylopectin

Both polysaccharides contain residues of α-(1,4)-linked D-glucose, and these are connected via α-(1,6)-glucosidic linkages, hence forming the branches.

In spite of high moisture permeability and poor mechanical properties, the starch molecule still acts as an efficient oxygen barrier. Various properties of starch-based film are dependent on the amylose content, method of preparation, moisture content, etc. [27]. According to the research, it is revealed that, when water content is less than 15%, the barrier property is better [28].

1.3 Starch Nanocrystals

Two main uses of starch nanocrystals are listed below:

1. Emulsifiers in food industries
2. Bioplastics materials

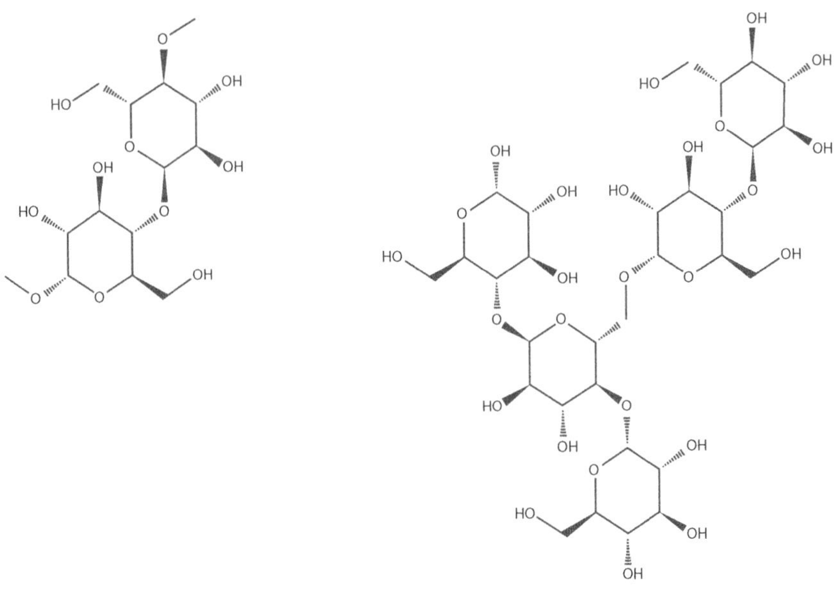

Figure 1.3 The chemical structure of amylopectin and amylose.

Table 1.1 Dependence of structural properties on factors—temperature and acid treatment.

Source	Corn starch	Pea starch	Waxy corn starch	Potato starch
Acid Concentration ($mol/L\ H_2SO_4$)	2.77	3.19	3.15	3.16
Temperature (°C)	45	40	40	40
Time Duration	7 days	5 days	5 days	5 days
Dimensions (L = length; T = thickness; W = width)	L = 50 nm	L = 70 nm W = 25 nm	T = 8 nm L = 90 nm	L = 55 nm W = 12 nm

The crystalline part of the starch is obtained by the process including disarrangement of structure of starch by an acid and hydrolysis under controlled environment (speed and hydrolysis duration). To understand the dependence of the structural properties on various factors, data have been provided in Table 1.1, indicating various factors like temperature and acid treatment.

A recent study shows how glucoamylase pretreatment creates cavities or pores due to disintegration of the α-(1,4) and α-(1,6) links in the polysaccharides, which subsequently increase the diffusion of the acid used for hydrolysis. The property of reinforcement is the reason why starch nanocrystals are very often preferred. One such experiment [29] done with starch-based composites along with jute fibers. The result was increase in degradation period of the composite film. In addition to that, it also showed increase in mechanical strength. Similarly, the water resistance and tensile properties can also be altered by starch-flax cellulose nanocrystal-based nanocomposite films [30].

1.4 Nanocomposites Based on Protein

1.4.1 Soy Protein

It is a composite macromolecule that consists of 37% non-polar and non-reactive amino acid residues and 59% polar and residues of reactive amino acid. They have good t prospective as food packaging material as they have high oxygen barrier property but films made from soy protein have poor mechanical properties, also they low water vapor barrier properties, therefore the need to get blend with other biodegradable polymers that have high hydrophobic properties, e.g., lipids that increase moisture barrier properties [31]. Recently an active nanocomposite packaging material that was based on soy protein isolate was developed in which clove essential oil was used to preserve fillets of sea food like blue fin tuna [32]. Freshness was maintained up to 17 days at 2°C by this packaging without any growth of microbes. A nanocomposite material was prepared with microcrystalline cellulose and metal nanoclusters to improve the tensile strength of soy protein. It has been studied that water capacity and tensile properties of soy protein can be enhanced by nano reinforcements.

1.4.2 Gluten

The most commonly used edible biopolymer and is employed in the form coatings and films is wheat gluten *(Triticum aestivum)*, which is composed of gliadin (a prolamin protein) and glutenin (a glutelin protein), and are known to improve the viscous properties, and strength and elasticity respectively [33]. Thus, the ratio of gliadin and glutenin is responsible for the varying properties of film. The main sources of gluten are wheat, rye, barley, triticale, and oats. Wheat gluten films possess lower permeability as compared to other films [34]. Carboxylated cellulose nanocrystals were utilized as nanofiller material for the growth of nanocomposite packaging film based on gluten [35]. C6 hydroxyls of cellulose were changed to carboxyl groups by oxidization reaction that induces a repulsive force between the nanocrystals of cellulose, ensuring the complete dispersion of nanocrystals of cellulose within the gluten matrix. This improved tensile strength for nanocomposite as compared to gluten film without any filler material. Breaking elongation of nanocomposite films and decreased reduction in water vapor permeability was also observed. TiO_2 nanoparticle was used along Carboxylated cellulose nanocrystals to coat commercial unbleached Kraft sheets [36]. Breaking strength and burst index were increased by 56% and 53% when three layers of the nanocomposite coating were applied. Properties such as antimicrobial against *Saccharomyces cerevisiae, Escherichia coli* and *Staphylococcus aureus* is also improved by coated paper.

1.4.3 Gelatin

The water-soluble protein that is derived from fractional hydrolysis of an insoluble structural protein collagen is known as gelatin. Collagen is primarily present in fibrous tissues of vertebrate and invertebrate animals like skin, tendons, ligaments, and bones and gelatin is abundantly derived from pig, cow, buffalo, and fish scales [37]. Usually Gelatin show higher strength flexibility, gas barrier properties [38]. But properties depend on collagen characteristics and method of processing while as acid-treated gelatin retains a higher isoelectric point and lowers hydroxyl groups with residues of aspartate and glutamate compared to those of alkaline treated gelatin [37]. The potential application of gelatin as a biopolymer for nanocomposite preparation was evaluated by several researchers [39]. Silver and organoclay reinforced gelatin nanocomposite film was developed by [40] they proved that the films exhibited stronger antimicrobial and tensile properties and stated that the gelatin-based nanocomposites benefit to improve the quality of the food. Correspondingly, chitin nanofiber gelatin-based nanocomposite was prepared and proved that nanofiber and gelatin exhibited great compatibility and showed improved mechanical properties [41].

1.5 Food Packaging Matrix with Nano Reinforcements

Nanomaterials that are distributed in a continuous polymer phase are known as nano reinforcements. They are classified depending upon their source, dimensions, or geometry, and they can be organic or in organic [37]. Polysaccharides such as starch, chitosan, chitin, and cellulose, antimicrobials like nisin, and carbon nanofillers such as fullerenes, graphene, and carbon nanotubes belong to organic nano reinforcements while, as inorganic nano reinforcements, include clays (montmorillonite, mica, sepiolite, halloysite, and laponite) [42], metals (gold, silver, copper, iron, and palladium), and metal oxides (silica, zinc oxide, titanium dioxide, magnesium oxide, silver oxide, and cupric oxide). Depending upon the number of dimensions, nano reinforcements are further categorized into three types [43].

a) Nanolayers that are one nano-sized in dimension.
b) Nanotubes also known as nano whiskers are two nano-sized in dimensions.
c) Nanoparticles having three nano-sized dimensions.

Depending upon geometry, nano reinforcements can be divided into more types layered, spherical, or acicular particles. Many properties of biopolymers can be often improved by the use of nano reinforcements [44].

Quantity of the added filler used is the main factor on which the property of nanocomposite depends but added filler can act as limited factor if added amount of filler is utilized beyond a the threshold value. A total of 1% to 5% (w/w) of bacterial cellulose nanocrystals were added to gelatin in which the tensile strength were enhanced gradually up to 4% having the maximum value of 108.6 Mpa but with same neat gelatin showed only 83 Mpa. Some factors such as aspect ratio, flexibility, size, and shape are also some important factors on which properties of nano reinforcement depend.

1.5.1 Nano Reinforcements and Their Effects on Properties of Nanocomposite

When 5% (w/w) of crystalline nanocellulose is added to pectin, glass transition temperature and melting temperature are increased, while absorption of moisture is reduced by 5.72%, but melting temperature will decrease when 7% (w/w) of crystalline nanocellulose is added, and moisture absorption will be decreased by 10.13% [45]. According to Ghelejlu *et al.* (2016), 3% (w/w) of sodium montmorillonite decreases water solubility, moisture content, and permeability when added to chitosan by 30.11%, 26.71%, and 39.42%, respectively [46]. Similarly, halloysite is added to starch at 0.1%, which affects its tensile strength, value of transparency, Young's modulus also, which decreases its elongation at break by 30.01% [47]. Wheat gluten is reinforced by filler cellulose nanocrystals and titanium dioxide by 12.5% and 1%, respectively, that increase the moisture content by 3% and contact angle by 74%. Montmorillonite also increases the puncture force by 99%, and puncture deformation is decreased by 10.65% when added to amaranth flour; also, tensile strength is increased by 33%, and solubility is also reduced by 6.32% [48]. It is also added to soy protein at 10% due to which rheological characteristics of film forming solution and viscosity at shear rate of 60 s^{-1} is increased by 570.11% and 242.1% correspondingly [32]. Moisture content is decreased by 31.23% when filler chitosan nanoparticles are added to fish gelatin at 8% [49]. Zolfi *et al.* (2014) studied that adding 5% (w/w) montmorillonite to kefiran-isolated whey protein decreases the opacity and whiteness index by 47.1% and 6.3%, respectively, while roughness parameters are increased by 55.75% [50]. Nanotitanium dioxide (5%) is also added to kefiran-isolated whey protein that increases both whiteness index as well as roughness parameters. Silver nanoparticles are added at 1% concentration to PLA, which reduces both oxygen transmission rate by 20% and water vapor permeability by 4%, while as oxygen transmission rate is decreased by 46% when both cellulose and silver nanocrystals are added to PLA at the rate of 1% and 5% [51]. Bacterial cellulose nanocrystals (*Gluconacetabacterxylinus*) when added at 4% (w/w) to gelatin increases the tensile strength, glass transition temperature, elastic modulus, and degradation temperature by 30.11%, 5%, 8%, and 5%, respectively.

1.5.2 Antimicrobial Properties of Nano Reinforcements

Selection of appropriate antimicrobial agent depends on its effect on the food stuff and the microorganism that is associated with the spoilage of that food product. Thus, nano reinforcement is used to upsurge the antimicrobial properties of the film and, therefore, prevents the food product from deterioration. Besides, when antimicrobial agents are reinforced in nano form, the polymer matrix possesses improved activity, owing to their high surface-to-volume ratio [52]. Essential oils are incorporated with other reinforcements to improve the antimicrobial properties. These oils may affect different properties of the film such as mechanical, physical, and barrier properties. Apart from antimicrobials, there are few inorganic nanofillers that possess antimicrobial properties such as zinc oxide, copper oxide, silver, and copper [53].

1.5.3 Antimicrobial Nanocomposite Based on Silver

There are many unique properties of silver such as thermal conductivity, catalytic activity, and chemical stability, but they also exhibit antimicrobial and antifungal properties which discover their application in various fields like textiles, medical applications, and consumer products [54]. Nanocomposite film was made by combining hydroxypropyl methylcellulose with particles of silver that resulted in good mechanical and barrier properties and also possessed antimicrobial property against *E. coli* culture [48]. Thus, it is revealed that silver nanoparticle acts as an antimicrobial agent and escalates the mechanical properties of the film that is why silver nanoparticle has a potential application in the arena of food packaging. Development of nanostructured low-density polyethylene and silver oxide film bags decreased microbial spoilage in apple slices. Even after 12 days, quality of apple slices stored at 5°C in an low-density polyethylene and silver oxide bag was acceptable but those that were packed in conventional bags indicated visible deterioration low-density polyethylene after 6 days [55]. Some more antimicrobial reinforcements were produced by combination of two or more materials to develop the synergistic effect of microbial inactivation. Nanocomposites made up of less density polyethylene comprising zinc and silver oxide were beneficial to inactivate *Lactobacillus plantarum* in orange juice at 4°C. Zinc oxide nanoparticles showed low antimicrobial activity against *Lactobacillus plantarum* as compared to silver nanoparticles particularly in longer storage time period, that is why orange juice filled in nanocomposites of low-density polyethylene containing nanosilver and titanium oxide had longer shelf life without any deterioration. Some researchers developed nanosilver with combination of gelatin, carboxymethyl, and chitosan hydrogel by a green and simple fabrication method that revealed that hydrogels showed an antibacterial effect on *Escherichia coli* [55].

1.6 Antimicrobial Nanocomposite Based on Zinc Oxide

The most widely used inorganic compound in different applications is zinc oxide as zinc oxide has been regarded as a generally recognized as safe (GRAS) material by the Food and Drug Administration. Therefore, it is mainly used as food additive. Zinc oxide is known to be highly toxic against many microorganisms among all metals. Zinc oxide and polyvinyl chloride were used for inactivation of foodborne pathogens and were proved to be effective against *E. coli* and *S. aureus* [56]. Zinc oxide used nanoparticles against two pathogens that are *Salmonella typhimurium* and *Staphylococcus aureus* and were very highly effective against both of them. Calcium alginate was loaded with zinc oxide nanoparticles and were used for active packaging application against the foodborne pathogens present in ready-to-eat poultry meat.

1.6.1 Antimicrobial Nanocomposite Based on Titanium Dioxide

It is a multipurpose material with remarkable features, e.g., biocompatible, stability to chemicals, high reactivity, electrochemical properties, economic, and harmless production. It is utilized as a photocatalyst for water-dye degradation and also used as a white colorant in the food and pharmaceutical industries. Moreover, it shows other properties such as UV protection and antimicrobial properties. It is utilized as physical cross-linking agent to develop the technological and functional properties of polysaccharide-based materials. Soybean polysaccharide biocomposite reinforced with Titanium dioxide displayed good antimicrobial activity against *Escherichia coli* and *Staphylococcus aureus* [57]. There is the generation of hydroxyl radicals (_OH) and reactive oxygen species (ROS) on the titanium dioxide surface due to photocatalytic reaction of titanium dioxide, resulting in the oxidation of polyunsaturated phospholipids present in cell membranes of microorganisms, which results inactivation of microorganism blending of titanium dioxide nanoparticles with suitable packaging material to increase its efficiency [58]. Whey protein film was also blended with titanium dioxide nanoparticles to obtain nanocomposite food package and the enhanced mechanical strength of the film. The variation of titanium dioxide concentration had an effect on the tensile strength and water barrier properties of the film [55]. Concentration of nano-TiO_2 and bacteria populations at initial stage and UV exposure time effects on nano-titanium dioxide photocatalytic disinfection against *Salmonella typhimurium* and *Listeria monocytogenes* was studied—the two most typical pathogens found in meat products. Antibacterial activity of the titanium dioxide-polyethylene film against *Escherichia coli* and *Staphylococcus aureus* found that the nanocomposite film exhibited more effective antibacterial activity for *Staphylococcus aureus* [59].

1.6.2 Barrier Properties of Nano Reinforcements

The key parameters in selecting the packaging material for food packaging application is barrier property. These properties signify gas permeability such as O_2, CO_2, and N_2, water vapor, and different aroma compounds that determine the quality of the food. Processes such as corrosive phenomena, oxidations, and modification of organoleptic properties degrade the permeability to oxygen facilitates [60]. Diffusion and permeation behaviors are very significant in order to design novel packaging material. Mainly, diffusion of gas/water vapor between the packaging films is influenced by the following factors [61]:

a) Structure of film
b) Permeability properties to a specific substance of film
c) Physical properties of the film, e.g., thickness and area
d) Environmental conditions of the packaging material such as temperature

e) Change in pressure and concentration gradient across the film
f) Theoretically, diffusion model is used to define the permeation of gas across the films used for packaging of foods.

Henry's law and Fick's law correlated the rate of permeation with the thickness as well as area of the packaging film. Figure 1.4 defines the permeation mechanism of gas across the film in terms of thickness (l), pressure (p), and concentration (c).

Fick's first law described the permeation of gas/vapor across the film (one-dimensional diffusion) and is given by following equation:

$$J = -D \times DC$$

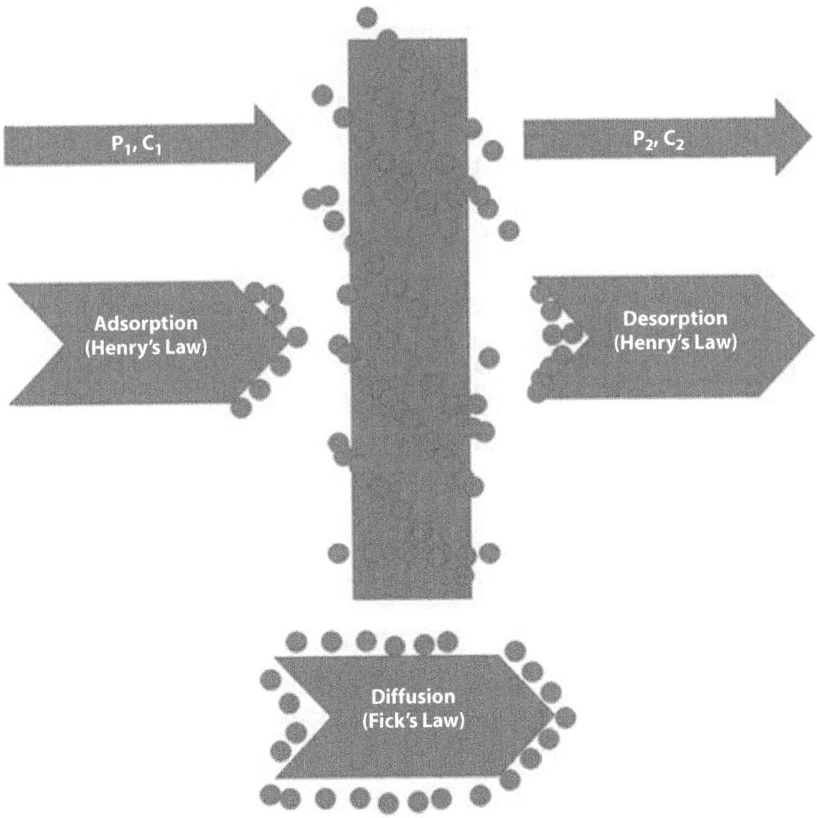

Figure 1.4 The permeation of gas/vapor mechanism across the packaging film.

where

J denotes diffusion flux (mol cm^{-2} s^{-1});

D denotes diffusion coefficient (cm^2 /s);

DC denotes concentration gradient (mol/cm^3).

When the diffusion is in stable state, then the equilibrium relationship between the concentration of sorbed gas at the interfaces and partial pressure will be considered through Henry' s law [62]:

$$C = kp$$

where

C denotes solubility of a gas at a fixed temperature in a particular solvent;

k denotes Henry's law constant;

p denotes partial pressure of the gas;

In recent years, materials used for bio-based packaging such as starch, cellulose, chitin, and polymers such as polycaprolactone and PLA are getting more attention. However, due to their poor barrier properties and weak mechanical properties, the application of these natural polymers in the packaging industry is restricted [1]. By incorporating nano reinforcements in the polymeric film, mechanical and barrier properties are improved, which lead to the growth of nanocomposite packaging material [63]. The nanofillers that are multiwalled carbon nanotubes are mainly used in nanocomposite technology such as nanoclays and chitin nanoparticles. These have been proved as an effective technique to increase the mechanical and barrier properties as well as to enhance and maintain the thermal stability [64]. Also, barrier properties are found to increase in starch nanocomposites, in which cellulose nanofibers are added [65]. Starch or zinc oxide nanocomposite films are used to get protection from UV rays and also decrease the water vapor permeability, chitosan-based nanocomposite reinforced films against UV rays, and lipid oxidation in the system of food packaging [66]. Apart from the protection against UV lights, these nanocomposite films also possess moisture barrier properties and are available in the visible transparent range [67]. Nano reinforcement incorporation improves the different barrier properties of packaging materials. Different nanoclays have been introduced in the field of packaging system such as montmorillonite, and these are used because of their nanosized thickness and macrosized length. Hemicelluloses and montmorillonite-based packaging films with chitin and polyvinyl alcohol as filers were found to have improved and better barrier properties than simple hemicelluloses montmorillonites films [68].

1.7 Aspects of Food Packaging

Because of efficient potential application of nanocomposites, they have been attracted as an important field in research sciences particularly in defense system, pharmaceuticals, food packaging, cosmetics, etc. It has been estimated that, by the year 2022, the global market of nanocomposites will reach \$7,475 million from \$5,100 million in year 2020. Mainly, global player in the field of nanocomposite clay-based nanocomposites [69]. To reach the highest levels in the intelligent packaging, biodegradable and bio-based plastic matrices have the potential to change the packaging concepts from conventional to new packaging with biodegradation. Figure 1.5 provides the information about the bioplastics application in different sectors. Nearly, 51% of global market share belongs to starch in year 2007 followed by PLA. Starch-based bioplastic market will reach USD 561 million by 2023, and this was reported by global opportunity analysis and the annual compound interest rate will increase 3.7% from 2017 to 2023 [69]. The market and importance of bioplastics will increase and is expected to reach 2.44 million tonnes in 2022.

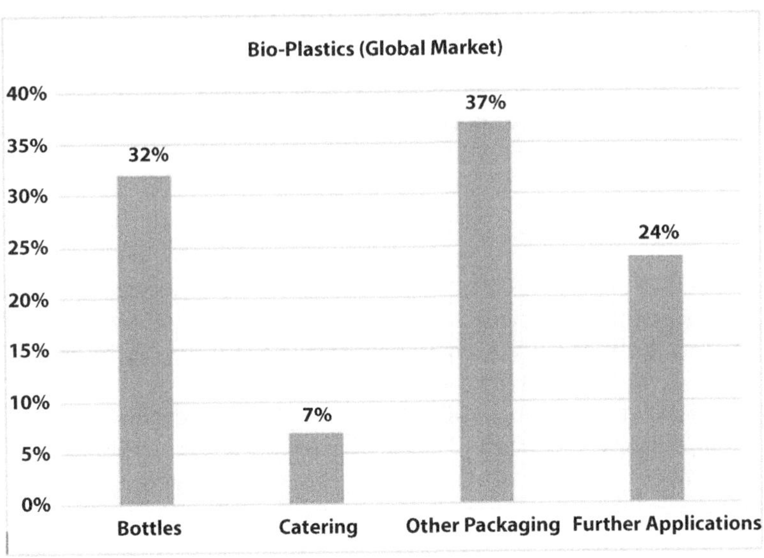

Figure 1.5 The application of bioplastics in the global market.

1.8 Conclusion

The utilization of biopolymers has faced possibility problems that are linked mostly to their extraordinary cost and overall performance in the food industry as equated to those of synthetic polymers, but with the workable development, the price of production is reduced that allowed biopolymer-based materials to progressively develop. The growing industrial and scientific interest of nanopackaging in food sector led to the improvement of different reinforcement material so that there is the growth in the utilization of recyclable and eatable packaging films; meanwhile, adding the nano reinforcements has been found to improve the complete performance of biopolymers, which include improving their different properties. Therefore, nanoparticles have vital role to enhance the feasibility of biopolymers for different application mainly in food sector as they protect food stuffs from the external damage and help in maintaining the superiority of food inside. Nanocomposites protect the food products from different environmental factors, as well as incorporate several active properties to the packaging material so as to escalate the shelf life of food and increasing the stability of foods. However, the nanocomposite provides the many benefits in food packaging sectors with the validated results from the recent studies, but due to their nanosize, they enter into the food material from packaging material. Therefore, safety regarding the possibility of migration/diffusion of packaging matrix to food is still arising and needs future studies and more accurate information regarding all properties of nanocomposite packaging material so that our global environment remains safe from the chronic exposure.

References

1. Rhim, J.W. and Ng, P.K.W., Natural biopolymer-based nanocomposite films for packaging applications. *Crit. Rev. Food Sci. Nutr.*, 47, 4, 411–433, 2007.
2. de Azeredo, H.M.C., Nanocomposites for food packaging applications. *Food Res. Int.*, 42, 9, 1240–1253, 2009.
3. Dufresne, A. and Vignon, M.R., Improvement of starch film performances using cellulose microfibrils. *Macromolecules.*, 31, 8, 2693–2696, 1998.
4. Sorrentino, A., Gorrasi, G., Vittoria, V., Potential perspectives of bionanocomposites for food packaging applications. *Trends Food Sci. Technol.*, 18, 2, 84–95, 2007.

5. Azeredo, H.M.C., Miranda, K.W.E., Rosa, M.F., Nascimento, D.M., de Moura, M.R., Edible films from alginate-acerola puree reinforced with cellulose whiskers. *LWT - Food Sci. Technol.*, 46, 1, 294–297, 2012.
6. Bawa, R., Bawa, S.R., Maebius, S.B., Flynn, T., Wei, C., Protecting new ideas and inventions in nanomedicine with patents. *Nanomed. Nanotechnol. Biol. Med.*, 1, 2, 150–158, 2005.
7. Bhushani, A. and Anandharamakrishnan, C., Food-grade nanoemulsions for protection and delivery of nutrients. In: *Nanoscience in Food and Agriculture 4. Sustainable Agriculture Reviews*, Springer International Publishing AG, Switzerland, 2017.
8. Prashanth, K.V.H. and Tharanathan, R.N., Chitin/chitosan: Modifications and their unlimited application potential-an overview. *Trends Food Sci. Technol.*, 18, 3, 117–131, 2007.
9. Fernandez-Saiz, P., Lagaron, J.M., Hernandez-Muñoz, P., Ocio, M.J., Characterization of antimicrobial properties on the growth of S. aureus of novel renewable blends of gliadins and chitosan of interest in food packaging and coating applications. *Int. J. Food Microbiol.*, 124, 1, 13–20, 2008.
10. De Moura, M.R., Mattoso, L.H.C., Zucolotto, V., Development of cellulose-based bactericidal nanocomposites containing silver nanoparticles and their use as active food packaging. *J. Food Eng.*, 109, 3, 520–524, 2012.
11. Xu, Y., Ren, X., Hanna, M.A., Chitosan/clay nanocomposite film preparation and characterization. *J. Appl. Polym. Sci.*, 99, 4, 1684–1691, 2006.
12. Rahman, P.M., Mujeeb, V.M.A., Muraleedharan, K., Thomas, S.K., Chitosan/nano ZnO composite films: Enhanced mechanical, antimicrobial and dielectric properties. *Arab. J. Chem.*, 11, 1, 120–127, 2018.
13. Youssef, A.M., El-Sayed, S.M., El-Sayed, H.S., Salama, H.H., Dufresne, A., Enhancement of Egyptian soft white cheese shelf life using a novel chitosan/carboxymethyl cellulose/zinc oxide bionanocomposite film. *Carbohydr. Polym.*, 151, 9–19, 2016.
14. Vlacha, M., Giannakas, A., Katapodis, P., Stamatis, H., Ladavos, A., Barkoula, N.M., On the efficiency of oleic acid as plasticizer of chitosan/clay nanocomposites and its role on thermo-mechanical, barrier and antimicrobial properties - Comparison with glycerol. *Food Hydrocoll.*, 57, 10–19, 2016.
15. Rubilar, J.F., Candia, D., Cobos, A., Díaz, O., Pedreschi, F., Effect of nanoclay and ethyl-Nα-dodecanoyl-l-arginate hydrochloride (LAE) on physico-mechanical properties of chitosan films. *LWT - Food Sci. Technol.*, 732, 206–214, 2016.
16. Kovacic, P. and Jacintho, J., Mechanisms of Carcinogenesis Focus on Oxidative Stress and Electron Transfer. *Curr. Med. Chem.*, 8, 7, 773–796, 2012.
17. Ahmed, J. and Varshney, S.K., Polylactides-chemistry, properties and green packaging technology: A review. *Int. J. Food Prop.*, 14, 1, 37–58, 2011.
18. Sin, L.T., Rahmat, R.R., Rahman, W.A.W.A., *Polylactic Acid: PLA Biopolymer Technology and Applications*, William Andrew, Oxford, 2012.

19. Chang, J.H., An, Y.U., Sur, G.S., Poly(lactic acid) nanocomposites with various organoclays. I. Thermomechanical properties, morphology, and gas permeability. *J. Polym. Sci. B: Polym. Phys.*, 2002.
20. Nakayama, N. and Hayashi, T., Preparation and characterization of poly(l-lactic acid)/TiO2 nanoparticle nanocomposite films with high transparency and efficient photodegradability. *Polym. Degrad. Stab.*, 41, 1, 94–103, 2007.
21. Busolo, M.A., Fernandez, P., Ocio, M.J., Lagaron, J.M., Novel silver-based nanoclay as an antimicrobial in polylactic acid food packaging coatings. *Food Addit. Contam. Part A Chem. Anal. Control Expo. Risk Assess.*, 92, 7, 1255–1264, 2010.
22. Auras, R., Harte, B., Selke, S., An overview of polylactides as packaging materials. *Macromol. Biosci.*, 4, 9, 835–864, 2004.
23. Clark, A.J., *Biodegradation of cellulose: enzymology and biotechnology*, Technomic Publishing Co. INC., Lancaster-Basel, 1996.
24. Cheng, Q.Y., An, X.P., Li, Y.D., Huang, C.L., Zeng, J.B., Sustainable and biodegradable superhydrophobic coating from epoxidized soybean oil and ZnO nanoparticles on cellulosic substrates for efficient oil/water separation. *ACS Sustain. Chem. Eng.*, 5, 12, 11440–11450, 2017.
25. Moon, R.J., Martini, A., Nairn, J., Simonsen, J., Youngblood, J., Cellulose nanomaterials review: Structure, properties and nanocomposites. *Chem. Soc. Rev.*, 40, 7, 3941–3994, 2011.
26. Morton, J.F., Adhesives from renewable resources. *Econ. Bot.*, 1993.
27. Mali, S., Grossmann, M.V.E., García, M.A., Martino, M.N., Zaritzky, N.E., Barrier, mechanical and optical properties of plasticized yam starch films. *Carbohydr. Polym.*, 56, 2, 129–135, 2004.
28. Forssell, P., Lahtinen, R., Lahelin, M., Myllärinen, P., Oxygen permeability of amylose and amylopectin films. *Carbohydr. Polym.*, 7, 2, 125–129, 2002.
29. Wang, P., Chen, F., Zhang, H., Meng, W., Sun, Y., Liu, C., Large-scale preparation of jute-fiber-reinforced starch-based composites with high mechanical strength and optimized biodegradability. *Starch/Staerke*, 69, 11–12, 2017.
30. Cao, X., Chen, Y., Chang, P.R., Muir, A.D., Falk, G., Starch-based nanocomposites reinforced with flax cellulose nanocrystals. *Express Polym. Lett.*, 2, 7, 502–510, 2008.
31. Gupta, P. and Nayak, K.K., Characteristics of protein-based biopolymer and its application. *Polym. Eng. Sci.*, 55, 3, 485–498, 2015.
32. Echeverría, I., López-Caballero, M.E., Gómez-Guillén, M.C., Mauri, A.N., Montero, M.P., Active nanocomposite films based on soy proteins-montmorillonite- clove essential oil for the preservation of refrigerated bluefin tuna (Thunnus thynnus) fillets. *Int. J. Food Microbiol.*, 266, 142–149, 2018.
33. Khatkar, B.S., Bell, A.E., Schofield, J.D., The dynamic rheological properties of glutens and gluten sub-fractions from wheats of good and poor bread making quality. *J. Cereal Sci.*, 22, 1, 29–44, 1995.

34. Films and Coatings Produced from Biopolymers and Composites, in: *Biopolymer Engineering in Food Processing*, 2020.
35. Rafieian, F., Shahedi, M., Keramat, J., Simonsen, J., Mechanical, thermal and barrier properties of nano-biocomposite based on gluten and carboxylated cellulose nanocrystals. *Ind. Crops Prod.*, 53, 282–288, 2014.
36. El-Wakil, N.A., Hassan, E.A., Abou-Zeid, R.E., Dufresne, A., Development of wheat gluten/nanocellulose/titanium dioxide nanocomposites for active food packaging. *Carbohydr. Polym.*, 124, 337–346, 2015.
37. Ramos, M., Jiménez, A., Garrigós, M.C., Active nanocomposites in food contact materials. In: *Nanoscience in Food and Agriculture 4. Sustainable Agriculture Reviews*, Springer International Publishing AG, Switzerland, 2017.
38. Shankar, S., Reddy, J.P., Rhim, J.W., Kim, H.Y., Preparation, characterization, and antimicrobial activity of chitin nanofibrils reinforced carrageenan nanocomposite films. *Carbohydr. Polym.*, 117, 468–475, 2015.
39. Rouhi, J., Mahmud, S., Naderi, N., Raymond Ooi, C.H., Mahmood, M.R., Physical properties of fish gelatin-based bio-nanocomposite films incorporated with ZnO nanorods. *Nanoscale Res. Lett.*, 8, 1, 1–67, 2013.
40. Kanmani, P. and Rhim, J.W., Properties and characterization of bionanocomposite films prepared with various biopolymers and ZnO nanoparticles. *Carbohydr. Polym.*, 106, 190–199, 2014.
41. Chen, C., Wang, Y., Yang, Y., Pan, M., Ye, T., Li, D., High strength gelatin-based nanocomposites reinforced by surface-deacetylated chitin nanofiber networks. *Carbohydr. Polym.*, 195, 387–392, 2018.
42. Sanchez, C., Julián, B., Belleville, P., Popall, M., Applications of hybrid organic-inorganic nanocomposites. *J. Mater. Chem.*, 15, 35–36, 3559–3596, 2005.
43. Murray, K.A., Killion, J.A., Major, I., Geever, L.M., Thermal degradation of bio-nanocomposites. In: *Thermal Degradation of Polymer Blends, Composites and Nanocomposites*, Springer International Publishing AG, Switzerland, 2015.
44. Mishnaevsky, L., Pontefisso, A., Dai, G., Hierarchical nanoreinforced composites: Computational analysis of damage mechanisms. *IOP Conference Series: Materials Science and Engineering*, 2016.
45. Chaichi, M., Hashemi, M., Badii, F., Mohammadi, A., Preparation and characterization of a novel bionanocomposite edible film based on pectin and crystalline nanocellulose. *Carbohydr. Polym.*, 157, 167–175, 2017.
46. Beigzadeh Ghelejlu, S., Esmaiili, M., Almasi, H., Characterization of chitosan-nanoclay bionanocomposite active films containing milk thistle extract. *Int. J. Biol. Macromol.*, 86, 613–621, 2016.
47. Wahab, R.A., Gowon, J.A., Elias, N., On the renewable polymers from agro-industrial biomass: A mini review. *J. Indones. Chem. Soc*, 2, 1, 24, 2019.
48. Diéguez, M.C.V., Pelissari, F.M., Sobral, P.J. d. A., Menegalli, F.C., Effect of process conditions on the production of nanocomposite films based on amaranth flour and montmorillonite. *LWT - Food Sci. Technol.*, 61, 1, 70–79, 2015.

49. Hosseini, S.F., Rezaei, M., Zandi, M., Farahmandghavi, F., Preparation and characterization of chitosan nanoparticles-loaded fish gelatin-based edible films. *J. Food Process Eng.*, 39, 5, 521–530, 2016.
50. Zolfi, M., Khodaiyan, F., Mousavi, M., Hashemi, M., The improvement of characteristics of biodegradable films made from kefiran-whey protein by nanoparticle incorporation. *Carbohydr. Polym.*, 109, 118–125, 2014.
51. Fortunati, E., Peltzer, M., Armentano, I., Jiménez, A., Kenny, J.M., Combined effects of cellulose nanocrystals and silver nanoparticles on the barrier and migration properties of PLA nano-biocomposites. *J. Food Eng.*, 118, 1, 117–124, 2013.
52. Llorens, A., Lloret, E., Picouet, P.A., Trbojevich, R., Fernandez, A., Metallic-based micro and nanocomposites in food contact materials and active food packaging. *Trends Food Sci. Technol.*, 24, 1, 19–29, 2012.
53. Mallakpour, S. and Behranvand, V., Nanocomposites based on biosafe nano ZnO and different polymeric matrixes for antibacterial, optical, thermal and mechanical applications. *Eur. Polym. J.*, 84, 377–403, 2016.
54. González, A. and Alvarez Igarzabal, C.I., Nanocrystal-reinforced soy protein films and their application as active packaging. *Food Hydrocoll.*, 43, 777–784, 2015.
55. Zhou, Y., Zhao, Y., Wang, L., Xu, L., Zhai, M., Wei, S., Radiation synthesis and characterization of nanosilver/gelatin/carboxymethyl chitosan hydrogel. *Radiat. Phys. Chem.*, 81, 5, 553–560, 2012.
56. Espitia, P.J.P., Soares, N., d. F.F., Coimbra, J.S. d. R., de Andrade, N.J., Cruz, R.S., Medeiros, E.A.A., Zinc oxide nanoparticles: Synthesis, antimicrobial activity and food packaging applications. *Food Bioprocess Tech.*, 5, 5, 1447–1464, 2012.
57. Anaya-Esparza, L.M., Ruvalcaba-Gómez, J.M., Maytorena-Verdugo, C.I., González-Silva, N., Romero-Toledo, R., Aguilera-Aguirre, S., Pérez-Larios, A., Montalvo-González, E., Chitosan-Tio2: A versatile hybrid composite. *Materials*, 13, 4, 811, 2020.
58. Vejdan, A., Ojagh, S.M., Adeli, A., Abdollahi, M., Effect of TiO2 nanoparticles on the physico-mechanical and ultraviolet light barrier properties of fish gelatin/agar bilayer film. *LWT - Food Sci. Technol.*, 71, 88–95, 2016.
59. Long, M., Wang, J., Zhuang, H., Zhang, Y., Wu, H., Zhang, J., Performance and mechanism of standard nano-TiO2 (P-25) in photocatalytic disinfection of foodborne microorganisms - Salmonella typhimurium and listeria monocytogenes. *Food Control*, 39, 68–74, 2014.
60. López, O.V., Castillo, L.A., García, M.A., Villar, M.A., Barbosa, S.E., Food packaging bags based on thermoplastic corn starch reinforced with talc nanoparticles. *Food Hydrocoll.*, 43, 18–24, 2015.
61. Siracusa, V., Food packaging permeability behaviour: A report. *Int. J. Polym. Sci.*, 2012.
62. Li, N.N. and Henley, E.J., Permeation of gases through polyethylene films at elevated pressures. *AIChE J.*, 15, 1, 43–80, 1964.

63. Khan, A., Khan, R.A., Salmieri, S., Le Tien, C., Riedl, B., Bouchard, J., Chauve, G., Tan, V., Kamal, M.R., Lacroix, M., Mechanical and barrier properties of nanocrystalline cellulose reinforced chitosan based nanocomposite films. *Carbohydr. Polym.*, 90, 4, 1601–1608, 2012.
64. Paszkiewicz, S., Szymczyk, A., Janowska, I., Jedrzejewski, R., Linares, A., Ezquerra, T.A., Wagner, H.D., Tenne, R., Rosłaniec, Z., Comparative study on the properties of poly(trimethylene terephthalate) -based nanocomposites containing multi-walled carbon (MWCNT) and tungsten disulfide (INT-WS2) nanotubes. *Polym. Adv. Technol.*, 28, 6, 645–657, 2017.
65. Babaee, M., Jonoobi, M., Hamzeh, Y., Ashori, A., Biodegradability and mechanical properties of reinforced starch nanocomposites using cellulose nanofibers. *Carbohydr. Polym.*, 132, 1–8, 2015.
66. Hosseini, S.F., Rezaei, M., Zandi, M., Farahmandghavi, F., Fabrication of bio-nanocomposite films based on fish gelatin reinforced with chitosan nanoparticles. *Food Hydrocoll.*, 44, 172–182, 2015.
67. Babaei-Ghazvini, A., Shahabi-Ghahfarrokhi, I., Goudarzi, V., Preparation of UV-protective starch/kefiran/ZnO nanocomposite as a packaging film: Characterization. *Food Packag. Shelf Life*, 16, 103–111, 2018.
68. Chen, G.G., Qi, X.M., Li, M.P., Guan, Y., Bian, J., Peng, F., Yao, C.L., Sun, R.C., Hemicelluloses/montmorillonite hybrid films with improved mechanical and barrier properties. *Sci. Rep.*, 5, 1, 1–12, 2015.
69. Idumah, C.I., Hassan, A., Ihuoma, D.E., Recently emerging trends in polymer nanocomposites packaging materials. *Polym.-Plast. Tech. Mat.*, 58, 10, 1054–1109, 2019.
70. Zahiri Oghani, F., Tahvildari, K., Nozari, M., Novel antibacterial food packaging based on chitosan loaded ZnO nano particles prepared by green synthesis from Nettle leaf extract. *J. Inorg. Organomet. Polym. Mater.*, 31, 43–54, 2021.
71. Smirnov, M.A., Nikolaeva, A.L., Vorobiov, V.K., Bobrova, N.V., Abalov, I.V., Smirnov, A.V., Sokolova, M.P., Ionic conductivity and structure of chitosan films modified with lactic acid-choline chloride NADES. *Polymers, 12*, 2, p.350, 2020.
72. Rodrigues, C., Francieli, D., Pier, M.D.L., Natan, P., Fiori, M.A., Cíntia, S., Gracher, R.H., Mechanical, thermal and antimicrobial properties of chitosan-based-nanocomposite with potential applications for food packaging. *J. Polym. Environ.*, 28, 4, 1216–1236, 2020.
73. Mohamad, N., Mazlan, M.M., Tawakkal, I.S.M.A., Talib, R.A., Kian, L.K., Fouad, H., Jawaid, M., Development of active agents filled polylactic acid films for food packaging application. *Int. J. Biol. Macromol.*, 163, 1451–1457, 2020.
74. Huang, Y., Wang, Y., Li, Y., Luo, C., Yang, C., Shi, W., Li, L., Covalent immobilization of polypeptides on polylactic acid films and their application to fresh beef preservation. *J. Agric. Food Chem.*, 68, 39, 10532–10541, 2020.

2

Polymer-Based Nanostructures in Nanopackaging

Apoorva Sood, Manpreet Kaur and Reena Gupta*

Department of Biotechnology, Himachal Pradesh University, Summerhill, Shimla, India

Abstract

Nanotechnology has gained prominence as a potential area of innovation and research for various disciplines such as medicine, agriculture, food, environment biotechnology, electronics, and engineering. A recently emerging application of nanotechnology is nanopackaging. Over the decades, assorted nanomaterials and nanocomposites have been developed, which can be used as nanopackaging materials. Nanopackaging has brought about a profound change in all industrial domains, notably in the food and electronic industries. Development of such "smart" and "active" packaging has improved the shelf life, quality, and traceability of foods. Polymers being the most versatile class of biomaterials are quite suitable for forming nanomaterials. Polymeric nanomaterials have unique features including nanoscale size, a large surface area-to-mass ratio, and strong reactivity, making them excellent for food packaging. The objective of this chapter is to go through the many types of polymer-based nanostructures, as well as their different types and applications in various industries, with a focus on nanopackaging.

Keywords: Nanotechnology, polymer-based nanostructures, nanocomposites, nanosensors, polymeric nanomaterials

**Corresponding author*: reenagupta_2001@yahoo.com

Annu, Tanima Bhattacharya and Shakeel Ahmed (eds.) Nanotechnology in Intelligent Food Packaging, (25–58) © 2022 Scrivener Publishing LLC

2.1 Introduction

Nanotechnology deals with the study and use of materials with structures having one dimension or more than one dimension (length/width/height) in the nanometer scale (10^{-9} m that is one billionth of a meter). It is a completely new branch of science and technology, which has revolutionized basically every aspect of our lives through its myriad applications. It has led to a deeper understanding of materials, new devices, and advancement in technology. Nanotechnology can be best explained as the designing, fabrication, characterization, and use of materials and devices at the nanoscale by customizing their form and size. Because of their physico-chemical characteristics that differ from bulk materials, nanomaterials have played an important role in research. This interesting property of a material emerges from the organization or pattern of random or well-ordered nanopatterns and the resultant material is referred to as nanostructure or nanostructured material. The increased surface-volume ratio and the quantum-confinement of these materials is the reason for these unique characteristics [1, 2].

For symmetrical objects like spherical particles, the surface-to-volume (S/V) ratio and the radius are inversely related, while for a cube, S/V ratio is inversely related to the length of the side of the cube. As the S/V ratio increases, the atoms or molecules of the nanomaterial become more exposed to the environment, and more of these "dangling bonds" become exposed on the particle surface, which makes them more chemically active. As a result, noble metals such as Ag, Au, Pt, and others act as very powerful catalysts in their nanoparticulate state. Graphene is ideal for sensing chemicals and gases due to the huge amount of unsaturated "dangling bonds" present on its surface, along with high S/V ratio. The low melting point of nanocrystals is also due to the high S/V ratio [1, 3, 4].

The other reason for unique properties of nanomaterials is quantum confinement. This occurs as a result of the material's size reduction, which causes the wave functions of electrons to become very tightly contained; this results in changes in the nanomaterial's electrical and optical characteristics. Changes also occur in the material's "effective mass" and "electron mobility", as well as its relative dielectric constant and optical characteristics [1, 5].

2.2 Properties of Nanomaterials

2.2.1 Optical and Electronic Properties

Electronic structures of the nanomaterials are significantly dependent on surface atoms and hence differ for various morphologies. Surface atoms have unusual localized electronic states that are frequently important in influencing optical characteristics. As the number of surface atoms in bulk materials is very less as compared to the inner atoms, their characteristics are totally governed by the interior atoms. However, the contribution of surface atoms rises as the particle size decreases in nanoparticles and the surface atoms impact nanoparticle characteristics significantly.

The optical characteristics of noble metal nanomaterials are size dependent. They exhibit an exceptional UV-visible spectrum, which is not present in the bulk material's spectrum. The localized surface plasma resonance is an excitation band that is displayed when the frequency of the input photon is consistent with the combined excitation of the non-localized electrons. The maximal wavelength of LSPR (localized surface plasmon) spectrum depends upon the size, shape, and interparticle spacing. These unbound (free) electrons present on the surface of nanoparticles (d-electrons of gold and silver) can be transported easily across the nanomaterial. Ag and Au have a mean free path of around 50 nm; this is greater than the nanoparticle size of both the materials. Because of this, no scattering is predicted from the bulk when an interaction between the light and the material occurs. Instead, the light resonates and standing waves are formed, and this is the cause of LSPR in nanomaterials [2, 7].

2.2.2 Magnetic Properties

Nanomaterials have a unique magnetic characteristic due to their unequal electron distribution. The methods utilized to make nanomaterials, like solvo-thermal, co-precipitation, microemulsion, thermal breakdown, and flame-spray synthesis, all influence these characteristics.

Electrons in atoms have distinct energy levels or bands. The electrons in the outermost shell in a bulk metal occupy energy bands that are continuous. Nanoparticles of less than 10 nm in size and comprising of roughly 50,0000 atoms exhibit their quantum character because of the discrete energy levels of the outermost electrons. This can be explained by the fact that, for a macroscopic metal particle, only a small fraction of the total atoms is exposed on the surface atomic layer, approximately less than 1 in 10 million. For a nanoparticle, on the other hand, this proportion increases to 50%. This behavior can distort the characteristics of the whole nanoparticle since the fundamental magnetic behavior of the particles is size-dependent. An example for this novel behavior is seen in the fact a better way of producing more potent magnetic materials is producing thin magnetic films from magnetic nanoparticles. Magnetic nanoparticles have applications in a myriad of disciplines such as catalysis (heterogeneous and homogeneous), biomedicine, use as magnetic fluids, data storage solutions, medical diagnostics [magnetic resonance imaging (MRI)], and environmental remediation (soil and water cleanup) [6, 7].

2.2.3 Mechanical Properties

Nanomaterials possess such mechanical properties that microparticles and their bulk components do not. In order to ascertain the actual mechanical characteristics of nanomaterials, several mechanical characteristics, *viz.*, elastic modulus, stress and strain, hardness, friction, and adhesion, are considered. The mechanical characteristics of nanoscale grains, like as strength, can be substantially enhanced. The mechanical properties of materials are determined by a number of basic characteristics. The amount a material deforms in response to a force is determined by its strength. Hardness is determined by the quantity of material that can be penetrated by another body as well as wear resistance. The grain structure of the material and the sliding of grains past each other govern these characteristics. Materials of nanoscale grain size can be produced utilizing a variety of processes, including nanoparticle deposition, electrodeposition, and specific low temperature milling methods.

The unique mechanical characteristics of nanoparticles have made them useful in a variety of disciplines, including tribology, surface engineering, nanofabrication, and nanomanufacturing. This usually necessitates a thorough understanding of the fundamentals of nanoparticle mechanical properties and their size dependency. A limitation of nanostructured materials is that they break at relatively low strains. This problem can be addressed by improvements in processing [1, 3, 7].

2.2.4 Chemical Properties

Chemical reactivity is also a size-dependent property of nanoparticles. A typical example of this is gold which is essentially a noble, inert metal in bulk and does not corrode or tarnish and thus is highly valued. Due to this noble behavior, gold is useless as a catalyst for chemical or biochemical reactions. Gold nanoparticles having a diameter of less than 5 nm, on the other hand, is a strong catalyst. For carbon monoxide (CO) oxidation, gold nanoparticles have been employed as a strong catalyst. Chemical reactivity not only is size dependent but also depends significantly on the material on which they are supported. Catalysts based on nanoparticles have made a breakthrough in the chemical industry and have become a popular research topic [6, 8].

2.2.5 Interaction of Nanoparticles With Living Systems

As chemical reactivity is enhanced in nanoparticles, their interaction with biological systems is also affected and becomes dependent on particle size. Materials become much more effective and reactive in the form of nanoparticles because a substantial amount of their surface/volume fraction is directly in contact with their surroundings. Nanomaterials being nanoscopic are able to penetrate past cell membranes and other cellular. Particles ranging from 1 to 10 nm can adhere to cell membranes and disrupt processes such as breathing, or they can penetrate the outer membrane and destroy the cell DNA.

Silver nanoparticles, for example, have long been recognized to have an inhibitory effect on a variety of bacterial species, and "silver-based compounds" have been widely employed for their bactericidal effects. Nanoparticle-based antimicrobials are a great alternative in the age of increasing antibiotic resistance in bacteria [28]. Nanoparticles having diameter in the range 1–10 nm adhere to the viral surface like that of HIV-1 virus in an ordered array. Silver nanoparticles envelope the virus and prevent it from adhering to host cells by attaching to the glycoprotein "knobs" in a regular pattern. Gold and silver nanoparticles are proposed for use as "carriers" in cosmetic and drug industry, where even nano or pico amounts of compounds are bound to these nanoparticles which transport them effectively into deeper layers [6, 9].

2.3 Classification of Nanomaterials

Nanomaterials are classified based on their characteristics into several categories. A generalized classification of nanomaterials is shown in Figure 2.1.

2.3.1 Carbon-Based Nanomaterials

Carbon-based nanomaterials are carbon nanoparticles in their purest form. The different types of carbon-based nanomaterials are explained in Table 2.1. Nanosized activated carbon can also be included in this category [7, 10].

2.3.2 Metal Nanomaterials

Metal precursors are used to make metal nanomaterials. They exhibit specific opto-electrical properties. Chemical, electrochemical, and photochemical

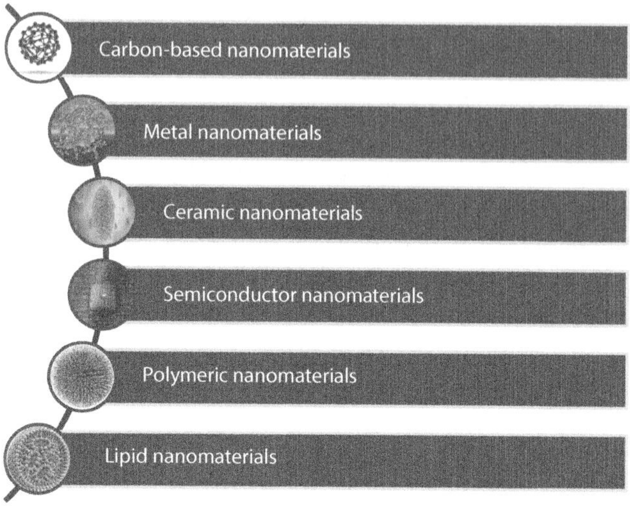

Figure 2.1 Classification of nanomaterials.

techniques are used to make these nanomaterials due to their well-known (LSPR) properties. Chemical reducing agents are used to reduce the metal-ion precursors in solution, resulting in metal nanomaterials. These have a high surface energy and the capacity to adsorb tiny molecules. Metal nanomaterials may be used in various fields, including research, detecting and imaging biomolecules, as well as bioanalytical and environmental applications, attributable to their superior optical characteristics, capacity to adsorb tiny molecules, and high surface energy. For example, gold nanoparticle coating is frequently employed in SEM sampling to improve the electron-stream, which aids in getting SEM images of high quality [7].

2.3.3 Ceramic Nanomaterials

These are nonmetallic inorganic materials composed of carbides, oxides, phosphates, and carbonates that are produced by their repeated heating and cooling. They exhibit a variety of shapes and sizes, *viz.*, amorphous, polycrystalline, dense, porous, and hollow. Ceramic nanomaterials have been utilized in photocatalysis, dye photodegradation, imaging, and drug delivery due to their excellent heat resistance and chemical inertness [7].

Table 2.1 Different types of carbon-based nanomaterials.

Carbon-based nanomaterials	Key properties
Fullerenes (C-60) buckminsterfullerene	• They are allotropes of carbon having a hollow-cage structure containing 60 or more than 60 C-atoms. • Carbon atoms are bound together by sp2 hybridization to form a spherical carbon molecule. • They have C-atoms arranged in a pentagonal and hexagonal arrangement. • They are economically viable due to their structure, high strength and electrical conductivity
Graphene	• It is a 2D planar carbon alloptrope having a hexagonal or honeycomb lattice made up of C- atoms. • Thickness is ~1 nm.
Carbon nanotubes (CNTs)	• They are graphene-nanofoil with a "honeycomb lattice" rolled into hollow cylinders of diameter ~0.7 nm. • They are 100× stronger than steel and used for in reinforcement of structures. • Single-walled C-nanotubes (SWCNTs) and multi- walled C-nanotubes (MWCNNTs) are the two types of CNTs. • Their unique characteristic is their thermal conductivity along the length of the tube and non-conductivity across the tube.
Carbon nanofibers	• Carbon nanofiber is made from graphene nanofoils forming a cone or a cup form instead of a cylindrical tube.
Carbon black	• Carbon in the form of an amorphous substance. • They are spherical in form with size ranging from 20 to 70 nm. • The particles are bound into agglomerates of about 500-nm agglomerates since the interaction between them is so high.

2.3.4 Semiconductor Nanomaterials

Semiconductor materials belong to II-VI, III-V, and IV-VI groups of the periodic table and exhibit characteristics that are halfway between metals and nonmetals. They have large bandgaps and show substantial changes in characteristics when bandgap tuned. They have been discovered to be effective in water splitting applications and are essential materials in photocatalysis, photo optics, and electrical devices. GaN, GaP, InP, and InAs are group III-V semiconductors; ZnO, ZnS, CdS, CdSe, and CdTe are group II-VI semiconductors, while Si and Ge are group IV semiconductors [7].

2.3.5 Polymeric Nanomaterials

Organic nanomaterials are referred to as polymeric nanomaterials. They possess nanocapsular or nanospherical structures depending on the technique of production. The active chemicals and polymers are evenly distributed in a matrix of nanospheres. The functional chemicals components are contained and encapsulated by a polymer shell. Controlled release, drug molecule protection, the potential to combine therapy with imaging, biocompatibility and biodegradability, and precise targeting are all benefits of polymeric nanomaterials, making them ideal for drug delivery and diagnostics [7].

2.3.6 Lipid Nanomaterials

Lipid-based nanomaterials have lipid moieties that are usually spherical in form and have a diameter of 10 to 100 nm. They have a solid core that is lipid-based and a matrix consisting of soluble lipophilic molecules. The exterior core is stabilized by the surfactants and emulsifiers. In cancer therapy, lipid nanomaterials have been utilized as a medication carrier and delivery system, as well as for RNA release [7].

2.4 Synthesis of Nanomaterials

Several methods have been developed to synthesize nanoparticle-based materials. Nanomaterials can be prepared by the use of a variety of techniques including biological, chemical, hybrid, and physical approaches. The main methods by which nanomaterials can be synthesized are demonstrated in Figure 2.2 [12].

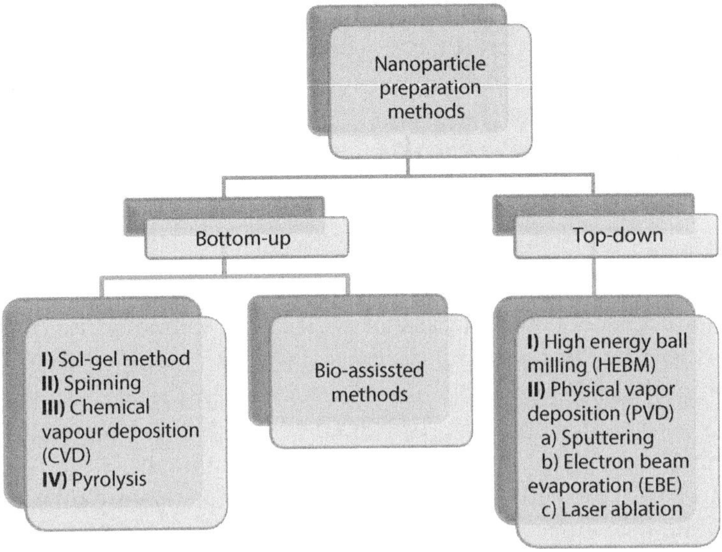

Figure 2.2 Methods for nanomaterial synthesis.

2.4.1 Bottom-Up Approach

It is a building up process where nanomaterials are synthesized from relatively simpler substances. It is a constructive method where materials are built-up from atoms to clusters and finally to nanomaterials.

2.4.1.1 Sol-Gel

Since it is simple and can produce a variety of nanomaterials, sol-gel is one of the most often used techniques. In this procedure, two components are used: the sol and the gel. Sol is a colloidal solution of particles floating in a liquid phase, while a solid macro-molecule immersed into a solvent is known as a "gel". "Sol" particles are connected in a liquid to form a network of distinguished particles or network of polymers. Hydrolysis and condensation are the two common stages in the sol-gel process. Water is utilized to dissolve the precursor's bonds during condensation. This is the initial stage in the gel phase's development. The excessive water is extracted after hydrolysis, and the eventual structure of the nanomaterial is identified, and it is created. In the "sol-gel" process, chlorides and metal oxides are generally employed as precursors, which are disseminated in a liquid by processes like stirring, shaking and sonication, resulting in a system of a liquid and a solid phase. For recovering the nanomaterials, phase separation procedures like sedimentation, filtering, and centrifugation are used. The moisture is removed further by drying [10, 11].

2.4.1.2 *Spinning*

The production of nanomaterials takes place in a spinning-disc reactor or SDR. A rotating disc of an SDR is housed within a chamber. To eliminate oxygen and therefore inhibit oxidation processes, the reactor is purged with N_2 or another inert gas. The liquid i.e., precursor along with water, is injected through the spinning disc. While spinning, the atoms or molecules fuse with each other and precipitate. After that, they are collected and dried. The properties of nanomaterials synthesized using SDR are determined by many operational factors like as temperature, flow-rate of the liquid, rotation speed of disc, ratio of liquid/precursor, position of feed, disc-surface and so on.

2.4.1.3 *Chemical Vapor Deposition*

Chemical processes occurring at very high temperatures deposit solids in a vaporized phase onto a substrate in chemical vapor deposition (CVD). When the mixed gases and the heated substrate interact, a chemical reaction takes place, resulting in development of a thin product layer on the surface of substrate. After this, the film is retrieved and utilized. The substrate temperature is a major influencing factor for CVD. CVD produces nanomaterials that are homogeneous, pure, and strong. The process' drawbacks are the need for specialized instrument and equipment and the severe toxicity of the gaseous by-products [11].

2.4.1.4 *Pyrolysis*

Pyrolysis is the thermal breakdown of materials in an inert environment at extreme temperatures, resulting in a change in chemical composition. It is a widely used technology for producing nanomaterials on a big scale. It is a high-yielding, easy, efficient, cost-effective, and a continuous process. A high pressure liquid or a vaporized precursor is delivered into the furnace by an orifice, where it is burned. Molecular breakdown happens when molecules are stimulated over the dissociation threshold, followed by chemical processes. The result is the formation of condensable and/or volatile products. When a laser is used for pyrolysis, the heating is highly targeted and quicker (leading to rapid nucleation) and the particle development is quenched sooner. This technique produces nanomaterials with an average diameter of 5 to 60 nm [10, 11].

2.4.1.5 Biosynthesis or Bio-Assisted Methods

"Green" synthesis is one of the biosynthetic approaches because it is an ecologically friendly, less toxic, cheaper, and efficient way for making nanomaterials. For the production of nanoparticles of metals and their oxides, these approaches use biological systems like bacteria, fungus, viruses, yeast, actinomycetes, and plant extracts [2].

The three types of bio-assisted techniques are as follows:

i. Microorganism-based bio-genic synthesis
ii. Plant extract-based biogenic synthesis
iii. Biomolecules as templates for bio-genic synthesis

For the production of nanomaterials, microorganisms such as bacteria, fungus, yeast, actinomycetes, and algae are employed as bioreactors. Microorganisms absorb ions of interest from their surroundings and then use cellular enzymes to convert them to the element metal. Biological membranes have also been utilized as templates to build nanomaterials due to the existence of ultra-fine holes which make their structure. In the manufacture of Au nanoparticles, a rubber-based membrane produced from *Hevea brasiliensis*, a tree, was employed as a preservative. Viruses have also been recommended as a template for producing nanomaterials because they have hollow cavities in the center and may produce nanomaterials of consistent size and shape.

Nanomaterials may be synthesized using templates like biomolecules [nucleic acids (DNA/RNA)], viruses, diatoms, and biological membranes. DNA is a good bio-molecular template because it has a high affinity for transition metal ions. Cross-linking DNA hydrogels allows transition metal ions, for example, Au (III) ions, to be incorporated into DNA-macromolecules, resulting in Au nanoparticles.

Nanoparticles of noble metals, metal oxides, and bimetallic alloys have been synthesized using biomass or plant extracts, which is a very successful, fast, clean, non-toxic, and eco sustainable technique. The kinetics of nanoparticle photosynthesis has been found to be comparable to the rate of chemical pathways in certain situations and significantly higher than that of other biosynthetic techniques [11].

2.4.2 Top-Down Approach

For the production of nanomaterials, the top-down method uses a destructive approach. Larger molecules are taken and broken down into suitably smaller units using mechanical or chemical methods in order to form nanomaterials [2].

2.4.2.1 High-Energy Ball Milling

It is an efficient and a very commonly used technique for creating various types of nanomaterials, nanocomposites, and alloys. Tungsten carbide, zirconia, sintered corundum, silicon nitride, Cr-Ni steel, chrome steel, and plastic-polyamide are used as moving balls. Their movement gives kinetic energy to the "milled" material, causing chemical bonds to dissolve and the milled material to rupture into smaller particles with freshly formed surfaces. The sum of energy transferred between the material and the balls throughout the process is regulated by the milling medium, speed of milling, ball/powder weight ratio, type of milling employed: dry/wet, type of high-energy ball milling (HEBM): vibrator mill, planetary mill, attritor mill, tumbler ball mill, milling conditions, and milling time. Surfactant-aided HEBM, a newer technique, has also been utilized to synthesize nanoparticles with exact size and surface properties. Surfactant molecules inhibit particle agglomeration and cold welding, which can result in particle size increase due to electrostatic or steric forces. These factors help to keep milling particles stable and prevent uncontrolled fracture. Surfactants also reduce the surface energy of newly formed tiny particles by creating a thin organic layer and providing long-range capillary forces, which reduce the likelihood of breaking [11].

2.4.2.2 Physical Vapor Deposition

Physical vapor deposition (PVD) is a technique that uses vacuum-deposition, which is safe to the environment. PVD are a group of synthetic methods used to make nanoparticles by depositing thin layers of material with a thickness ranging from a few nanometers to several micrometers. PVD is made up of three basic steps:

- Material vaporisation from a solid-state (source).
- Transport of the vaporized matter.
- Nucleation and development to form thin films of nanoparticles.

2.4.2.2.1 Sputtering

Sputtering is a PVD process that takes place in a vacuum. It ejects particles off a surface by colliding with ions, resulting in nanoparticles being deposited on it. Sputtering is based on momentum transfer concept. The atoms from the target (that is, the material which are to be deposited) are expelled by ion bombardment. The form and size of the nanoparticle are determined factors such as the layer thickness, annealing temperature and time, and the type of substrate.

The following are the steps involved in sputtering:

(i) Electron collisions with gaseous molecules produce plasma of neutral gases, most frequently argon between the two electrodes.
(ii) By providing voltage between the two electrodes, ions in the plasma are propelled toward the target.
(iii) As a result, these ions with the right energy strike the target, causing the material to be ejected from the target.
(iv) The expelled material is subsequently transferred to the substrate and deposited [11].

2.4.2.2.2 Electron Beam Evaporation

Thin films and nanoparticles are made with EBE. A vacuum unit, an electron beam source, and substrates or target materials make up the EBE system. By flowing current through an electron beam source, it is heated, which results in the creation of an electron beam. Using magnets, the produced beam of electrons is focused and guided toward the target materials. EBE has a high rate of deposition and may be used to deposit both conducting and insulating materials. Materials with a higher boiling point can be deposited using EBE. During the EBE process, optimal process parameters and circumstances may be selected to deposit nanoparticles of various sizes and shapes.

EBE includes the following steps:

(i) The target is struck by an electron beam, which heats the material.
(ii) When the temperature rises over the boiling point of the target substance (atoms), it evaporates.
(iii) The substance that has evaporated is transported.
(iv) Condensation onto the substrate [11].

2.4.2.2.3 Laser Ablation

LASiS or laser-ablation synthesis in solution is an efficient method for producing nanomaterials from different types of liquids. It uses a high-powered laser-beam to evaporate particles from a source (solid), which then condenses into a plasma plume, resulting in nanomaterials. To synthesize metal-based nanomaterials, laser ablation offers a substitute to the conventional chemical reduction of metals. It is an environment friendly method since it is a stable process that does not need stabilizing agents or any such chemicals [10, 11].

2.5 Polymer-Based Nanostructures

Polymer nanostructures are colloidal systems with sizes ranging between 5 and 10 nm and diameter up to 1000 nm. Polymers are the most versatile type of biomaterial, with applications in a myriad of fields. Paul Ehrlich and Ursula Scheffel pioneered the creation of polymer nanoparticles. Toward the end of 1960s and early 1970s, Peter Speiser and his colleagues at ETH Zurich did considerable research on polymeric nanomaterials [13]. Polymeric nanomaterials have been utilized in drug delivery systems (DDSs) for a variety of reasons, including controlled release, the capacity to shield drugs and other biologically active molecules from the environment, and the potential to increase bioavailability and therapeutic index.

2.5.1 Types of Polymer-Based Nanomaterials

Polymers used for nanomaterial formulation may be natural or synthetic. Nanomaterials may be formed by using any of the two types and more commonly as a composite of both natural and synthetic.

2.5.1.1 Natural Polymers

Natural polymers, also known as biopolymers, are generated from a range of living sources, including plants, animals, marine life, and microorganisms. H_2O and CO_2 are considered renewable resources since they are vulnerable to breakdown into inorganic compounds.

Natural polymers fall into two classes: polysaccharide-derived polymers and protein-derived polymers (Table 2.2) [14].

Table 2.2 Nanomaterials derived from natural polymers.

Polymer	Derived nanomaterial	Size (nm)
Alginate	Nanocapsules Nanospheres Nanoaggregates	78–667 64 157
Cellulose	Cellulose nanocrystals (CNCs) Cellulose nanofibrils (CNFs)	Length = 100–200, diameter = 2–50; Length = 5, width = 10–150
Chitin	Nanoparticles	150
Chitosan	Nanoparticles	185
Dextran	Nanoparticles	105–400
Inulin	Nanoparticles	112
Pectin	Nanoparticles	100–200
Starch	Nanocrystals	Length = 30, width = 7
Albumin	Nanoparticles	10
Collagen	Nanoparticles	693
Gelatin	Nanoparticles	200
Fibrin	Nanotubes Nanoparticles	150–300 250
Zein	Nanoparticles Nanofiber	365 200–450

2.5.1.2 *Synthetic Polymers*

Biodegradable or non-biodegradable synthetic polymers are available. Polyesters, polyethers, and polyamides are examples of synthetic polymer nanomaterials (Table 2.3). Polyesters have been utilized in drug delivery, medical device manufacturing, and tissue engineering scaffolding [14].

Table 2.3 Nanomaterials derived from synthetic polymers.

Polymer	Derived nanomaterial	Size (nm)
Polylactic acid (PLA)	Nanoparticles	137–156
PLGA	PLGA-Lipid Nanoparticles	215
Poly(ε-caprolactone)	Nanoparticles	500
Polyhydroxybutyrate	Nanoparticles	160
Polyethylene glycol	Nanoparticles	330
Polyethylene oxide (PEO)	Nanofibers	105–240
Poly(N-isopropylacrylamide) (PNIPAM)	Nanoparticles	82

2.5.2 Types of Polymer-Based Nanomaterial Structures

2.5.2.1 Polymeric Nanocarriers

2.5.2.1.1 Polymeric Micelles

When amphiphilic block copolymers and an aqueous solvent are mixed, they get self-assembled into polymeric micelles. These are nanosized molecules having a core-shell based. They are ideal for DDSs as they are biocompatible, less toxic. Their miceller association, morphology, nanoscale size, and high stability also support their use in the biomedical field. Most important property of polymeric micelles is their core-shell structure: core being hydrophobic carries the drug and protects it and the shell (hydrophilic) stabilizes the hydrophobic core under aqueous conditions and enhances their solubility in water.

Polymeric micelles are synthesized using different types of polymers which affect their make and properties. These are of three types, namely:

- Di-block co-polymers: poly-styrene and poly (ethylene glycol) (PEG).
- Tri-block co-polymers: poly ethylene oxide.
- Graft co-polymers: stearic acid with G-chitosan.

Based on the intermolecular forces, polymeric micelles are classified into three types:

- Ones prepared from hydrophilic interactions; e.g., poly(ethylene oxide) are the conventional type of micelles.

- Micelles prepared using electrostatic interactions between oppositely charged polymers; e.g., polyethyleneglycol are poly-ion complex micelles.
- Polymeric micelles with non-covalent connections are formed by polymer self-assembly; e.g., poly(4-vinyl pyridine).

2.5.2.1.2 Polymeric Nanomaterials

They are biodegradable and biocompatible colloidal solid particles having sizes between 10 and 1000 nm. Polymeric nanomaterials are small in size, biodegradable, water soluble, nontoxic, have longer shelf life, and are stable when stored which makes them suitable for delivery of molecules to the target sites which may be tissues or organs. These molecules may be drugs, proteins (peptides), and nucleic acid (DNA or genes).

2.5.2.1.3 Dendrimers

Dendrimers are 3D macromolecules a branched structure having size less than 15 nm. Branching occurs from a central core giving them tree-like appearance, hence the name. They are novel type of polymeric nanomaterial and are also known as cascade molecules and arborols. Dendrimers are functional and soluble in aqueous solutions due to their properties like size, structure, hyper-branching, and multivalency; these characteristics also make it easier to control their molecular weight. The merits of dendrimers are increased solubility of drugs which are poorly soluble on their own, better circulation half-life of the drug, and also can be employed for controlled release of the drug [15].

2.6 Polymer Nanocomposites

Composites are man-made or naturally occurring solid materials that are created by combining two or more distinct constituent elements to produce a new entity. If composites have at least one phase in nanoscale dimension, i.e., 10–100 nm, they are called nanocomposites. Nanocomposites are created by combining inorganic nanoclusters, clays, fullerenes, metals, oxides, or semiconductors with a variety of organic polymers, organic and organometallic chemicals, biomolecules, enzymes, and "sol-gel" generated polymers. Nanocomposites have unique characteristics due to their high S/V ratio and interfacial contact. Because of their ideal properties such as film-forming capabilities, activated functions, and dimensional diversity, polymer-based nanomaterials are extensively utilized [16, 17].

Polymeric nanomaterials are of two types based upon the nanofillers used: carbon nanotubes (CNTs) and layered silicates.

CNTs: To make a nanosized cylinder of C-atoms, sheets of these atoms are rolled into tubes. CNTs are both stronger and lighter than steel and aluminium. Because of their high thermal and chemical durability, they are utilized as catalysts, conductors, capacitors, and transistors. CNTs may be single-walled/double-walled or zigzag, armchair, and chiral on the basis of direction of rolling [19].

Layered silicate/silicate clay: They are hybrids of organic and inorganic materials with properties such as strength and dimensional stability, ease of availability, and cost effectiveness [15, 16].

2.7 Methods of Synthesis

The method of synthesis of polymeric nanomaterials depends upon the requirement or the application of the product. Dispersion of premade polymers and polymerization of monomers to produce polymer nanomaterials are the two most often utilized methods. Many different techniques have evolved from these techniques to mitigate the limitations they had. Products are derived as aqueous colloidal solutions [36].

2.7.1 Solvent Evaporation Method

It is the most widely utilized approach for making polymeric nanoparticles, particularly for delivering active compounds like medicines. The polymers of interest and target drug are solubilized into an organic solvent using an aqueous surfactant solution to create an emulsion. Solvents used are chloroform, dichloromethane, and the less toxic ethyl acetate [37]. The main methods for synthesis emulsion production for this approach are synthesis of oil in water (o/w) single emulsions and water in oil in water (w/o/w) double emulsions.

Steps involved are as follows:

a) Preparation of emulsion by ultrasonication or high-speed homogenization.
b) Evaporation of the organic solvent. This step may affect the particle size and morphology of nanoparticles.
c) Collection by ultracentrifugation.
d) Removal of surfactants from the nanoparticles by washing with distilled water.
e) Lyophilized or freezing for long term storage.

2.7.2 Salting-Out Method

A water-soluble solvent is removed from an aqueous solution via this technique, resulting in the creation of nanospheres. It differs from the other methods in the composition of the oil-water emulsion. A water-soluble polymer solvent (acetone), an aqueous phase gel, an agent for salting-out and a colloidal stabilizer are all included. Electrolytes like $MgCL_2$, $CaCl_2$, magnesium acetate, or non-electrolytes, like sucrose, are employed as salting-out agents.

Steps involved are as follows:

a) Due to the decreased miscibility of acetone and water, saturation of the aqueous phase permits the creation of an oil water emulsion from the other miscible phases.
b) The oil water emulsion is formed at room temperature with vigorous agitation.
c) The emulsion is then diluted with deionized water or an aqueous solution in the appropriate volume. This causes the organic solvent to diffuse to the exterior phase and, as a result, the polymer to precipitate into nanospheres.
d) To remove the residual solvent and the salting-out agent, cross-flow filtration method is used [36].

Heat sensitive molecules can be processed by this method since there is no heat evolved in this technique [37].

2.7.3 Nanoprecipitation or Solvent Displacement Method

This is a simple, rapid, and reproducible method. It is based on the deposition of a polymer between two surfaces using an organic solvent or salts. Rapid diffusion of the solvent into a non-solvent phase lowers the interfacial tension between the two phases, increasing surface area and resulting in the creation of nanoparticles. Polymer nanoparticles such as poly(D,L-lactic acid) (PLA), poly(D,L-lactic-co-glycolic acid) (PLGA), and poly(-caprolactone) are produced using this technique (PCL) [38]. The three basic components of this nanoprecipitation system include polymer (synthetic, semisynthetic, or natural); the polymer's organic solvent, such as acetone, should be miscible in water and easily removed by evaporation; and the water (non-solvent for the polymer) [36, 37].

Steps involved are as follows:

a) The polymer is dissolved in a polarity-intermediate water-miscible solvent.
b) The solution is constantly added to an aqueous solution while stirring.
c) Because the polymer solution diffuses quickly into the aqueous phase, nanoparticles develop almost instantaneously as they evade the water molecules.
d) As the solvent diffuses out of the nanodroplets, the polymer precipitates in the form of nanocapsules or nanospheres [37].

2.7.4 Dialysis

It is a basic and effective way to prepare small polymeric nanomaterials. Steps involved are as follows:

a) The polymer is dissolved in an organic solvent before being placed in a dialysis tube that is appropriate for the nanoparticle's molecular weight.
b) During dialysis, the solvent loses its solubility due to displacement.
c) This causes progressive aggregation of the polymer and a homogeneous suspension of nanoparticles is obtained [37].

2.7.5 Super Critical Fluid Technology

Fluids that do not change in phase despite the change of pressure are defined as supercritical fluids. Supercritical CO_2 has a critical state temperature, Tc = 31.1°C and critical pressure, Pc = 73.8 bar. Being nontoxic, non-flammable, and inexpensive, it is the most commonly used supercritical fluid. Supercritical fluid technology is an interesting and effective particle production technique, which mitigates many disadvantages of traditional methods. This method does not require any solvent and hence is an environment friendly technique with nanoparticles having nontoxic effects. The two most commonly utilized methods of this approach are rapid expansion of supercritical solution (RESS) and supercritical antisolvent (SAS) method.

a) RESS method: The active substance (drug) is dissolved in the organic solvent and then released to the supercritical fluid. The organic phase rapidly dissolves in the supercritical solvent, and remaining nanoparticles can be filtered back.

b) SAS method: The active substance and polymers are dissolved in the supercritical solvent at high pressure [37].

2.7.6 Monomer Polymerization

A low-water-solubility monomer, a water-soluble initiator, and a surfactant are combined at a certain temperature and agitated for a prolonged length of time in this approach. At the end of the process charged stabilized polymers are formed. It is a costly and a lengthy procedure because it is difficult to remove the stabilizing agents from the particles. This is an effective method to produce nanoparticles for drug delivery [38].

2.8 Characterization of Polymeric Nanomaterials

While dealing with any material, characterization is one the most essential step. Particle size, surface area, composition, zeta potential, polydispersity index (PDI), shape, crystallinity, and other factors are used to classify nanomaterials. The most widely utilized methods are electron microscopy, X-ray diffraction, electrophoresis, chromatography, dynamic light scattering (DLS), photon correlation spectroscopy (PCS), and near-infrared spectroscopy, which are all improving analytical techniques [37]. These robust techniques provide a better insight into the properties of polymer-based nanomaterials, which helps one understand the applicability of the said nanomaterial. Figure 2.3 shows the various methods used for the characterization of polymer-based nanomaterials.

2.9 Applications of Polymeric Nanomaterials

2.9.1 Biomedical Applications

Purity, crystallinity, electrical characteristics, surface reactivity, functionality, solubility, form, and aggregation are all distinct physicochemical features of nanoscale materials. Polymeric nanomaterials are comparable in size to biological entities, allowing them to interact easily with biomolecules on the cell surface as well as within the cell. In certain circumstances, the characteristics of nanomaterials are also the source of harmful effects on the body, and hence, they should be characterized when the cytotoxicity,

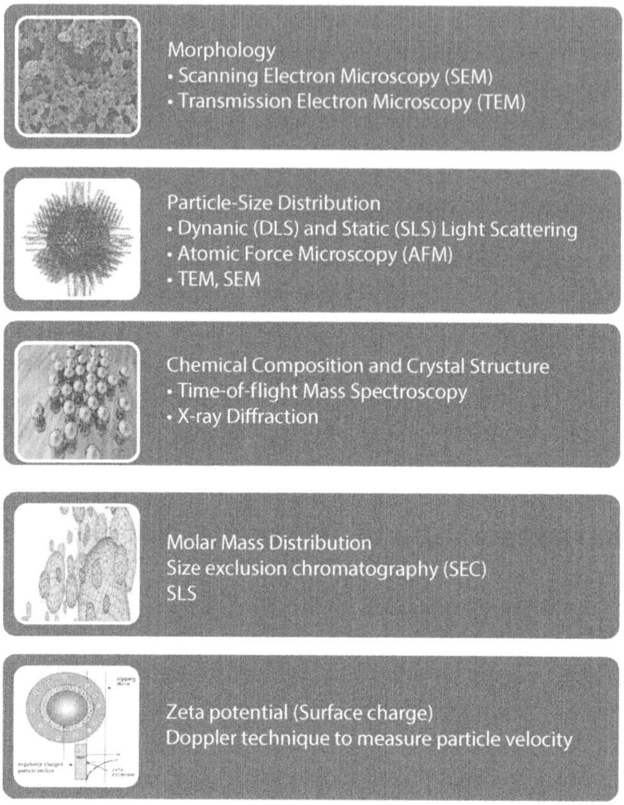

Figure 2.3 Techniques used to characterize polymer-based nanomaterials.

hematoxicity, carcinogenicity, teratogenicity, and immunogenicity are being defined [20].

The particular adsorption of proteins to the polymer surface determines its biocompatibility and the distribution of functional groups on it, which influence the cellular interactions with the surroundings too [21, 22].

2.9.1.1 Smart Drug Delivery Systems

A significant and promising usage of polymeric nanoparticles is as drug delivery systems (DDSs). A DDS is described as a device that permits the entry of any therapeutic substance (drug) into the body and increases efficiency and safety by controlling the rate, timing, and site of release of the said drug in the body according to the National Institutes of Health (NIH). Because of their compact size, these systems may pass through even the tiniest blood capillaries and escape being digested by phagocytes, extending their time in the circulation. Polymer nanostructures that are implemented in regulated drug delivery because their volume allows them to readily bypass tissue and gaps, reaching all types of target organs (spinal cord, lung, spleen, lymph, and liver). The biodegradability, temperature sensitivity, and pH of controlled-release polymers significantly influence their controlled-release characteristics [18]. Drugs or pharmaceutical materials need to be delivered to the patient in the most physiologically acceptable manner [24]. Nanofiber-based drug delivery techniques administer drugs in a tailored way, lowering overall drug dosage, and hence toxicity for non-target locations. This eliminates the need for alternative systemic therapies such as oral administration [24–26]. The release of drugs is influenced by factors such as porosity of the material, its surface area, and the time taken for its biodegradation [24, 25, 27].

2.9.1.2 *Vaccine Adjuvants*

A vaccine that activates both the innate and adaptive immune systems is considered successful. New age vaccines use cellular components such as nucleic acids, polysaccharides, and proteins, which are safer than live or attenuated pathogens but less immunogenic. Adjuvants, which are immune-stimulating signal molecules, are utilized to stimulate a robust protective immunological response. Adjuvants containing aluminium have been utilized in the past, but newer adjuvants such as nucleotide polymers DNA, RNA, and their equivalents are currently being employed. They need to be protected from extracellular degradation and hence are encapsulated or bound with polymeric nanomaterials. Polymer nanomaterials, single-chained particles, polymersomes dendrimers, and DNA-based architectures all have inherent features that make them preferable for usage, such as biodegradability and biocompatibility, as well as simplicity of production and modification. Antigens and TLR7/8 agonists can be delivered together using PLGA which are difficult to deliver due to its hydrophobic nature. Stability and non-toxicity are main concerns while dealing with the choice of type of nanostructure, apart from possibility of controlling the morphology and tuneable payload release [23].

2.9.1.3 *Scaffolds for Tissue Engineering*

Tissue engineering is an interdisciplinary discipline that uses life science and engineering concepts to the development of biological replacements that maintain, repair, or improve tissue and entire organ function.

Tissue engineering scaffolds should have characteristics that promote oxygen and nutrient transport as well as the elimination of metabolic waste during tissue regeneration. High porosity, broad surface area, uniform size, structural integrity with tissue, biocompatibility, biodegradability, suitable mechanical qualities, and nontoxicity for cells are all characteristics of nanofibers utilized for scaffolding. Natural (alginate, collagen, chitosan, chitin, and gelatin) and synthetic [PLGA, PCL, and poly-L-lactic acid (PLLA)] biodegradable polymers, non-degradable polymers (silicone), and conducting polymers (polyaniline and polypyrrole) polymers have all been utilized [24].

2.9.1.4 Antimicrobial Action

Antimicrobial agents are substances that inhibit the development of bacteria or kill them in part while causing no damage to the surrounding tissues. Antibacterial drugs are crucial in the prevention and treatment of infectious illnesses. Antibiotic resistance is emerging as a problem world-over as a result of their misuse [18]. Nanoparticles can be used as efficient antibacterial agents due to their inherent and unique mechanical, chemical, magnetic, electrical, and optical properties, owing to their high surface area to volume ratio. *E. coli* growth and biofilm formation have been demonstrated to be inhibited by AgNPs-chitosan nanoparticles. The bacterium's negatively charged surface interacts with the positively charged chitosan to enhance AgNPs' binding to proteins on the cell walls, causing internal components and other proteins to permeate out and the bacteria to deactivate [30]. Adsorption to the nanoparticle surface, chemical conjugation, or physical encapsulation also can be used to load antimicrobial compounds onto nanoparticulate structures. Polymer-based nanocarriers can transport encapsulated antimicrobial chemicals into infected cells via adsorption, endocytosis, or contact release. Better solubility, precise site-targeted delivery, and prolonged release increase therapeutic efficiency [29].

2.9.1.5 Cell-Based Chips

Nanostructured conducting polymers are being widely studied and utilized for use in cell-based chips for improving the adhesion and ability to proliferate and differentiate various cell types especially neurons [45]. Use of cell-based chips is a potential substitute for animal experimental models as they violate animal rights and are costly too. They are time consuming and have less relevance to human biology. Cell-based chips can be also be utilized as biosensors which can monitor the effects of antitumor drugs or treatment and can be used to better predict drug efficacy and safety, since they provide better understanding of target-drug interaction at a cellular as well as organism level. To immobilize PC12 cells, an electrochemically conducting scaffold consisting of pyrrole N-hydroxyl succinimidyl ester and pyrrole (PPy-NSE) copolymer was modified with a nerve growth factor (NGF) copolymer. It was noted that cells grown in NGF-containing media grew neurites in the same way that cells grown in NGF-free media did [46].

2.9.2 Food Industry Applications

2.9.2.1 Food Packaging

Traditional packing methods have been extended, improved, and polished by modern food science and technology. Food safety and shelf life are top priorities in this age of globalisation, as is the necessity to monitor food safety and quality using international standards. Nanomaterial-based packaging, also known as nanopackaging, has improved performance and can help minimize the massive quantity of food wasted between the sites of production and consumption, which is estimated to be 30%–40% of the total. With the introduction of cellophane in the 1930s, polymers became popular as food packaging materials. Thermoplastics may be readily moulded into any shape required for packaging purposes, and they are also recyclable [31]. PNFPs (polymer nanoparticles for food packaging) provide additional benefits. They have improved barrier characteristics, as well as improved mechanical and flame resistance. In the 1990s, montmorillonite clay was combined with polymers such as polyethylene, nylon, polyvinyl chloride, and starch to create nanocomposites for food packaging [33].

a) Smart nanopackaging materials may include a nanosensor, which is a tiny device that can attach to a target and send back a signal when an abnormality is discovered, as well as convert that response into a signal or output that people can utilize. Physicochemical (sensors) and biological signals can be detected and responded to by these small sensors (biosensors). Nanosensors can detect microorganisms, toxins, and pollutants, as well as assure food freshness [32].
b) The capacity of intelligent packaging to monitor or record internal and exterior changes in the product's surroundings gives knowledge to the consumer. They monitor

both internal and external change and record any change throughout the supply change. Self-heating and self-cooling packaging have also been proposed but the major limitation is that the device occupies much space [33].

c) Bioactive packaging offer greater advantage by having a positive impact on the consumer's health in this age of human health consciousness. Techniques such as enzyme encapsulation, nanoencapsulation, microencapsulation, and enzyme immobilization. In this new packaging method, biopolymers that are known to preserve biopolymer characteristics are used [33].

2.9.3 Electronic Applications

2.9.3.1 Flexible Devices

Conjugated polymers (CPs) are promising materials for flexible and stretchable next-generation electronic applications which include light-emitting diodes, photovoltaics, transistors, photodetectors and bio-inspired devices. Chemical modification of CPs increases the mechanical stretchability of organic semiconductor (OSC) films and suppresses the formation of microscale cracks, which reduce the charge transport. Poly(3,4-ethylenedioxythiophene):poly(s tyrenesulfonate) (PEDOT:PSS), a well-known conductive CP, has high transparency in the visible range and superior electric conductivity. Non-ionic surfactants and ionic compounds have been used as additives to improve the electrical conductivity and mechanical stretchability of PEDOT:PSS films. Individual single nanowires provide flexible and stretchable electronic devices with nanoscale active channels. These one-dimensional CP nanomaterials are soft and flexible; hence, they are promising components of flexible and stretchable devices for wearable electronics and implantable biomedical applications. They can also be modified by geometric engineering, which is advantageous for flexible and stretchable electronic devices [35].

2.9.3.2 Chemical and Bio-Sensors

Conducting polymers have sensitive reactivity to certain chemical or biological species, which can be tuned by adding functional groups to the polymer chains or adding appropriate counter ions during the polymerization process. Nanomaterial-based sensors have been used for toxic as well as non-toxic gases, aroma, humidity, and even for microbes [39]. Food borne pathogens such as *B. cereus*, *Vibrio parahemolyticus*, and *Salmonella* sp. have been identified using conducting polymer nanoparticle sensors composed of carbon black and polyaniline [40].

2.9.3.3 *Nanogenerators*

One-dimensional polymer nanomaterials such as PTFE, fluorinated ethylene propylene, and poly vinyl chloride have been used in nanogenerators such as TENG (triboelectric nanogenerator). Nanogenerators with 1D PTFE nanowires have been proposed for use in monitoring the cardiovascular working and applied in throat-attached anti-interference voice recognition [34].

2.9.3.4 *Semiconductors*

Because of their unique size and characteristics, semiconductor nanoparticles have garnered a lot of interest. Since semiconductor-polymer hybrid materials integrate the merits of polymers with the optical, electrical, and catalytic properties of semiconductor nanoparticles, these are particularly valuable in nanoscience. These hybrids are used in environmental, optico-electronic, biomedical sectors, and so on, because they combine the unique characteristics of organic and inorganic components in a single material [47].

2.9.3.5 *Lithium-Ion Batteries*

Lithium-ion batteries (LIBs) have several advantages, including high efficiency, minimal maintenance, and low self-discharge. An anode, a polymer porous membrane as a separator, a cathode, and an electrolyte comprising dissociated lithium salts are among the battery's components. These can be made using electrospun polymer nanoparticles. Due to their porous structure, which reduces the diffusion distance of lithium (Li+) ions and tends to increase the area of contact of electrode-electrolyte for elevated Li+ ion flux, electrospun PVP and vanadium (IV) acetylacetonate fabricated to porous V2O5 nanotubes can be utilized as a cathode for LIBs and show an inherently reversible capacity and good cycling performance along with excellent rate capacity. For LIB separators, polymer nanomaterials such as polyimide (PI), PAN, PVDF, and their copolymers have been produced, which give better ionic conductivity, thermal stability, and electrolyte affinity [34].

2.9.4 Environmental Applications

2.9.4.1 Biosensors

To assess toxic gases in the atmosphere and hazardous soluble chemicals in water, biosensors are widely utilized in environmental pollution control. They transform a biological event into an electrical signal. The most used biosensors involve co-immobilization of an enzyme specific for the target molecule on a conducting polymer, namely, polypyrrole and polyaniline. They also offer advantages such as thermal stability and better shelf life. Polymers with a tunable and appropriate free volume paired with a certain molecular mass or affinity provide effective biosensors. Weak intermolecular interactions such as hydrogen bonding and van der Waal forces provide significant value for a speedy reaction and recovery period since no strong interaction between the functionalized polymers and the analyte occurs. The linked metal oxide nanoparticles on the polymer matrix may boost the sensor's sensitivity, owing to their substantial surface area and conductance [42]. Using a quartz crystal microbalance, polymer nanostructures based on ZnO/siloxane have been synthesized for gas-phase detection of dimethylmethylphosphonate (DMMP). By forming hydrogen bonds, hydrophilic polymers like siloxane or carboxilane improved target analyte binding [43]. For iron detection in wastewater containing ferrous and ferric ions, the ether/thioether-functionalized network polymer (PAF1–ET) was developed. Thermochemical stability, a wide surface area of up to 5,600 m2/g, and specificity for complex and trace analytes are all advantages of the porous polymers, which provide a feasible active site for adsorption of molecules as well as ions [44]. Environmental, sensor, solar cell, and medicinal applications are just a few of the domains where semiconductor-polymer nanocomposites can be used. Polyaniline/ZnO nanocomposites are used in chemical vapor sensing [48], while hybrid polymer/TiO_2 are used in solar cells [49].

2.9.4.2 Water Purification

Heavy metals such as mercury can be adsorbed from contaminated water bodies using polymer nanostructured membranes having a magnetic core. HMO-001, a composite adsorbent produced by inoculating nano-sized hydrous manganese dioxide (HMO) on the porous polystyrene cation exchanger resin (D-001), can be used to adsorb lead [41].

2.9.4.3 *Catalytic or Redox Degradation of Pollutants*

The nanoparticles can be utilized for catalytic or redox degradation of toxins from water and air due to high and specific surface area, superior reactivity, and optical, electronic, and catalytic properties, which are shape-dependent. Polychlorinated biphenyls (PCBs), azo dyes, halogenated aliphatics, organochlorine pesticides, halogenated herbicides, and nitroaromatics have all been shown to be degraded by nanoparticles such as nano-TiO_2, ZnO, CdS, and WO3, zero-valence metals (such as Fe, Cu, and Zn), and bimetallic nanoparticle. These can be stabilized and made more effective by fabricating with polymers such as chitosan, alginate, resin, PHB, polyaniline (PANI) and polyethylene (PE) [41].

2.10 Conclusion

Nanotechnology offers a promising technology for the development of new food products and new packaging options, which, in turn, can benefit both the consumers and the industry. Nanotechnology is advantageous as it helps to improve the functional properties and advanced characteristics of the packaging materials. Food packaging materials with improved health functionalities, longer shelf life, enhanced transportability, low costs, and environment-friendly properties can be produced by employing using nanotechnology. Nanostructures possess the improved properties like thermal, mechanical, and biodegradable functionalities, which make them useful to be used in food packaging that is active and intelligent.

Because of the smaller particle size of nanoparticles and the differences in physical and chemical characteristics between nanoparticles and their macro-counterparts, using nanostructures in food packaging may be difficult. Nanoparticles may enter the food stuff and can cause severe health issues. So, risk assessment of nanostructures needs to be studied and clinical trials must be carried out so as to limit the side effects of these nanostructures on human health as well as on environment. Both pros and cons of nanostructures should be acknowledged properly before they are applied for any application.

Acknowledgements

The fellowship granted to Ms. Manpreet Kaur (CSIR Fellowship-09/237(0170)/2018-EMR-I) in the form of SRF from Council of Scientific and Industrial Research CSIR is thankfully acknowledged. The financial support from Department of Biotechnology, Ministry of Science and Technology, Govt. of India to Ms. Apoorva Sood in the form of studentship and to Department of Biotechnology, Himachal Pradesh University, Shimla, India, is also thankfully acknowledged.

References

1. Sengupta, A. and Sarkar, C.K., Introduction, in: *Introduction to Nano Basics to Nanoscience and Nanotechnology*, A. Sengupta and C.K. Sarkar (Eds.), pp. 1–3, Springer, Berlin, Heidelberg, 2015.
2. *Nanotechnology: Synthesis to Applications*, CRC Press, Boca Raton, 2018.
3. Bhushan, B. (Ed.), *Springer Handbook of Nanotechnology*, Springer, Berlin, Heidelberg, 2010.
4. Poole, C.P., Jr and Owens, F.J., *Introduction to Nanotechnology*, Wiley, New York, 2003.
5. Ratner, M.A., Ratner, D., Pearson, U.K., Nanotechnology: A gentle introduction to the next big idea. *Mater. Today*, 6, 50, 2002.
6. Binns, C., *Introduction to Nanoscience and Nanotechnology*, John Wiley & Sons, Inc., Hoboken, NJ, USA, 2010.
7. Khan, I., Saeed, K., Khan, I., Nanoparticles: Properties, applications and toxicities. *Arab. J. Chem.*, 12, 908–931, 2019.
8. Lopez, N., Janssens, T.V.W., Clausen, B.S., Xu, Y., Mavrikakis, M., Pedersen, T.B., Nørskov, J.K., On the origin of the catalytic activity of gold nanoparticles for low-temperature CO oxidation. *J. Catal.*, 223, 232–235, 2004.
9. Morones, J.R., Elechiguerra, J.L., Camacho, A., Holt, K., Kouri, J.B., Ramírez, J.T., Yacaman, M.J., The bactericidal effect of silver nanoparticles. *Nanotechnology*, 16, 2346–2353, 2005.
10. Ealia, A.M.S. and Saravanakumar, M.P., A review on the classification, characterisation, synthesis of nanoparticles and their application. *IOP Conf. Ser.: Mater. Sci. Eng*, vol. 263, p. 032019, 2017.
11. Dhand, C., Dwivedi, N., Loh, X.J., Ying, A.N.J., Verma, N.K., Beuerman, R.W., Lakshminarayanan, R., Ramakrishna, S., Methods and strategies for the synthesis of diverse nanoparticles and their applications: a comprehensive overview. *RSC Adv.*, 5, 105003–105037, 2015.
12. Rane, A.V., Kanny, K., Abitha, V.K., Thomas, S., Methods for synthesis of nanoparticles and fabrication of nanocomposites, in: *Synthesis of Inorganic Nanomaterials*, S.M. Bhagyaraj, O.S. Oluwafemi, N. Kalarikkal, S. Thomas (Eds.), pp. 121–139, Woodhead Publishing, Sawston, Cambridge and Elsevier, 2018.

13. Lu, X.Y., Wu, D.C., Li, Z.J., Chen, G.Q., Polymer nanoparticles, in: *Progress in Molecular Biology and Translational Science*, A. Villaverde (Ed.), Academic Press, Elsevier, Cambridge, 2011.
14. Ramachandraiah, K. and Hong, G.P., Polymer based nanomaterials for strategic applications in animal food value chains. *Food Rev. Int.*, 7, 1–30, 2020.
15. Yadav, H.K.S., Almokdad, A.A., Shaluf, S.I.M., Debe, M.S., Polymer-based nanomaterials for drug-delivery carriers, in: *Nanocarriers for Drug Delivery*, Elsevier, Amsterdam, Netherlands, 2019.
16. Arunkumar, L. and Venkataraman, A., Polymer nanocomposites. *Resonance-J. Sci. Educ.*, 10, 49–60, 2005.
17. Vauthier, C. and Bouchemal, K., Methods for the preparation and manufacture of polymeric nanoparticles. *Pharm. Res.*, 26, 1025–1058, 2009.
18. Han, J., Zhao, D., Li, D., Wang, X., Jin, Z., Zhao, K., Polymer-based nanomaterials and applications for vaccines and drugs. *Polymers*, 10, 31, 2018.
19. Mitchell, L.A., Lauer, F.T., Burchiel, S.W., McDonald, J.D., mechanisms for how inhaled multiwalled carbon nanotubes suppress systemic immune function in mice. *Nat. Nanotechnol.*, 4, 451–456, 2009.
20. Malinova, V. and Meier, W., Polymer materials for biomedical applications, in: *Nanoscience & Nanotechnology Series*, P. Broz, (Ed.), pp. 3–15, Royal Society of Chemistry, Cambridge, 2010.
21. Ahad, I.U., Bartnik, A., Fiedorowicz, H., Kostecki, J., Korczyc, B., Ciach, T., Brabazon, D., Surface modification of polymers for biocompatibility via exposure to extreme ultraviolet radiation. *J. Biomed. Mater. Res.*, 102A, 3298–3310, 2014.
22. Webb, G.A., Introduction. *Annu. Rep. Prog. Chem. Sect. C: Phy. Chem.*, 109, 9–9, 2005.
23. Comberlato, A., Paloja, K., Bastings, M.M.C., Nucleic acids presenting polymer nanomaterials as vaccine adjuvants. *J. Mater. Chem. B.*, 7, 6321–6346, 2019.
24. Pawłowska, S., Kowalewski, T.A., Pierini, F., Fibrous polymer nanomaterials for biomedical applications and their transport by fluids: An overview. *Soft Matter*, 14, 8421–8444, 2018.
25. Zhang, Y., Lim, C.T., Ramakrishna, S., Huang, Z.-M., Recent development of polymer nanofibers for biomedical and biotechnological applications. *J. Mater. Sci: Mater. Med.*, 16, 933–946, 2005.
26. Verreck, G., Chun, I., Rosenblatt, J., Peeters, J., Dijck, A.V., Mensch, J., Noppe, M., Brewster, M.E., Incorporation of drugs in an amorphous state into electrospun nanofibers composed of a water-insoluble, nonbiodegradable polymer. *J. Control. Release*, 92, 349–360, 2003.
27. Fonte, P., Araújo, F., Silva, C., Pereira, C., Reis, S., Santos, H.A., Sarmento, B., Polymer-based nanoparticles for oral insulin delivery: Revisited approaches. *Biotechnol. Adv.*, 33, 1342–1354, 2015.

28. Huh, A.J. and Kwon, Y.J., "Nanoantibiotics": A new paradigm for treating infectious diseases using nanomaterials in the antibiotics resistant era. *J. Control. Release*, 156, 128–145, 2011.
29. Gumustas, M., Sengel-Turk, C.T., Gumustas, A., Ozkan, S.A., Uslu, B., Effect of polymer-based nanoparticles on the assay of antimicrobial drug delivery systems, in: *Multifunctional Systems for Combined Delivery, Biosensing and Diagnostics*, Elsevier, Amsterdam, Netherlands, 2017.
30. Sanpui, P., Murugadoss, A., Prasad, P., Ghosh, S., Chattopadhyay, A., The antibacterial properties of a novel chitosan–Ag-nanoparticle composite. *Int. J. Food Microbiol.*, 124, 142–146, 2008.
31. Silvestre, C. and Cimmino, S., *Ecosustainable polymer nanomaterials for food packaging: innovative solutions, characterization needs, safety and environmental issues*, CRC Press, Boca Raton, 2016.
32. Kour, H., Ahmad, N., Wani, T., Malik, A., Kaul, R., Chauhan, H., Gupta, P., Bhat, A., Advances in food packaging – a review. *Stewart Postharvest Rev.*, 9, 1–7, 2013.
33. Majid, I., Ahmad Nayik, G., Mohammad Dar, S., Nanda, V., Novel food packaging technologies: Innovations and future prospective. *J. Saudi Soc. Agric. Sci.*, 17, 454–462, 2018.
34. Huang, L.B., Xu, W., Hao, J., Energy device applications of synthesized 1D polymer nanomaterials. *Small*, 13, 1701820, 2017.
35. Lee, Y., Zhou, H., Lee, T.W., One-dimensional conjugated polymer nanomaterials for flexible and stretchable electronics. *J. Mater. Chem. C*, 6, 3538–3550, 2018.
36. Zielińska, A., Carreiró, F., Oliveira, A.M., Neves, A., Pires, B., Venkatesh, D.N., Durazzo, A., Lucarini, M., Eder, P., Silva, A.M., Santini, A., Souto, E.B., Polymeric nanoparticles: production, characterization, toxicology and ecotoxicology. *Molecules*, 25, 3731, 2020.
37. Derman, S., Katmıs, A., Fide, S., Karaismailoglu, S., Synthesis and characterization methods of polymeric nanoparticles. *Characterization Appl. Nanomaterials (CAN)*, 1, 60–66, 2018.
38. Lu, X.Y., Wu, D.C. *et al.*, Polymer nanoparticles, in: *Progress in Molecular Biology and Translational Science*, D.B. Teplow, (Ed.), pp. 299–323, Academic Press, Elsevier, Cambridge, 2011.
39. Oh, W.K., Kwon, O.S., Jang, J., Conducting polymer nanomaterials for biomedical applications: cellular interfacing and biosensing. *Polym. Rev.*, 53, 407–442, 2013.
40. Idumah, C.I., Zurina, M., Ogbu, J., Ndem, J.U., Igba, E.C., A review on innovations in polymeric nanocomposite packaging materials and electrical sensors for food and agriculture. *Compos. Interfaces.*, 27, 1–72, 2020.
41. Sahoo, T.R., Polymer Nanocomposites for Environmental Applications, in: *Properties and Applications of Polymer Nanocomposites: Clay and Carbon Based Polymer Nanocomposites*, B.P. Sahoo, and D.K. Tripathy, (Eds.), Springer, Berlin Heidelberg, 2017.

42. Fadillah, G., Saputra, O.A., Saleh, T.A., Trends in polymers functionalized nanostructures for analysis of environmental pollutants. *Trends Environ. Anal. Chem.*, *26*, e00084, 2020. https://doi.org/10.1016/j.teac.2020.e00084.
43. Das, R., Bandyopadhyay, R., Pramanik, P., Stereo-regulated Schiff Base Siloxane Polymer Coated QCM Sensor for Amine Vapor Detection. *Mater. Chem. Phys.*, 226, 214–219, 2019.
44. Suhail, M.H., Abdullah, O.Gh., Kadhim, G.A., Hydrogen Sulfide Sensors Based on PANI/f-SWCNT Polymer Nanocomposite Thin Films Prepared by Electrochemical Polymerization. *J. Sci. Adv. Mater. Devices*, *4*, 143–149, 2019.
45. Yea, C.-H., Lee, B., Kim, H.-H., Kim, S.-U., El-Said, W.A., Minc, J.-H., Oh, B.-K., Choi, J.-W., The immobilization of animal cells using the cystine-modified RGD oligopeptide. *Ultramicroscopy*, 108, 1144–1147, 2019.
46. El-Said, W.A., Abdelshakour, M., Choi, J.-H., Choi, J.-W., Application of Conducting Polymer Nanostructures to Electrochemical Biosensors. *Molecules*, 25, 307, 2020.
47. Kango, S., Kalia, S., Thakur, P., Kumari, B., Pathania, D., Semiconductor–polymer hybrid materials, in: *Organic-Inorganic Hybrid Nanomaterials*, S. Kalia and Y. Haldorai (Eds), pp. 283–311, Springer, Cham, 2014.
48. Kondawar, S.B., Patil, P.T., Agrawal, S.P., Chemical Vapour Sensing Properties of Electrospun Nanofibers of Polyaniline/ZnO Nanocomposites. *Adv. Mater. Lett.*, 5, 389–395, 2014.
49. Yoriya, S. and Grimes, C.A., Self-Assembled Anodic TiO 2 Nanotube Arrays: Electrolyte Properties and Their Effect on Resulting Morphologies. *J. Mater. Chem.*, 21, 102–108, 2011.

3

Role of Green Nanocomposites in Smart/Active Food Packaging

Samrat K.[1], Sharath R.[2*], Chandraprabha M. N.[1], Hari Krishna R.[3†] and Kumaraswamy H. M.[4]

[1]*Department of Biotechnology, M.S. Ramaiah Institute of Technology, Bangalore, Karnataka, India*
[2]*Department of Studies in Food Technology, Davangere University, Shivagangotri, Davanagere, Karnataka, India*
[3]*Department of Chemistry, M.S. Ramaiah Institute of Technology, Bangalore, Karnataka, India*
[4]*Department of PG Studies and Research in Biotechnology, Kuvempu University, Shivamogga, Karnataka, India*

Abstract

Major concern in food packaging industry is to produce biodegradable materials for packaging with enhanced mechanical, barrier, functional and thermal properties to prevent food products from microbial spoilage, loss of nutrition, and increase shelf life. To meet this requirement, numerous biopolymers and composite materials have been exploited. The usage has been restricted because of poor barrier and mechanical properties. Advent of nanotechnology has provided significant opportunity to develop high strength packaging materials by incorporation of nanosized fillers to biopolymers forming bionanocomposites, which are eco-friendly alternative to petroleum-based polymers. In this regard, present chapter provides insight about different types of biopolymers, bionanocomposites/green nanocomposites classification and innovative packaging applications in particular about active packaging and intelligent packaging systems.

***Keywords*:** Biopolymers, bionanocomposites, barrier properties, shelf life, food products

**Corresponding author*: sharathsarathi@gmail.com
†*Corresponding author*: rhk.chem@gmail.com

Annu, Tanima Bhattacharya and Shakeel Ahmed (eds.) Nanotechnology in Intelligent Food Packaging, (59–78) © 2022 Scrivener Publishing LLC

3.1 Introduction

Plastics being versatile material have been widely used in packaging applications. Currently, 320 million tons of plastics per year are produced globally to meet the increasing demand, because of widespread application in different fields [1]. Preparation of plastics is solely dependent on fossil resources, and there increasing prices has influenced on the economy of plastic market. Polymers that are commonly used for food packaging applications are polypropylene (PP), polystyrene (PS), polyvinylchloride (PVC), and polyethylene (PE) [2]. They are commonly preferred due to their excellent mechanical and barrier properties, abundant availability, and cost effectiveness. However, these materials cause environmental burden due to non-biodegradable and non-recyclable nature along with greenhouse gas emissions (i.e., methane and CO_2) after disposal [3]. Therefore, increasing demand for innovative packaging materials in food industry is of great importance that could rapidly and completely degrade in environment [4]. In addition, materials used for packaging of food must provide barrier against oxygen, volatile compounds like flavors, carbon dioxide, and water vapor permeability [5, 6]. In this regard, biopolymers have been found be one of the alternatives into eco-friendly food packaging materials because of biodegradability nature. Biopolymers used for food packaging purpose get degraded once disposed, leaving behind biomass, CO_2, and water as by-products [4].

Unfortunately, biopolymers used for food packaging applications have disadvantages like brittleness, poor barrier properties, and low heat distortion temperature compared to petroleum-based synthetic polymers [7]. To overcome the drawbacks to improve and enhance biopolymer properties, nano-fillers like hybrid nanoclay and nano cellulose are incorporated forming bionanocomposites [8]. Bionanocomposites possess greater potential in food packaging technology to extend packed shelf life of food products and enhance mechanical, thermal, and barrier properties. In this regard, present chapter provides the insight on preparation of nanocomposites, classification of packaging and its functions, and food packaging applications of bionanocomposite material.

3.2 Bionanocomposite/Green Nanocomposites

Bionanocomposites consist of reinforced nanofillers in the biopolymer matrix [9]. They can also be called as nanobiocomposites, green composites, biohybrids, or green nanocomposites [10]. The nanofillers act as structural support, and they play an important role in augmenting the mechanical and barrier properties of the biopolymer matrix [4].

3.3 Biopolymers

Biopolymers are polymers made of covalently bonded chain like molecules of monomeric units. They are biodegradable, abundant, and renewable materials, with the capability of degrading through the action of microorganism resulting in formation of organic by-products like water and carbon dioxide that are safe to the environment [4].

3.3.1 Classification of Biopolymers

Biopolymers used for green nanocomposites preparation are classified and summarized as follows [11–13]:

1. **Biopolymers of natural origin:** carbohydrates like cellulose, starch, chitosan, and agar; proteins such as corn zein, soy protein isolate, wheat gluten, and wheat protein isolate.
2. **Synthetic biopolymers:** polyvinyl alcohol, polycaprolactone, polylactic acid, and polybutylene succinate.
3. **Microbial origin biopolymers:** polyhdroxybutrate and polyhydroxyalkanoates.

3.4 Nanofillers

Biopolymers performance has been enhanced by using nanofillers. Two classes of nanofillers are widely used for food packaging applications based on source as (i) organic nanofillers and (ii) inorganic nanofillers. Organic nanofillers include chitin nanofibrils, lignin, cellulose nanofibrils, starch nanocrystals, cellulose nanowhiskers, chitin nanocrystals, clay (i.e., Montmorillonite), and cellulose nanoparticles. Inorganic nanofillers commonly used for preparation of green nanocomposites are graphene oxide nanoparticles, titanium oxide nanoparticles, zinc oxide nanoparticles, gold nanoparticles, and sulfur nanoparticles. Apart from this, inorganic-inorganic and organic-inorganic hybrids are also used as nanofillers [14]. Addition of nanofillers into biopolymer matrix provides large interface due to nanosize and improves mechanical properties, relaxation behavior, and barrier and thermal properties of green nanocomposites [15, 16]. They also exhibit desired functions like antimicrobial property and oxygen scavenger in packaging materials, resulting in improvement of shelf life of food products [11].

3.5 Types of Green Nanocomposites

Green nanocomposites are prepared using different biopolymers of natural origin with inorganic or organic nanofillers. Incorporation of inorganic nanofillers (i.e., nanoparticles) enhances their mechanical and barrier properties, owing to high aspect ratio and surface area of nanoparticles [11].

3.5.1 Carbohydrate-Based Green Nanocomposites

Cellulose: Cellulose is a robust natural polymer and building block of fibrous cells. It is eco-friendly, recyclable, cost-effective, and ubiquitous. During synthesis of nanomaterials, cellulose is used as supporting materials because cellulose application along with nanoparticles leads to increase in surface area and their by enhance the activity. Therefore, cellulose nanofibers are considered as important class of nanomaterials due to additive nature [17]. Nanoreinforced cellulose improves elastic modulus of polymer, thermal properties, and strength [18, 19]. Polymer films incorporated with nanoreinforced cellulose enhanced the moisture barrier properties [20, 21]. Cellulose films impregnated with 5% chitosan nanoparticles inhibited the *E. coli* growth by 85% [22].

Chitosan: Chitosan, a heteropolysaccharide obtained by chitin deacetylation at alkaline conditions, consists of D-glucosamine and N-acetyl-D-glucosamine units connected by β-1,4-glycosidic linkages. The deacetylation degree depends on ratio of the two units [23, 24]. Soy protein isolates that are incorporated with chitin whiskers composite significantly enhanced tensile strength and water absorption resistance of the matrix [25]. In 2005, U.S. Food and Drug Administration listed chitosan under GRAS (generally recognized as safe) category materials and used as food preservatives due to non-toxic and biocompatible nature for animals and human beings [26, 27]. Chitosan is known to possess antimicrobial property due to polycationic nature against wide variety of bacteria and fungi [28]. Significant antibacterial activity against *L. monocytogenes* and *E. coli* was exhibited by sulfur nanoparticles incorporated with chitosan nanocomposite with enhanced properties [29].

Starch: Starch is the widely used plant-based polysaccharide in the applications of food packaging because of biodegradability, biocompatibility, low cost, abundance, non-toxic, and easy availability [30–32]. Water absorption resistance has been improved by addition of synthetic and inorganic materials with starch films [33]. Presence of positive charge on the antimicrobial agents attributes to its antimicrobial activity. Therefore, metals impregnated polysaccharide exhibits increased antimicrobial activity due to enhanced surface area [28].

3.5.2 Protein-Based Green Nanocomposites

Gelatin: Gelatin is produced from water soluble protein called collagen by partial hydrolysis. Animal age, collagen type, method of processing and source influences the gelatin properties [34]. Potential sources of gelatin production are cattle bones, bovine hide and pig skin. Gelatin is consists of 6.8% hydrogen, 25.2 % oxygen, 7% nitrogen, and 50.5% carbon [35]. Gelatin is a tasteless and odorless substance processed in the form of powder or granules. Due to biodegradability nature, less cost, abundant availability, and film forming ability, it is considered as one of the promising protein biopolymer [36]. In food industry, gelatin is used as coating and packaging materials and in dairy products, beverages, bakery and confectionery for emulsification, stabilization, gelling, and texturization [37]. Gelatin matrix incorporated with nanoparticles greatly improves overall properties like mechanical and physical properties, biodegradability, and biocompatibility of nanocomposites [38]. In addition, organic polymers like cellulose nanocrystals, starch, chitosan, and chitin have been used for gelatin bionanocomposite film preparation with complete biodegradable nature for food packaging applications [39, 40]. Other commonly used protein biopolymers for preparation of green nanocomposites are soy protein [41].

3.6 Green Nanocomposite Preparation Methods

The basic step involved to obtain nanocomposites with improved properties is homogeneous dispersion of nanofillers into polymer matrix. Nanofillers possess increased self-agglomeration tendency due to presence of hydroxyl groups on the nanofillers surface. To minimize the tendency of agglomeration, different methods have been used during formation of nanocomposite films [42]. They are described as follows:

1. **Solvent casting method:** Organic solvent or water is used for dissloving polymer under room temperature or heating conditions. Post dissolution, particular size particles are added to the solution. Then, thin films are formed by casting film forming solution on to glass plate and evaporating solvent via drying. As a result, composite films containing particles with polymer is obtained. Three steps are followed for formation of films: solubilization, casting, and drying [43].
2. ***In situ* intercalative polymerization method:** In this approach, specific monomer is mixed with clay mineral physically. Then, within the interlayer, monomer undergoes intercalation leading to delamination. Polymerization is induced via numerous method to form matrix of cross-linked or liner polymer. Before the initiation, swelling of clay mineral is carried out to achieve dispersion, exfoliation, and intercalation of long-chain alkyl ammonium ion of clay during intercalative polymerization method [42].
3. **Thermocompression method:** Polymer is compressed thermally between preheated aluminium foiled stainless steel plates. For 3–5 min, pressure of about 1,000 psi is applied. The film obtained after cooling in the air is removed from the aluminium foil layer [42].
4. **Solution intercalation method:** Polymer and pre-swollen clay are dissolved separately in the solvent. Afterward, solutions are mixed leading to intercalation of polymer chains by displacing solvent from clay interlayer. Low or no polarity polymers are suitable for obtaining polymer-clay intercalated layer of thin films by this method. Water-soluble biopolymers are frequently used to produce nanocomposites by intercalation [10].

3.7 Green Nanocomposites for Applications of Food Packaging

The major role of any packaging materials are food product protection from microbial spoilage and prevent damage from outside via external force, moisture, oxygen, extended shelf life with storage, transportation, handling, distribution, reuse, recovery, or disposal with maximizing value for consumer and to provide nutritional needs with proper food components for consumer

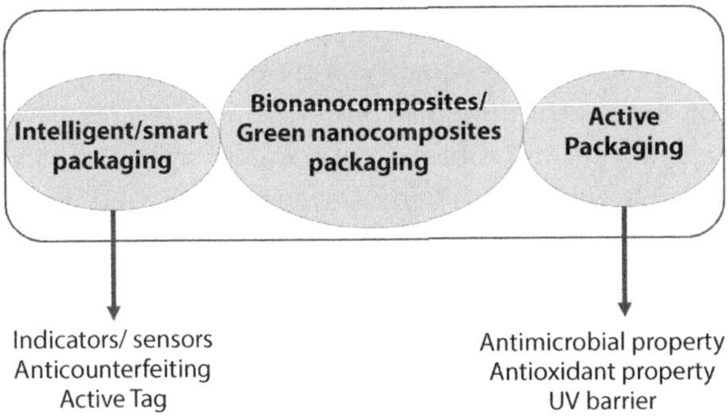

Figure 3.1 Innovative food packaging applications and its functions.

[8]. Further, advent of bionanocomposite/green nanocomposite systems via nanotechnological principles has resulted in enhancement of convenience, protection and preservation, containment, communication, and marketing compared to traditional food packaging systems [17]. The green nanocomposite applications toward food packaging can be described via two classifications like smart/intelligent packaging and active packaging (Figure 3.1).

3.7.1 Active Packaging

Active packaging provides protection and preservation of food product by interacting with active components incorporated into food packaging

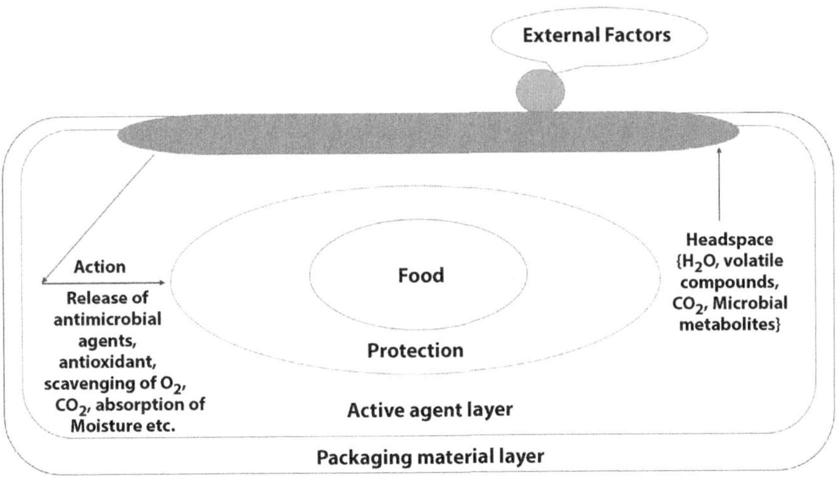

Figure 3.2 Conceptual repetition of active packaging system.

activated by intrinsic or extrinsic factors leading to stimulating action, thereby increase the shelf life, quality, and safety of food product [44]. The most common active components combined with in green composites are antioxidant agents, antimicrobial agents, water vapor and oxygen absorbers, carbon dioxide emitter, ethylene scavengers, etc. [45]. Figure 3.2 shows the conceptual representation of active packaging system exhibiting protection role against internal and external factors.

3.7.1.1 *Antimicrobial Packaging*

The major reason for food spoilage and deterioration quality of food is microbial growth. This results in the nutritional loss, shelf-life reduction, risk of food born-illness, changes in texture, bad odor and development of off-flavor, and discoloration [46]. Development of green nanocomposites with antimicrobial properties by incorporation of nanofillers has shown promising results in food packaging applications than bulk materials. Due to high surface reactivity and high volume-to-surface ratio of nanofillers at nanometer scale [47]. Antimicrobial packaging films are prepared by mixing antimicrobial agents (i.e., inorganic nanomaterials) with polymer matrix during processing of polymers. The most commonly used nanomaterials for enhancing antimicrobial property of food packaging materials are metal and metal oxides like copper, silver, zinc oxide, silicon dioxide, and titanium dioxide [48]. Among them, silver nanoparticles incorporated with biopolymers such as chitosan [49], gelatin [50] and agar [47] forming

Table 3.1 Examples of antimicrobial packaging of green nanocomposites.

Nanocomposite	Antibacterial activity against	References
Chitosan/silica (SiO_2)	*S. Typhimurium, L. Monocytogenes, S. aureus,* and *E. coli*	[52]
Cellulose nanofibrils/silver	*L. monocytogenes and E. coli*	[53]
Chitosan/silver	*P. citrinum, S. aureus, E. coli,* and *A. niger*	[54]
Starch/ZnO	*E. coli*	[55]
Whey protein isolate/ cellulose nanofiber/TiO_2	*Pseudomonas* spp., *S. aureus, E. coli, Enterobacteriaceae,* and *L. monocytogenes*	[56]

nanocomposites exhibited significant antimicrobial activity against both gram-positive and gram-negative bacteria. Antimicrobial agents function either by slow migration and reaction with organics of food or by direct contact with the surface of food. The mechanism with which nanoparticles exhibit antimicrobial property is by ROS generation, resulting in oxidative stress, enzyme inhibition, impairment of membrane integrity, and direct interaction with microbial cells causing disruption of electron transport chain, leading to cell components leakage and death [51]. Examples of antimicrobial property of green nanocomposites are tabulated in Table 3.1.

3.7.1.2 *Antioxidant Agents*

One of the important functions of active packaging is to enhance oxidative stability of foods by the antioxidant activity of packing materials. Residual oxygen in packing results in reduction of food product shelf life via toxic aldehydes formation, oxidative rancidity, lipid oxidation, enzymatic browning, and loss of nutrition loss [57]. The major factor that could extend shelf oxygen in packing results in reduction of food product shelf life via toxic life and reduce spoilage and reactions that cause deterioration of food is by controlling oxygen level in packaging [58]. Therefore, antioxidant agents have been employed in the advanced packaging systems to protect lipids, proteins and nucleic acids from oxidative damage by reactive free radicals. Antioxidant agents preserve food materials from oxidative damage no matter what is the reaction mechanism [59, 60]. Coating, sachets, immobilization and labels from antioxidant agents (natural or synthetic origin) have been impregnated successfully into active packaging systems [61]. Antioxidant activity in packaging films is determined by assessing radical scavenging capacity via DPPH and TBARS assay. Due to reluctance by consumer to use synthetic

Table 3.2 Examples of antioxidant packaging of green nanocomposites.

Nanocomposite	Function	References
Chitosan/pine needle extract	Antioxidant property	[63]
Chitosan/flavanols (myricetin, kaempferol, and quercetin)	Antioxidant property	[64]
Polylactic acid/polycaprolactone loaded carvacrol or thymol	Antioxidant property	[65]
Chitosan/zinc oxide nanoparticles	Antimicrobial/antioxidant property	[66]

antioxidants, ascorbic acid, tocopherol, plant essential oils, carvacrol, and plant extracts like natural-based antioxidants are commonly used as component of food packaging material [57]. Luzi *et al.* [62] reported that polyvinyl alcohol-based nanocomposite incorporated with chitosan, CNC, and carvacrol exhibited significant antioxidant and scavenging activity. Antioxidant properties of few green nanocomposites are tabulated in Table 3.2.

3.7.1.3 UV-Barrier

Substantial photochemical reactions caused by UV radiations ranging from 315 to 400 nm (UVA) and 280 to 315 nm (UVB) lead to formation of free radicals. These free radicals results in vitamins, proteins, lipids oxidation, antioxidants degradation, and texture and color change of food products. Hence, UV radiation exposure causes reduction in food quality and shelf life, organoleptic, and nutritional loss [67]. This provides the radical scavenging importance in the applications of food packaging [68, 69]. Cazon *et al.* [70] prepared PVA/bacterial cellulose/glycerol films exhibited excellent UV-blocking property. Apart from this, κ-carrageenan/CNC nanocomposite films showed better mechanical properties as well as barrier to UV radiations [71].

3.7.1.4 Oxygen Scavengers

The oxygen presence in food packaging favors microbial growth, modification of color, development of off-flavor, and nutrient loss and also significantly affects the respiration rate in vegetables and fruits and production of ethylene. Passive barrier packaging materials such as barrier nanocomposite and copolymers of ethylene vinyl alcohol are used to pack oxygen sensitive foods [72]. However, complete oxygen removal from passive method is not possible and remains in head space. Such drawbacks are solved by using bionanocomposites/green nanocomposites [42, 73]. Various polymeric matrices like PP carbonate with PE glycol [74], collagen [75], and kappa carrageenan [76] with cellulose nanocrystals forming green nanocomposites have exhibited excellent oxygen scavenging properties.

Apart from this, various other applications of active packaging include moisture regulators/barriers, carbon dioxide emitters/absorbers, ethylene scavengers, emitter, and sulfur dioxide emitter.

3.7.2 Smart/Intelligent Packaging

Smart or intelligent packaging utilizes nanotechnology-based sensors or indicators impregnated with packaging materials register and conveys

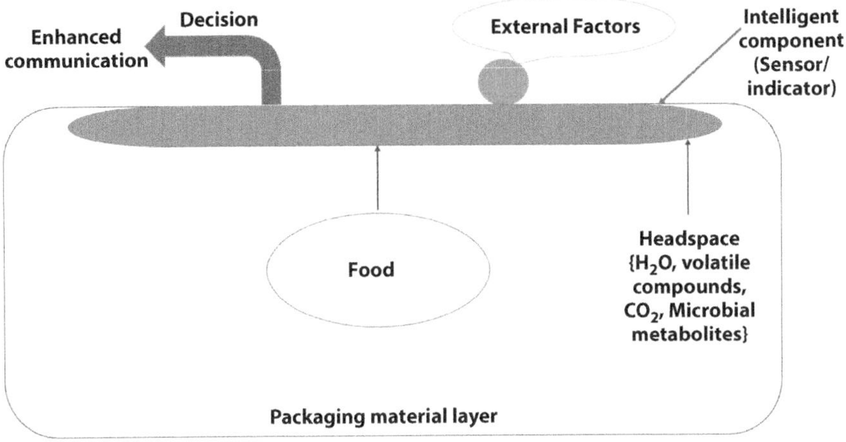

Figure 3.3 Conceptual framework of intelligent packaging systems.

information by interacting with internal and external environmental factors about food product condition, safety, and digestibility [77, 78]. Figure 3.3 shows the conceptual framework of intelligent packaging systems, exhibiting enhanced communication role.

3.7.2.1 Time-Temperature Indicators

During storage, transportation, and processing of food, one of the important factors that affects quality and safety of different food products is temperature [79]. Effects of temperature on food products are monitored and recorded by time-temperature indicators (TTIs), thus providing information about shelf life and food quality [80]. Therefore, quality of food product information is obtained by embedding TTIs with food packaging and monitoring for specified period and temperature. Based on the color changes of the dye incorporated nanocomposites, TTIs depict the food product stability via dye degradation [81]. Rachmelia and Imawan [82] studied about of chitosan/anthocyanin nanocomposite as TTIs at temperature range of 10°C to 40°C and found excellent stability between 10°C and 25°C.

3.7.2.2 Freshness Indicators

Due to high consumer demand for fresh and healthy foods, freshness indicators development increased from the last decade. These indicators

Table 3.3 Examples of green nanocomposites as freshness indicators.

Nanocomposite	Function	Food	References
Chitosan/polyvinyl alcohol/anthocyanin from jabuticaba fruit or purple sweet potato peels	Freshness indicator	Meat	[87]
Chitosan/oxidized chitin nanocrystals/ red cabbage anthocyanin	Freshness indicator	Hairtail/ Shrimp	[88]
Oxidized chitin/gelatin/ black rice bran anthocyanin	Freshness indicator	Fish	[89]
Polyvinyl alcohol/ chitosan/xylan/ hydroxyapatite	Freshness indicator	Fish	[90]

provide the information by monitoring quality of food products during storage and transportation in addition to chemical changes or microbial growth. Loss of food product freshness is usually caused because of exposure to expiration of shelf life and adverse conditions [83]. Meat and sea food products' freshness is assessed depending on total volatile basic nitrogen content (TVB-N), because they release volatile amines once food spoils. Freshness of food product is usually detected by two methods like pH variations [84] and conductometric [85]. Ezati *et al.* [86] reported cellulose-based nanocomposites incorporated with chitosan and alizarin as pH-sensitive indicator to track freshness of minced beef via color changes during storage. Few examples of green nanocomposites as freshness indicator are shown in Table 3.3.

Other smart/intelligent applications include gas indicators, integrity indicators, radiofrequency identification tags, and bar codes.

3.8 Conclusion

Green nanocomposites/bionanocomposites have provided a new insight for preparation of innovative packaging films by incorporating inorganic and organic nanomaterials. These composites have exhibited good barrier and thermal and mechanical properties. In addition, intelligent packaging systems such as different types of indicators alert the customer about the food product quality. In conclusion, still, extensive research needs to be carried out toward green nanocomposites usage for variety of food products packaging and marketability.

References

1. Paletta, A., Filho, W.L., Balogun, A.L., Foschi, E., Bonoli, A., Barriers and challenges to plastics valorisation in the context of a circular economy: Case studies from Italy. *J. Clean. Prod.*, 241, 118149, 2019.
2. Luzi, F., Torre, L., Kenny, J.M., Puglia, D., Bio-and fossil-based polymeric blends and nanocomposites for packaging: Structure–property relationship. *Materials.*, 12, 471, 2019.
3. Asgher, M., Qamar, S.A., Bilal, M., Iqbal, H.M.N., Bio-based active food packaging materials: Sustainable alternative to conventional petrochemical-based packaging materials. *Int. Food Res. J.*, 137, 109625, 2020.
4. Othman, S.H., Bio-nanocomposite materials for food packaging applications: types of biopolymer and nano-sized filler. *Agric. Agric. Sci. Procedia.*, 2, 296, 2014.
5. Marsh, K. and Bugusu, B., Food packaging—Roles, materials, and environmental issues. *J. Food Sci.*, 72, 39, 2007.
6. Lee, S.J. and Rahman, A.T.M.M., Intelligent packaging for food products, in: *Innovations in Food Packaging*, 2nd, J.H. Han (Ed.), pp. 171–212, Elsevier Academic Press, California, 2014.
7. Kuswandi, B., Nanotechnology in food packaging, in: *Nanoscience in Food and Agriculture*, 1st, S. Ranjan (Ed.), pp. 151–183, Springer, Switzerland, 2016.
8. Moustafa, H., Youssef, A.M., Darwish, N.A., Abou-kandil, A.I., Eco-friendly polymer composites for green packaging: Future vision and challenges. *Compos. B. Eng.*, 172, 16, 2019.
9. Youssefa, M. and El-Sayed, S.M., Bionanocomposites materials for food packaging applications: Concepts and future outlook. *Carbohydr. Polym.*, 193, 19, 2018.
10. Pande, V.V. and Sanklecha, V.M., Bionanocomposite: A review. *Austin J. Nanomed. Nanotechnol.*, 5, 1045, 2017.
11. Rhim, J.W., Park, H.M., Ha, C.S., Bio-Nanocomposites for Food Packaging Applications. *Prog. Polym. Sci.*, 38, 1629, 2013.
12. Adeyeye, O.A., Sadiku, E.R., Reddy, A.B. *et al.*, The use of biopolymers in food packaging, in: *Green Biopolymers and their Nanocomposites*, D. Gnanasekaran (Ed.), pp. 137–158, Springer, Switzerland, 2019.

13. Rahman, R., Sood, M., Gupta, N., Bandral, J.D., Hameed, F., Ashraf, S., Bioplastics for food packaging: a review. *Int. J. Curr. Microbiol. App. Sci.*, 8, 2311, 2019.
14. Shankar, S. and Rhim, J.W., Bio-nanocomposites for food packaging applications, in: *Encyclopedia of Renewable and Sustainable Materials*, vol. 5, S. Hashmi (Ed.), pp. 29–41, Elsevier, Netherlands, 2020.
15. De Oliveira, A.D. and Beatrice, C.A.G., Polymer nanocomposites with different types of nanofiller, in: *Nanocomposites - Recent Evolutions*, S. Sivasankaran (Ed.), pp. 103–128, IntechOpen, London, 2019.
16. Cakmak, H., Sogut, E., Gutierrez, T.J., Functional biobased composite polymers for food packaging applications, in: *Reactive and Functional Polymers*, vol. 1, pp. 95–136, Springer, Switzerland, 2020.
17. Sharma, C., Dhiman, R., Namita, R., Panwar, H., Nanotechnology: An untapped resource for food packaging. *Front. Microbiol.*, 8, 1735, 2017.
18. Zhang, B., Huang, C., Zhao, H., Wang, J., Yin, C., Zhang, L., Zhao, Y., Effects of cellulose nanocrystals and cellulose nanofibers on the structure and properties of polyhydroxybutyrate nanocomposites. *Polymers.*, 11, 2063, 2019.
19. Kumar, N. and Kumbhat, S., *Essentials in Nanoscience and Nanotechnology*, John Wiley & Sons, New Jersey, 2016.
20. Wang, J., Gardner, D.J., Stark, N.M., Bousfield, D.W., Tajvidi, M., Cai, Z., Moisture and oxygen barrier properties of cellulose nanomaterial-based films. *ACS Sustain. Chem. Eng.*, 6, 49, 2018.
21. Jung, B.N., Jung, H.W., Kang, D.H., Kim, G.H., Shim, J.K., Synergistic effect of cellulose nanofiber and nanoclay as distributed phase in a polypropylene based nanocomposite system. *Polym.*, 12, 2399, 2020.
22. Romainor, A.N.B., Chin, S.F., Pang, S.C., Bilung, L.M., Preparation and characterization of chitosan nanoparticles-doped cellulose films with antimicrobial property. *J. Nanomater.*, 2014, 130, 2014.
23. Bedian, L., Villalba-Rodriguez, A.M., Hernandez-Vargas, G., Parra-Saldivar, R., Iqbal, H.M.N., Bio-based materials with novel characteristics for tissue engineering applications - A review. *Int. J. Biol. Macromol.*, 98, 837, 2017.
24. Verlee, A., Mincke, S., Stevens, C.V., Recent developments in antibacterial and antifungal chitosan and its derivatives. *Carbohydr. Polym.*, 164, 268, 2017.
25. Ahari, H., Anvar, A.A., Ataee, M., Naeimabadi, M., Employing nanosilver, nanocopper, and nanoclays in food packaging production: A systematic review. *Coatings.*, 11, 509, 2021.
26. Xing, Y., Xu, Q., Li, X., Chen, C., Ma, L., Li, S., Che, Z., Lin, H., Chitosan-based coating with antimicrobial agents: preparation, property, mechanism, and application effectiveness on fruits and vegetables. *Int. J. Polym. Sci.*, 2016, 4851730, 2016.
27. Garg, U., Chauhan, S., Nagaich, U., Jain, N., Current advances in chitosan nanoparticles based drug delivery and targeting. *Adv. Pharm. Bull.*, 9, 195, 2019.

28. Arora, D., Sharma, N., Sharma, V., Abrol, V., Shankar, R., Jaglan, S., An update on polysaccharide-based nanomaterials for antimicrobial applications. *Appl. Microbiol. Biotechnol.*, 100, 2603, 2016.
29. Shankar, S. and Rhim, J.W., Preparation of sulfur nanoparticle-incorporated antimicrobial chitosan films. *Food Hydrocoll.*, 82, 116, 2018b.
30. Wesley, S.J., Raja, P., Raj, A.A., Tiroutchelvamae, D., Review on nanotechnology applications in food packaging and safety. *Int. J. Eng. Res.*, 3, 645, 2014.
31. Hassan, B., Chatha, S.A.S., Hussain, A.I., Zia, K.M., Akhtar, N., Recent advances on polysaccharides, lipids and protein based edible films and coatings: A review. *Int. J. Biol. Macromol.*, 109, 1095, 2018.
32. Nesic, A., Cabrera-Barjas, G., Dimitrijevic-Brankovic, S., Davidovic, S., Radovanovic, N., Delattre, C., Prospect of polysaccharide-based materials as advanced food packaging. *Molecules.*, 25, 135, 2020.
33. Abdul Khalil, H.P.S., Chong, E.W.N., Owolabi, F.A.T., Asniza, M., Tye, Y.Y., Rizal, S., Nurul Fazita, M.R., Mohamad Haafiz, M.K., Nurmiati, Z., Paridah, M.T., Enhancement of basic properties of polysaccharide-based composites with organic and inorganic fillers: A review. *J. Appl. Polym. Sci.*, 136, 47251, 2019.
34. Milovanovic, I. and Hayes, M., Marine gelatine from rest raw materials. *Appl. Sci.*, 8, 2407, 2018.
35. Junuzovic, H., Selimovic, A., Begic, S., Salkic, M., Osmic, S., Removal of lead and zinc ions from their monocomponent and two-component aqueous solutions using sodium hydroxide. *Technologica Acta*, 12, 1, 2019.
36. Said, N.S., Howell, N.K., Sarbon, N.M., A review on potential use of gelatin-based film as active and smart biodegradable films for food packaging application. *Food Rev. Int.*, 2021.
37. Ramos, M., Valdes, A., Beltran, A., Garrigos, M.C., Gelatin-based films and coatings for food packaging applications. *Coatings.*, 6, 41, 2016.
38. Qiang, X., Zhou, S., Zhang, Z., Quan, Q., Huang, D., Synergistic effect of halloysite nanotubes and glycerol on the physical properties of fish gelatin films. *Polymers.*, 10, 1258, 2018.
39. Noorbakhsh-Soltani, S.M., Zerafat, M.M., Sabbaghi, S., A comparative study of gelatin and starch-based nano-composite films modified by nanocellulose and chitosan for food packaging applications. *Carbohydr. Polym.*, 189, 48, 2018.
40. Yanat, M. and Schroen, K., Preparation methods and applications of chitosan nanoparticles; with an outlook toward reinforcement of biodegradable packaging. *React. Funct. Polym.*, 161, 104849, 2021.
41. Gautam, S., Sharma, B., Jain, P., Green natural protein isolate based composites and nanocomposites: a review. *Polym. Test.*, 99, 106626, 2021.
42. Shankar, S. and Rhim, J.W., Polymer Nanocomposites for food packaging applications, in: *Functional and Physical Properties of Polymer Nanocomposites*, A. Dasari (Ed.), pp. 29–55, John Wiley & Sons, New Jersey, 2016.

43. Suhag, R., Kumar, N., Petkoska, A.T., Upadhyay, A., Film formation and deposition methods of edible coating on food products: A review. *Food Res. Int.*, 136, 109582, 2020.
44. Drago, E., Campardelli, R., Pettinato, M., Perego, P., Innovations in smart packaging concepts for food: An extensive review. *Foods.*, 9, 1628, 2020.
45. Vilela, C., Kurek, M., Hayouka, Z., Rocker, B., Yildirim, S., Antunes, M.D.C., Freire, C.S.R., A concise guide to active agents for active food packaging. Trends in food science & technology. *Trends Food Sci. Technol.*, 80, 212, 2018.
46. Malhotra, B., Keshwani, A., Kharkwal, H., Antimicrobial food packaging: Potential and pitfalls. *Front. Microbiol.*, 6, 611, 2015.
47. Rhim, J.W., Wang, L.F., Hong, S.I., Preparation and characterization of agar/silver nanoparticles composite films with antimicrobial activity. *Food Hydrocoll.*, 33, 327, 2013.
48. Bikiaris, D.N. and Triantafyllidis, K.S., HDPE/Cu-nanofiber nanocomposites with enhanced antibacterial and oxygen barrier properties appropriate for food packaging applications. *Mater. Lett.*, 93, 1, 2013.
49. Ediyilyam, S., George, B., Shankar, S.S., Dennis, T.T., Wacławek, S., Cerník, M., Padil, V.V.T., Chitosan/Gelatin/Silver nanoparticles composites films for biodegradable food packaging applications. *Polymers.*, 13, 1680, 2021.
50. Kanmani, P. and Rhim, J.W., Physical, mechanical and antimicrobial properties of gelatin based active nanocomposite films containing AgNPs and nanoclay. *Food Hydrocoll.*, 35, 644, 2014.
51. Omerovic, N., Djisalov, M., Zivojevic, K., Antimicrobial nanoparticles and biodegradable polymer composites for active food packaging applications. *Compr. Rev. Food Sci. F.*, 20, 2428, 2021.
52. Bi, F., Zhang, X., Liu, J., Yong, H., Gao, L., Liu, J., Development of antioxidant and antimicrobial packaging films based on chitosan, D-α-tocopheryl polyethylene glycol 1000 succinate and silicon dioxide nanoparticles. *Food Packag. Shelf Life.*, 2, 100503, 2020.
53. Yu, Z., Wang, W., Kong, F., Lin, M., Mustapha, A., Cellulose nanofibril/silver nanoparticle composite as an active food packaging system and its toxicity to human colon cells. *Int. J. Biol. Macromol.*, 129, 887, 2019.
54. Wu, Z., Huang, X., Li, Y.C., Xiao, H., Wang, X., Novel chitosan films with laponite immobilized Ag nanoparticles for active food packaging. *Carbohydr. Polym.*, 199, 210, 2018.
55. Ni, S., Zhang, H., Dai, H., Xiao, H., Starch-based flexible coating for food packaging paper with exceptional hydrophobicity and antimicrobial activity. *Polym.*, 10, 1260, 2018.
56. Sani, M.A., Ehsani, A., Hashemi, M., Whey protein isolate/ cellulose nanofiber/TiO_2 nanoparticle/rosemary essential oil nanocomposite film: Its effect on microbial and sensory quality of lamb meat and growth of common food-borne pathogenic bacteria during refrigeration. *Int. J. Food Microbiol.*, 251, 8, 2017.

57. Ahmed, I., Lin, L., Zou A comprehensive review on the application of active packaging technologies to muscle foods. *Food Control.*, 82, 163, 2017.
58. Criado, P., Fraschini, C., Hossain, F., Lacroix, M., Cellulose nanocrystals in food packaging, in: *Reference Module in Food Science*, G.W. Smithers (Ed.), pp. 1–13, Elsevier, Netherlands, 2019.
59. Amorati, R., Foti, M.C., Valgimigli, L., Antioxidant activity of essential oils. *J. Agric. Food Chem.*, 61, 10835, 2013.
60. Sanches-Silva, A., Costa, D., Albuquerque, T.G., Trends in the use of natural antioxidants in active food packaging: A review. *Food Addit. Contam. Part Chem. Anal. Control Expo. Risk Assess.*, 31, 374, 2014.
61. Shankar, S. and Rhim, J.W., Amino acid mediated synthesis of silver nanoparticles and preparation of antimicrobial agar/silver nanoparticles composite films. *Carbohydr. Polym.*, 130, 353, 2015.
62. Luzi, F., Fortunati, E., Giovanale, G., Mazzaglia, A., Torre, L., Balestra, G.M., Cellulose nanocrystals from Actinidia deliciosa pruning residues combined with carvacrol in PVA_CH films with antioxidant/antimicrobial properties for packaging applications. *Int. J. Biol. Macromol.*, 104, 43, 2017.
63. Kadam, A.A., Singh, S., Gaikwad, K.K., Chitosan based antioxidant films incorporated with pine needles (*Cedrus deodara*) extract for active food packaging applications. *Food Control.*, 124, 107877, 2021.
64. Zhang, N., Bi, F., Xu, F., Yong, H., Bao, Y., Jin, C., Liu, J., Structure and functional properties of active packaging films prepared by incorporating different flavonols into chitosan based matrix. *Int. J. Biol. Macromol.*, 165, 625, 2020.
65. Lukic, I., Vulic, J., Ivanovic, J., Antioxidant activity of PLA/PCL films loaded with thymol and/or carvacrol using scCO2 for active food packaging. *Food Packag. Shelf Life.*, 26, 100578, 2020.
66. Souza, V.G.L., Rodrigues, C., Valente, S., Pimenta, C., Pires, J.R.A., Alves, M.M., Santos, C.F., Coelhoso, I.M., Fernando, A.L., Eco-friendly Zno/chitosan bionanocomposites films for packaging of fresh poultry meat. *Coatings.*, 10, 110, 2020.
67. Ahankari, S.S., Subhedar, A.R., Bhadauria, S.S., Dufresne, A., Nanocellulose in food packaging: A review. *Carbohydr. Polym.*, 255, 117479, 2021.
68. Carrizo, D., Taborda, G., Nerın, C., Bosetti, O., Extension of shelf life of two fatty foods using a new antioxidant multilayer packaging containing green tea extract. *Innov. Food, Sci, Emerg.*, 33, 534, 2016.
69. Yildirim, S., Rocker, B., Pettersen, M.K., Nilsen-Nygaard, J., Ayhan, Z., Rutkaite, R., Radusin, T., Suminska, P., Marcos, B., Coma, V., Active packaging applications for food. *Compr. Rev. Food Sci. Food Saf.*, 17, 165, 2018.
70. Cazon, P., Velazquez, G., Vazquez, M., Characterization of mechanical and barrier properties of bacterial cellulose, glycerol and polyvinyl alcohol (PVOH) composite films with eco-friendly UV-protective properties. *Food Hydrocoll.*, 99, 105323, 2020.

71. Yadav, M. and Chiu, F.C., Cellulose nanocrystals reinforced κ-carrageenan based UV resistant transparent bionanocomposite films for sustainable packaging applications. *Carbohydr. Polym.*, 211, 181, 2019.
72. Gaikwad, K.K., Singh, S., Lee, Y.S., Oxygen scavenging films in food packaging. *Environ. Chem. Lett.*, 16, 523, 2018.
73. Bajpai, V.K., Kamle, M., Shukla, S., Mahato, D.K., Chandra, P., Hwang, S.K., Kumar, P., Huh, Y.S., Han, Y.K., Prospects of using nanotechnology for food preservation, safety, and security. *J. Food. Drug Anal.*, 26, 1201, 2018.
74. Jiang, G., Zhang, M., Feng, J., Zhang, S., Wang, X., High oxygen barrier property of poly(propylene carbonate)/polyethylene glycol nanocomposites with low loading of cellulose nanocrystals. *ACS Sustain. Chem. Eng.*, 5, 11246, 2017.
75. Long, K., Cha, R., Zhang, Y., Cellulose nanocrystals as reinforcements for collagen-based casings with low gas transmission. *Cellulose.*, 25, 463, 2018.
76. Azizi, S. and Mohamad, R., Mechanical and barrier properties of kappa-carrageenan/cellulose nanocrystals bio-nanocomposite films. *IOP Conf. Ser.: Mater. Sci. Eng*, vol. 368, p. 012013, 2018.
77. Mihindukulasuriya, S.D.F. and Lim, L.T., Nanotechnology development in food packaging: A review. *Trends Food Sci. Technol.*, 40, 149, 2014.
78. Majid, I., Nayik, G.A., Dar, M.S., Nanda, V., Novel food packaging technologies: Innovations and future prospective. *J. Saudi Soc Agric. Sci.*, 17, 454, 2016.
79. Muller, P. and Schmid, M., Intelligent packaging in the food sector: A brief overview. *Foods.*, 8, 16, 2019.
80. Tsironi, T., Ronnow, P., Giannoglou, M., Taoukis, P., Developing suitable smart TTI labels to match specific monitoring requirements: The case of Vibrio spp. Growth during transportation of oysters. *Food Control.*, 73, 51, 2017.
81. Bhargava, N., Sharanagat, V.S., Mor, R.S., Kumar, K., Active and intelligent biodegradable packaging films using food and food waste-derived bioactive compounds: A review. *Trends Food Sci. Technol.*, 105, 385, 2020.
82. Rachmelia, D. and Imawan, C., Time temperature indicator label using black corn extract and chitosan matrix. *J. Phys. Conf. Ser.*, 1120, 012041, 2018.
83. Ghaani, M., Cozzolino, C.A., Castelli, G., Farris, S., An overview of the intelligent packaging technologies in the food sector. *Trends Food Sci. Technol.*, 51, 1, 2016.
84. Jayus, B.K., Oktaviana, R., Abdullah, A., Heng, L.Y., A novel on package sticker sensor based on methyl red for real-time monitoring of broiler chicken cut freshness. *Packag. Technol. Sci.*, 27, 69, 2014.
85. Heising, J.K., Van Boekel, M.A.J.S., Dekker, M., Simulations on the prediction of cod (*Gadus morhua*) freshness from an intelligent packaging sensor concept. *Food Packag. Shelf Life.*, 3, 47, 2015.

86. Ezati, P., Tajik, H., Moradi, M., Fabrication and characterization of alizarin colorimetric indicator based on cellulose-chitosan to monitor the freshness of minced beef. *Sensor Actuat B Chem.*, 285, 519, 2019.
87. Capello, C., Trevisol, T.C., Pelicioli, J., Terrazas, M.B., Monteiro, A.R., Valencia, G.A., Preparation and characterization of colorimetric indicator films based on chitosan/polyvinyl alcohol and anthocyanins from agrifood wastes. *J. Polym. Environ.*, 29, 1616, 2021.
88. Chen, M., Yan, T., Huang, J., Zhou, Y., Hu, Y., Fabrication of halochromic smart films by immobilizing red cabbage anthocyanins into chitosan/oxidized-chitin nanocrystals composites for real-time hairtail and shrimp freshness monitoring. *Int. J. Biol. Macromol.*, 179, 90, 2021.
89. Ge, Y., Li, Y., Bai, Y., Yuan, C., Wu, C., Hu, Y., Intelligent gelatin/oxidized chitin nanocrystals nanocomposite films containing black rice bran anthocyanins for fish freshness monitorings. *Int. J. Biol. Macromol.*, 155, 1296, 2020.
90. Vadivel, M., Sankarganesh, M., Raja, J.D., Rajesh, J., Mohanasundaram, D., Alagar, M., Bioactive constituents and bio-waste derived chitosan/xylan based biodegradable hybrid nanocomposite for sensitive detection of fish freshness. *Food Packag. Shelf Life.*, 22, 100384, 2019.

4

Polymer Nanocomposites as Engineered Food Packaging Materials

Tugbahan Yilmaz

Nuclear Technology and Radiation Safety, Vocational School of Technical Sciences, Konya Technical University, Konya, Turkey

Abstract

The environmental pollution concerns provoked by non-biodegradable packaging materials along with guaranteed safety demand of the consumer, food products of high quality, ready-to-eat concept, and extended shelf life have played an important role to develop eco-friendly food packaging materials using polymer nanocomposites. In addition, antioxidant and antimicrobial properties of these materials are very related to food packaging applications. Over the last decades, polymer nanocomposite materials alternative to non-biodegradable plastic packaging materials open an opportunity in terms of biodegradability, biocompatibility, non-toxicity, high performance, and cost-effective nanocomposite materials. In this chapter, an overview of the application of polymer nanocomposites used most widely in food packaging along with the properties of these materials is reviewed.

Keywords: Food packaging, polymer nanocomposite, biodegradable, mechanical properties, barrier properties, antimicrobial and antibacterial activity

4.1 Introduction

According to the United States Environmental Protection Agency data in 2018 about plastic waste, over 18% of plastics ended up in the landfill that about 26.97 million tons of plastic waste. Additionally, plastics are recycled only 4.47%, combusted 16.27% corresponding to 3.09 million tons and 5.62 million tons, respectively [1]. If the whole world is considered, then it will

Email: tgbhn.ylmz@gmail.com

Annu, Tanima Bhattacharya and Shakeel Ahmed (eds.) Nanotechnology in Intelligent Food Packaging, (79–112) © 2022 Scrivener Publishing LLC

be inevitable to reach an enormous number. Although these results include all plastic waste, they may show why food packaging is important due to the plastics and their derivatives commonly used in food packaging applications. Furthermore, the environmental pollution concerns have become alarming through the excessive use of synthetic materials, the use of fossil resources, and the recycling processes of synthetic packaging materials. As a result of these concerns triggered by these non-biodegradables (petroleum-based) packaging materials, eco-friendly, biodegradable, and sustainable packaging materials were introduced. Hence, biodegradable and sustainable packaging materials have increased in connection with the advancement of materials science and improving consumer demand. Food packaging materials are of great importance not only for high-quality food but also assured food safety. In addition, they also have the capability to prevent adverse effects such as chemical and microbial contaminants, oxygen, moisture, and light. These materials show excellent properties, including gas barrier, vapor barrier and aroma barrier against oxygen (O_2), carbon dioxide (CO_2), water vapor, and volatile compounds (such as taints and flavors) [2–7]. Figure 4.1 is presented the properties of food packaging materials.

Polymer nanocomposites are considered to be excellent biodegradable food packaging materials. There has been an increase in the use of polymer nanocomposites for applications in food packaging due to their advantages such as biodegradability, biocompatibility, non-toxicity, functionality, lightweight, high performance, and cost-effectiveness [2, 8–13].

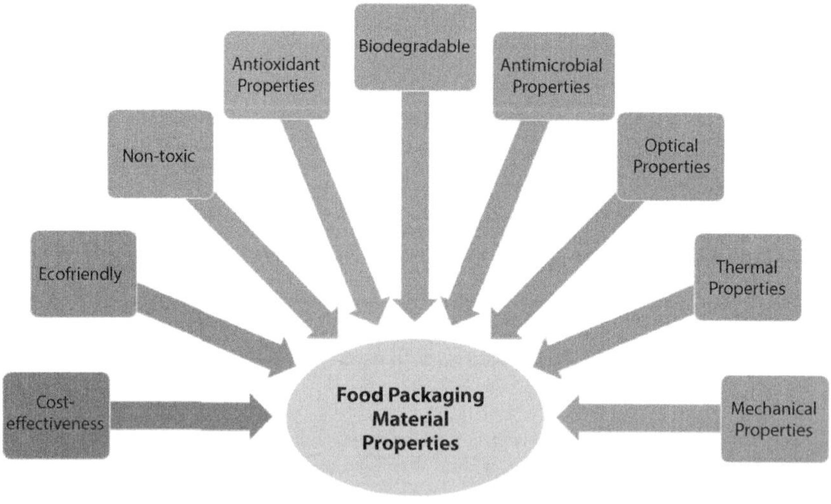

Figure 4.1 The properties of food packaging materials.

As seen in Figure 4.2a, there has been a slightly increasing attention from 2000 to 2010, whereas a dramatic increase has been observed between 2010 and 2020 in the number of publications per year. It is observed that the number of citations has increased with the increasing number of publications, as shown in Figure 4.2b (both publication- and citation-based data were collected from Web of Science with the keywords "polymer nanocomposites and food packaging"). Increasing the number of publications per year related to polymer nanocomposite materials in the application of food packaging clearly shows that the application of polymer nanocomposites has an important role as food packaging materials.

Bionanocomposite materials can be classified as polymer and ceramic. Polymer nanocomposites are classified into three main groups based on their resources and synthesis process as (i) chemically synthesized polymers, (ii) natural polymers, and (iii) produced by microorganisms, as shown in Figure 4.3 [14]. Conventionally used synthetic polymers are chemically synthesized from bioresources [e.g., polylactide acid (PLA) and poly(ethylene succinate) (PES)] and from petrochemical resources [e.g., poly(vinyl alcohol) (PVA), poly(butylene succinate) (PBS), poly(e-caprolactone) (PCL), and poly(glycolic acid) (PGA)]. Natural polymers are extracted from biomass resources from carbohydrate (e.g., agar, starch, chitosan, cellulose, and carrageenan), from plant and animal protein (e.g., collagen, corn zein, soy protein, casein, and wheat gluten gelatin), and from plant and animal lipids (e.g., fatty acids and wax). Further, using microbial polyesters [e.g., poly(b-hydroxybutyrate) (PHB) and poly(hydroxyalkanoates) (PHAs] and microbial polysaccharides (e.g., curdlan and pullulan), polymers are also produced by the fermentation process [14–17].

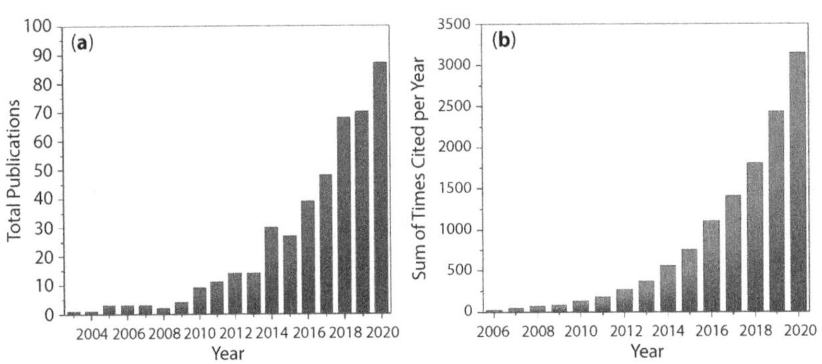

Figure 4.2 (a) Total publications and (b) sum of times cited per year on polymer nanocomposites for food packaging (search with "polymer nanocomposites and food packaging" as keywords by Web of Science on March 10, 2021).

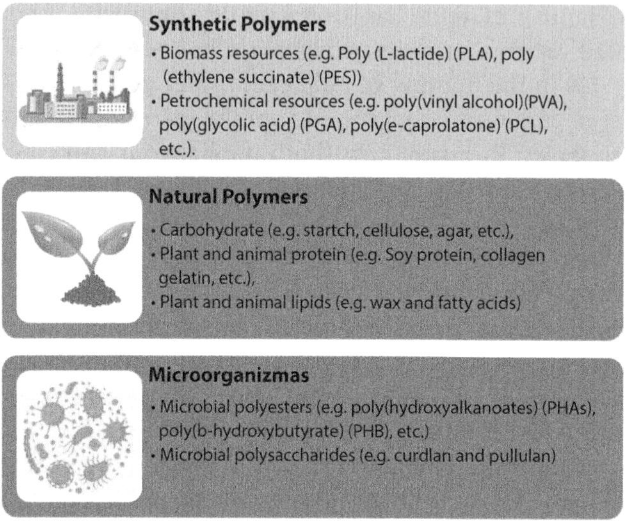

Figure 4.3 Classification of polymer nanocomposites according to their resources.

Polymer nanocomposite materials present a lack of mechanical properties, high gas and vapor permeability, and low heat distortion temperature. Nowadays, there is a number of studies based on polymer nanocomposites functionalized by nanofiller to overcome these limitations [18]. Polymer nanocomposites are basically obtained by mixing nanofiller into the polymer matrix. Using nanofiller in polymer nanocomposite is a new approach to improve antimicrobial properties, antioxidant properties, mechanical and barrier properties, thermal stability, and shelf life [19–21]. Most used as polymer nanocomposites in food packaging materials are reviewed in this chapter.

4.2 Synthetic Polymer Nanocomposites as Engineered Food Packaging Materials

Synthetic polymer nanocomposite materials including polylactic acid (PLA), polyvinyl alcohol (PVA), polycaprolactone (PCL), polyglycolic acid (PGA), and polybutylene succinate (PBS) are mostly investigated classes for food packaging materials to improve the properties of food packaging [22]. These synthetic polymer nanocomposites have been studied by various researchers for their properties such as flexibility, durability, clarity, and high gloss.

4.2.1 Polylactide Acid-Based Polymer Nanocomposites as Engineered Food Packaging Materials

PLA is a biodegradable, environmentally friendly polymer produced from biomass such as corn, wheat, straw, and sorghum [23, 24]. PLA can be utilized to develop barrier and mechanical properties for food packaging applications. In addition, PLA shows good biocompatibility, transparency, gloss, antibacterial, and water and oil resistance properties [25]. Hence, PLA has a wide range of application areas, including packaging materials, healthcare systems, and clothing [25–28]. Nowadays, PLA is used in food packaging as short shelf life bottles, blow-molded containers, and thermoformed films and food service ware [29, 30]. Using PLA in food packaging, however, has some limitations including low thermal stability, and poor barrier properties for oxygen and moisture. In this case, the addition of nanoclay has led to overcoming these limitations of PLA.

PLA-based polymer nanocomposites with montmorillonite (MMT), organomontmorillonite (OMMT), halloysite, bentonite, and cellulose nanocrystals (CNCs) show excellent barrier and mechanical properties. Many studies have been reported in the literature [31–35]. Kim *et al.* investigated the influence of clay type and MMT concentrations (Cloisite®20A, Cloisite®25A, Cloisite®30B, and Cloisite®93A) on the mechanical and barrier properties of PLA films [34]. They were coated PLA films using nanoclay/ink formulation. It was reported that the MMT/PLA composites showed not only lower oxygen and water permeability but also higher tensile strength. PLA film coated with Cloisite®30B indicated the best barrier and mechanical properties according to the other nanoclays. As shown in Figure 4.4a, the PLA film coated with Cloisite®30B presents the lowest

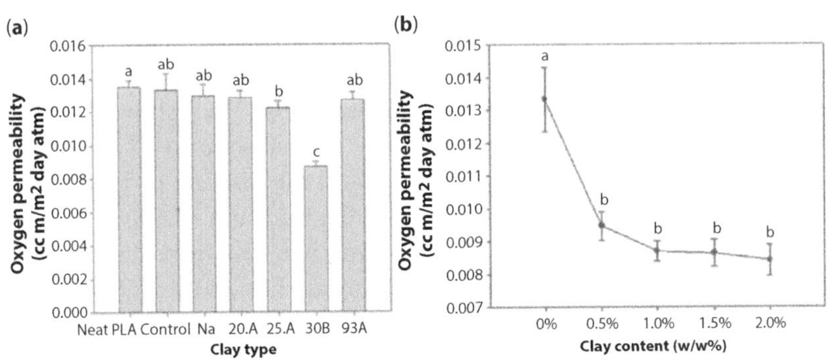

Figure 4.4 Oxygen permeability of PLA films coated by ink formulations: (a) clay type and (b) clay content. Reprinted with permission from Reference [34].

oxygen permeability compared with neat-PLA film, control film, and modified with Cloisite®20A, Cloisite®25A, and Cloisite®93A. Furthermore, the oxygen permeability of the non-modified PLA films (Cloisite Na[b]) was displayed a slightly higher value compared with modified ink with MMT (Cloisite®20A, Cloisite®25A, Cloisite®30B, and Cloisite®93A). Figure 4.4b shows the value of oxygen permeability of PLA films coated by ink with various concentration of Cloisite®30B. The value of the oxygen permeability of PLA decreased while the Cloisite®30B clay content increased (0%–2%). It can be said that the barrier and mechanical properties developed after the additive of MMT. In another work, Di Maio *et al.* investigated a commercial OMMT (Cloisite®30B) to observe the effect of Cloisite®30B on PLA nanocomposite within the range of 0.1–1 μm [36]. Compared with unfilled PLA films, the PLA films with Cloisite®30B showed a slightly rise in the surface wettability while dramatic decrease in the oxygen permeability. Nevertheless, there is no significant change in film transparency and chromatic perception for the PLA films with Cloisite®30B.

In a different work, the influence of nanoclay type on the oxygen barrier properties of PLA was investigated by Sengul and Dilsiz [37]. In order to improve the properties of the pristine PLA oxygen barrier, the PLA films containing bentonite and four different modify of OMMTs surface were utilized in this study. The value of oxygen permeability showed that the lowest value as a 13% for unmodified bentonite while dimethyl-dialkyl-tallow-amine modified OMMT reached the highest reduction value as 34%. The other three OMMTs with aminopropyl triethoxysilane, methyl dihydroxyethyl hydrogenated tallow ammonium, and distearyl dimethyl ammonium chloride also increased by 25%, 28%, and 29%, respectively. To increase the shelf life of processed meat products by enhancing their water barrier properties and lowering lipid oxidation, the PLA film incorporated with MMT-Na^+ was demonstrated by Vilarinho *et al.* [38]. Othman *et al.* studied the influence of MMT and halloysite with various concentrations (0–9 wt.%) on the oxygen barrier properties, mechanical, and transparency of PLA biopolymer [39]. They showed that the addition of 3 wt.% MMT and halloysite nanoclays into PLA polymers resulted in the optimum oxygen barrier and mechanical properties than that of pristine PLA film. Nevertheless, the addition of MMT and halloysite nanoclays (≥5 wt.%) into PLA film gives rise to reducing oxygen barrier and mechanical properties due to agglomeration of MMT and halloysite nanoclays. It was also observed that nanoclay types affected the polymer nanocomposite film, whereby PLA film incorporated with

MMT depicted a better barrier and mechanical properties than incorporated with Halloysite. In comparison to halloysite nanoclay, the structure of MMT nanoclay showed a lower fraction volume along with a higher surface area. Therefore, PLA/MMT nanocomposite showed strong interaction than PLA/halloysite nanocomposite. As seen in this study, the type of nanoclay and concentrations of nanoclay have an important role to improve the properties of polymer nanocomposite for applications in food packaging.

To investigate antibacterial properties as well as the barrier and mechanical properties of PLA with zeolites have performed by Bendahou *et al.* They demonstrated the PLA films incorporated with zeolites ($NaAlO_2$, SiO_2) in micro- and nano-scale to improve antibacterial properties, mechanical properties, water vapor permeability, and oxygen permeability [40]. In this study, using these two zeolites, antibacterial activity toward *Escherichia coli* (*E. coli*) was observed regardless of the zeolites scale. Additionally, both zeolites showed almost similar barrier properties. They also showed good mechanical properties with high tensile strength and young modulus when 1,000 g/mol for PEG (polyethylene glycol) used as a stabilizer. Moreover, PLA incorporated zeolites nanoclay was slightly decreased water vapor permeability while reduced oxygen permeability. In addition, several natural origins such as cellulose fiber CNCs, chitosan and chitin nanocrystals (ChNCs) can be used in PLA nanocomposite films as nanoclay materials to enhance the barrier and mechanical properties [41–45]. For instance, Vanitha and Kavitha demonstrated a composite film using natural cellulose fiber-based on PLA films to enhance the physical and chemical properties of food packaging film [46]. In this study, natural cellulose fiber extracted from palm sprouts depicted good interaction with PLA films. Hence, the results showed that the water absorption and mechanical properties could be enhanced after the addition of cellulose fiber in a PLA film. It can clearly say that the incorporation of cellulose fiber into PLA film can be used in food packaging applications. Another study performed by Long *et al.* [47]. They prepared composite films cellulose to incorporate with PLA. This study indicated that PLA/cellulose composite can be a good candidate for applications in food packaging on account of the enhancement of mechanical and water-barrier properties of PLA films. Kongkaoroptham *et al.* used chitosan nanoparticles in PLA films as additives to investigate antioxidant activity of chitosan NPs [48]. The results showed that the PLA composite incorporated with chitosan NPs could be used for food packaging application due to the increased antioxidant activity of chitosan NPs in PLA.

4.2.2 Polyvinyl Alcohol-Based Polymer Nanocomposites as Engineered Food Packaging Materials

PVA is biodegradability and non-toxic polymer with excellent physical properties, and chemical resistance despite its low mechanical strength [49, 50]. PVA is one of the most used polymers for applications in food packaging due to its high transparency, biodegradability, biocompatibility, and mechanical properties [51]. However, using PVA polymer in food packaging has limitations which are high water swelling, water solubility, and flammability. These limitations lead to restricting PVA film's mechanical integrity for high water swelling and water solubility during the application process. In addition, during storage and use in food applications, PVA is also accepted as a potential hazard because of its flammability [51]. In this case, reinforcing material nanoclays have been utilized to produce PVA polymer nanocomposite with good mechanical, thermal, and barrier properties [52–57].

For food packaging, oxygen permeability is one of the virtual parameters to improve the quality of food and prolong shelf life. Recently, *Ocimum tenuiflorium*-loaded chitosan and PVA-based films have been developed by Annu *et al.* for active food packaging with enhanced antioxidant potential [167]. Similarly, Li *et al.* prepared cellulose nanowhisker (CNW)- and chitosan-based PVA nanocomposite by an eco-friendly water-evaporation technique to enhance the oxygen barrier properties, mechanical properties, and antimicrobial activity [58]. It was observed that the PVA/CNW nanocomposite films' oxygen permeability had a higher oxygen permeability value than pristine PVA polymer film, while PVA/CNW/chitosan nanocomposite films had the lowest oxygen permeability value than pristine PVA polymer film, as seen in Figure 4.5a. Furthermore, increasing chitosan composition from 5% to 25% also slightly decreased the oxygen permeability of PVA/CNW/chitosan nanocomposite films.

It was proved that chitosan has a good oxygen barrier property. The results also presented that PVA/CNW/chitosan nanocomposite can be utilized as a natural packaging material to prevent oxidation of food. In terms of elongation at break, PVA/chitosan-15% composites had a lowest than pristine PVA films as shown in Figure 4.5b. According to elongation at break results, it is seen that adding chitosan into PVA-based nanocomposite reduces the flexibility of PVA film. The tensile strength and elongation at break of PVA/CNWs nanocomposite films increased compare to pristine PVA film. For these nanocomposites, the mechanical properties could be related to the increased interaction of PVA/CNWs. Compared with PVA/chitosan nanocomposite, PVA/CNW/chitosan nanocomposites showed a

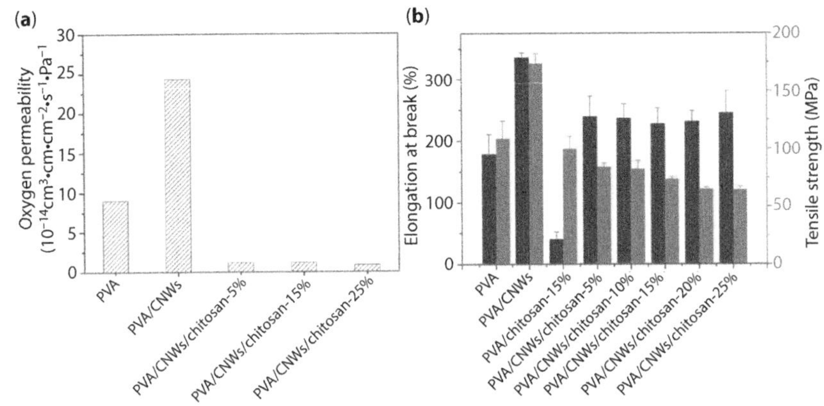

Figure 4.5 (a) Oxygen permeability and (b) elongation at break (black) and tensile strength (red) of cellulose nanowhisker (CNW)– and chitosan-based PVA nanocomposite. Reprinted with permission from Reference [58].

dramatic enhancement in flexibility. PVA/CNW/chitosan nanocomposites depicted a little change with the addition of chitosan for the elongation at break while a slight reduction for the tensile strength. Furthermore, the antimicrobial property was performed *toward E. coli* and *Staphylococcus aureus* (*S. aureus*) for PVA/CNW/chitosan nanocomposites. In terms of the antimicrobial property, there is no inhibition of *E. coli* and *S. aureus* for bare PVA and PVA/CNWs films, whereas PVA/chitosan and PVA/CNW/chitosan nanocomposite films showed the antimicrobial properties. It was observed that as the chitosan content increased, the antimicrobial properties of PVA/chitosan and PVA/CNW/chitosan also increased.

Sarwar *et al.* prepared a PVA-based nanocomposite film with nanocellulose (NC) and silver (Ag) nanocomposite films to observe antibacterial activity for food packaging [59]. PVA composite reinforced with NC showed increased tensile strength from 5.52 ± 0.27 MPa to 12.32 ± 0.61 MPa. PVA/NC/Ag nanocomposite films displayed antibacterial activity toward *E. coli* and *S. aureus*. This study clearly showed that increasing NC concentration in PVA films improves antimicrobial activity. Ben Shalom *et al.* prepared PVA-based CNC nanocomposite to enhancement mechanical properties of PVA/CNC nanocomposite film. In this study, they reported that the water absorption of PVA/CNC nanocomposite and cross-linked CNC films is dramatically reduced. However, PVA/CNC nanocomposite and cross-linked CNC films are remarkably enhanced as a cross-linker concentration and PVA function [60]. In addition to cellulose and chitosan, a number of studies have been reported PVA-based nanocomposite reinforced with starch [61–63], chitin [64, 65], and MMT [66] to improve barrier and mechanical properties and antimicrobial activity for applications in food packaging. For instance, Zhang *et al.* demonstrated a PVA/liquefied chitin composite in order to improve mechanical and barrier properties of blend films [65]. As a result, the PVA/liquefied chitin composite showed a good food preservation property. On the other hand, MMT incorporated with PVA composite films was demonstrated by Wu *et al.* to enhance thermal stability and mechanical properties [66]. Their result indicated that the PVA-based MMT composite can be a good candidate for food packaging applications.

4.2.3 Other Polymer Nanocomposites as Engineered Food Packaging Materials

There is a lot of research on polymer nanocomposites incorporated with synthetic polymers such as polyhydroxybutyrate (PHB), polyethylene terephthalate (PET), polypropylene (PP), PCL, PGA, and PBS for food packaging applications [67]. These polymeric materials provide a long shelf life, safety and quality [68, 69]. As mentioned in Section 4.2.1, there are lacking due to mechanical properties, barrier properties, and thermal properties of biopolymers. In order to come through these limitations, a number of nanoclays are utilized to form synthetic polymeric nanocomposites.

4.2.3.1 Polyhydroxybutyrate-Based Polymer Nanocomposites in Food Packaging

PHB is a water-insoluble polymer with biodegradability, biocompatibility permeability, processability, and high crystallinity [70]. PHB is a useful polymer for drug delivery application, tissue engineering, and food packaging application due to its biodegradability and biocompatibility [71]. Recently, Manikandan *et al.* reported the influence of PHB nanocomposite reinforced with graphene nano-platelets (Gr-NPs) on thermal stability, melting point, tensile strength, barrier properties, and shelf life [72]. The PHB nanocomposite incorporating different concentrations of Gr-NPs (0–1.3 wt.%) was developed using casting solution method. Since there is no agglomeration, the concentration of 0.7 wt.% Gr-NPs in PHB nanocomposite was considered to be ideal. Compared to the bare PHB, PHB reinforced Gr-NPs nanocomposite indicated tensile strength (by two times), water vapor permeability (by two times), oxygen reduction (by three times), thermal stability (by 10°C), and a higher melting point (by 10°C).

Furthermore, they were also showed that the shelf life improved four-fold using oxygen- and moisture-sensitive food items by a simulation study. Zobo *et al.* prepared PHB-based nanocomposite incorporating cellulose nanofibers (CNFs) and CNCs with different content (1, 3, and 5 wt.%) and investigated their mechanical and barrier properties [73]. Compared with pure PHB, PHB composites reinforced with CNCs showed increased elongation at break by 91.2% and Young's modulus by 18.4% while decreased tensile strength by 3.5%. It was found that the elongation at break of PHB/CNCs nanocomposite first increased with the increased CNCs content, then decreased. As increased CNCs content, tensile strength decreased while Young's modulus increased. PHB nanocomposite with CNFs is similar to PHB/CNC nanocomposite.

4.2.3.2 Polyethylene Terephthalate-Based Polymer Nanocomposites in Food Packaging

PET, a semicrystalline thermoplastic polyester, is commonly used in beverage bottles and other containers for food packaging due to its high resistance to carbon dioxide permeation, oxygen, water vapor, high clarity, and mechanical properties [74–77]. Lim *et al.* demonstrated PET-based nanocomposite coated with polyacrylic acid (PAA) and various compositions of montmorillonite (0, 1, 3, and 5 wt.%) and tested barrier properties of PET/PAA/MMT nanocomposite. It was reported that PET barrier property showed the lowest oxygen barrier and water vapor properties after PAA coating. Further, PET/PAA oxygen permeability decreased after coating with MMT of 1 wt.% and 3 wt.%, thereby PET/PAA/MMT nanocomposite is a good candidate for the application of food packaging, especially oxygen-sensitive food items [78]. In a recent study, Meng *et al.* depicted the improvement of the oxygen barrier and resistance of PET films coated with MMT/laponite (LAP)/PVA [79]. Kim *et al.* demonstrated a CNF-coated PET film in order to improve oxygen permeability and stability [80]. The results showed PET films coated with cellulose improved performance regarding mechanical, oxygen barrier, and water-resistant properties.

4.2.3.3 Polypropylene-Based Polymer Nanocomposites in Food Packaging

PP, as other thermoplastics which is mostly used, is preferred due to its mechanical properties, easy processability, and low cost [81]. For instance, Kalaj *et al.* demonstrated PP nanocomposite incorporated with various concentrations of OMMT (1, 2, and 4 wt.%) modified with different iron NPs content (0.05, 0.1, and 0.2 wt.%) to investigated their mechanical, morphological, thermal, and gas barrier properties [82]. In comparison with PP films, the property of water vapor and oxygen barrier of PP significantly reduced after 2 wt.% addition of OMMT, 33% and 22%, respectively. Moreover, after 0.2 wt.% addition of Fe-NPs in PP/OMMT nanocomposite, water vapor permeability and oxygen permeability also reduced by 77% and 55%, respectively. d'Eon *et al.* prepared PP-based nanocomposite with CNC to examined its mechanical and adhesion properties [83]. The oxygen permeability was tested 0.89 Barrers for PP film. Then, oxygen permeability of PP film increased by 12% a rate of 0.98 Barrers after exposed UV-ozone. It was found that, after UV ozone-exposed PP film after coating with CNC, oxygen permeability reduced by 39% a rate of from 0.98 to 0.53 Barrers. These results clearly showed that PP/CNC nanocomposite can be used in food packaging applications. A comparison study related to PP using an adhesive layer with polyurethane (PU) incorporating CNC and CNF was carried out by Wang *et al.* [84]. They prepared a multilayer film with the film structure of BOPP/PU/CNMs/PU/BOPP. Average tensile strength of PP/CNC20 and PP/CNC50 films was around 60 MPa, while PP/CNF20 and PP/CNF50 films were around 40 MPa and 110 MPa, respectively. These results could be related to the density of CNC and CNF. CNF films are softer than CNC films because of a lower crystallinity and density. The water vapor transmission rate of CNC and oxygen transmission of CNC reduced from 126 $cm^3/(m^2{\cdot}day)$ to 6.1 $cm^3/(m^2{\cdot}day)$ and from 516 $g/(m^2{\cdot}day)$ to 1.0 $g/(m^2{\cdot}day)$, respectively. Like CNC, the water vapor transmission rate of CNF and the oxygen transmission reduced. Zhang *et al.* reviewed the use of CNCs and CNFs to improve food packaging parameters, including mechanical, optical, barrier, antimicrobial, and antioxidant properties [85]. In addition, many research groups have been reported the life cycle assessment (LCA) of PET, polystyrene, and PLA, which is important for global warming impact and eco-efficiency [86–89].

4.2.3.4 Polycaprolactone-Based Polymer Nanocomposites in Food Packaging

PCL is another synthetic biodegradability polymer, which is commonly used in drug delivery applications, tissue engineering, and food applications [90–95]. Cesur *et al.* demonstrated PCL nanocomposite with various concentration of clay (0.4 and 3 wt.%), chitosan (25, 50, and 75 wt.%), and oleic acid or glycerol monooleate (5, 10, 20, and 30 wt.%) were used as a plastifer as well [91]. Prepared films developed to investigate biodegradable properties and antibacterial activity against *Bacillus cereus* (*B. cereus*), *E. coli*, *Candida albicans* (*C. albicans*), and *Pseudomonas aeruginosa* (*P. aeruginosa*). The results indicated that all PCL nanocomposite with and without chitosan were examined in point of antimicrobial activity, whereas there is no antimicrobial activity for neat PCL. However, none of PCL nanocomposite reinforced with clay, chitosan, and oleic acid or glycerol presented any antimicrobial activity against *B. cereus*. Besides, the PCL/chitosan nanocomposite biodegraded in 5 months; thus, they showed that their study associated with ASTM 6400. Their study also showed that increased chitosan content in PCL nanocomposite caused decreasing the biodegradation time of nanocomposite. Additionally, all PCL nanocomposite with chitosan indicated poor mechanical properties as well as neat chitosan in this study. Uzunlu and Niranjan demonstrated PCL nanocomposite with cinnamaldehyde (CNMA) and pomegranate to be used for antimicrobial activity toward *E. coli* and *S. aureus* [96]. Pomegranate was used in three different forms as pomegranate methanolic extract (PME), seed flour of pomegranate (SF), and freeze-dried arils of pomegranate (FDAP). Reinforced materials, CNMA, and pomegranate were used in various compositions (5 and 10 wt.%), as well. The antimicrobial activity tests were carried out under different temperature values (4°C, 20°C, and 37°C) and time (24 hours, 11 days, and 28 days). Their result indicated that the PCL nanocomposite with CNMA, and pomegranate showed antimicrobial activity. Thus, CNMA-based films could be used in cinnamon ingredients items, so its smell does not adversely affect the flavor of cinnamon ingredients items. Additionally, this study showed that pomegranate active compounds are considered that used in foods could be beneficial. Another study was performed by Correa *et al.* to figure out the effect of PHB/PCL nanocomposite on shelf life for food packaging [97]. In this study, PHB/PCL nanocomposites were prepared with organoclay materials, Cloisite®30B and Cloisite®10A to test tensile strength, water vapor permeability, and elongation at break. Even though prepared nanocomposites with Cloisite®30B and Cloisite®10A showed activity against microorganisms, they did not show antimicrobial activity. Therefore, prepared nanocomposites reinforced with nisin to analyzed antimicrobial activity. Their results showed that the addition of organoclays degradation temperature and increased barrier properties. PHB/PCL nanocomposite reinforced nisin depicted antimicrobial activity against *Lactobacillus plantarum* (*L. plantarum*, processed meat used as a spoilage bacteria model); thus, it has shown that the PHB/PCL nanocomposite reinforced nisin contributes to extending the shelf life.

4.3 Natural Polymer Nanocomposites as Engineered Food Packaging Materials

The priority is considered as food safety in food packaging applications. The food packaging factors are to be considered oxygen, temperature, moisture, humidity, carbon dioxide concentration, antibacterial and antimicrobial activity, as well as product attributes [98]. Natural polymers including cellulose, chitosan, starch, and their derivatives have been widely studied due to their biodegradability, non-toxic, low cost, and water soluble [99–102]. In this section, a comprehensive review of state-of-the-art natural origin polymer nanocomposite materials for food packaging applications is explained with some recent examples highlighting.

4.3.1 Cellulose-Based Polymers Nanocomposite as Engineered Food Packaging Materials

Cellulose is the most plenty polymer on the earth and shows excellent properties such as biodegradability, flexibility, non-toxic, low cost, hydrophilicity, water soluble, and high transparency [103, 104]. Cellulose is one of the valuable materials due to its excellent properties in food packaging application, pharmaceutical and cosmetic industries, and biomedical applications [105–107]. The use of cellulose in the composite can enhance the thermal and mechanical properties, whereas it can reduce water vapor permeability [108]. Cellulose can provide a good compatibility after formed as a nanocomposite. Hence, cellulose-based nanocomposites are considered to food packaging materials compared to petroleum-based food packaging materials. Cellulose is obtained from not only plants such as sugarcane, corn stalks, sorghum bagasse oats but also bacteria [109–112]. Cellulose could be used in used in three main forms as reinforcement in nanocomposites: (i) CNCs, (ii) CNFs, and (iii) CNWs [113]. Now that it would be difficult to explain both of three forms, recent studies are reviewed in this part.

Gonzales *et al.* prepared soy protein films reinforced with different CNF content (10, 20, and 40 wt.%) and investigated their mechanical properties [108]. Soy protein reinforced with CNF depicted after the addition of CNF elongation at break decreased while tensile strength increased from 1.2 to 4.8 MPa. The water vapor permeability of soy protein reinforced with CNF indicated a slightly decrease after the addition of CNF. Yu *et al.* developed PLA nanocomposite with NCs to improve mechanical and barrier properties [114]. In this study, the different shape of NC was used as cellulose nanospheres (CNSs), CNFs, and rod-like CNCs with various contents. Bare PLA, by contrast, elongation at break and tensile strength of PLA/CNS nanocomposites were increased by 140% and 130%, respectively. Moreover, tensile strength increased by 260% for PLA/CNF, and 210% for PLA/CNC, similarly Young's modulus increased by 350% for PLA/CNF, and 250% for PLA/CNC. It was showed that cellulose-based nanocomposite attributed to improving the mechanical properties of PLA films.

Cellulose-based nanocomposites with different polymers have been studied to examine antimicrobial activity and antibacterial activity as well as mechanical and barrier properties in food packaging [102, 115–120]. Fortunati *et al.* reported several studies to investigate the antibacterial activity of cellulose-based nanocomposite [118–120]. They prepared PVA-based cellulose and Ag NPs nanocomposite and analyzed their antimicrobial activity to *E. coli* and *S. aureus* [118]. They showed that *E. coli had greater* antibacterial activity than *S. aureus* regardless of incubation time and temperature in this study. Another study was also performed by Fortunati *et al.* with similar nanocomposite film through various contents of CNC and Ag apart from the previous works [120]. Besides, both studies presented similar results related to antimicrobial activity to *E. coli* and *S. aureus.* As a result, the nanocomposite of Ag NPs with CNCs in PLA polymers is promising a high potential to enhance the antibacterial activity for food packaging applications. In another study, they prepared PLA-based nanocomposite reinforced with surfactant-modified CNS (s-CNC) and zinc oxide nanoparticles (ZnO NPs) at various concentration values (at different concentrations, 0.1 and 0.5 wt.%) [119]. Increasing the content of ZnO NPs in PLA, the elongation at break showed decreased while stress at break increased. The elongation at break was monitored by PLA/s-CNC with ZnO NPs. In addition, *E. coli* and *Listeria innocua* (*L. innocua*) were also studied to analyze their antibacterial activity. However, there was no antibacterial activity against both bacteria at lower ZnO NPs content. These results were also supported by the literature [121]. Pantani *et al.* indicated that the PLA/ZnO NPs nanocomposite was not able to reduce the growth of bacteria, *E. coli*, and *S. aureus*. Fortunati *et al.* claimed the lack of antibacterial activity of ZnO NPs because of a lower Zn ions diffusion in PLA composite than that are supported by Shalumon *et al.* using ZnO NPs in a PVA instead of PLA. The antimicrobial activity was observed toward *E. coli* and *S. aureus* in this study [122]. Although ZnO NPs with CNCs in PLA have not shown antimicrobial activity, they can be used in food packaging because of their mechanical properties.

4.3.2 Chitosan-Based Polymer Nanocomposite as Engineered Food Packaging Materials

Chitosan, derivative from chitin, is one of the most used in applications for food packaging due to its non-toxic, biodegradable, biocompatible, and bio functionality properties [123]. Chitosan shows good barrier and mechanical properties aside from antimicrobial activity against an extensive range of yeast, foodborne filamentous fungi, and gram-negative and gram-positive bacteria [124–126]. Despite poor properties such as reactivity, solubility brittleness, and rigidity [127], chitosan can straight apply as a food packaging material. Thanks to its functional properties, it can be used in different forms such as nanofiber, nano- and micro- particle, and gel [128]. The mechanical properties of chitosan allow predictions of film behavior from shipping to packaged food storage. In order to obtain and maintain a high-quality food product, barrier properties have a critical role to ensure them [125].

Pandey *et al.* demonstrated chitosan-based PVA film in order to enhance antimicrobial activity toward *E. coli and Listeria monocytogenes* (*L. monocytogenes*) [101]. This study showed an effective antimicrobial activity of these bacteria species. Moreover, in this study, the addition of AgNPs in Chitosan/PVA composite showed significant antimicrobial activity against the tested bacteria species. Zhang *et al.* investigated the antimicrobial activity of chitosan-based titanium oxide (TiO_2) nanopowder nanocomposite film as food packaging material against bacteria, fungi, and molds under UV light [129]. They reported that the addition of TiO_2 nanopowder in chitosan film not only improved mechanical properties but also decreased light transmittance in the visible region of the chitosan film. Additionally, chitosan reinforced with TiO_2 nanocomposite showed antimicrobial activity against *Aspergillus niger (A. niger), C. albicans, E. coli*, and *S. aureus,* which are food-borne microbes after antimicrobial treatment for 12 hours. This study clearly showed that chitosan reinforced with TiO_2 nanocomposite could attribute shelf life due to its antimicrobial activity. Qin *et al.* prepared chitosan-based AgNPs nanocomposite films and studied their mechanical properties and antimicrobial activity against *E. coli, L. monocytogenes, Salmonella,* and *S. aureus*, which are food-borne microbes, as well [130]. Besides, the mechanical properties of chitosan film improved, water solubility, and water vapor permeability decreased after the addition of AgNPs. Moreover, AgNPs added chitosan film showed a higher antimicrobial property against food-borne microbes because of the interaction between AgNPs and enzymes, membrane protein, and nucleic acids. Recently, many studies have focused on the antimicrobial activity of copper oxide nanoparticles (CuO NPs) due to the low content of chitosan nanocomposite [131, 132]. Despite low content in chitosan film, it shows antimicrobial activity. However, the activity mechanism of CuO NPs is not figured out clearly. In order to understand this mechanism, a few studies were performed. Almasi *et al.* examined the influence of copper oxide nanoparticles (CuO NPs) on the antibacterial activity of chitosan nanofibers (CHNFs) films [133]. The antibacterial activity of CNF films reinforced with CuO NPs was observed against *E. coli*, *P. aeruginosa*, and *L. monocytogenes*. Furthermore, this study helped to understand the activity mechanism of CuO NPs. This mechanism could be explained with three different ways, which are (1) CuO NPs attached to the cell membrane, then interaction between CuO NPs and bacterial membrane, (2) copper ions leached, and then copper-peptide complex formed caused the integrity of the bacterial membrane to be disrupted (3) catalysis of reactive oxygen species and increased its production significantly led to cell death. As a result, chitosan, regardless of form, is a promising polymer for applications in food packaging. Further, CuO NPs can be used in food packaging due to their antimicrobial activity.

On the other hand, Riaz *et al.* investigated the antimicrobial properties of chitosan nanocomposite with different concentrations of apple peel polyphenols (APP) [134]. The water vapor permeability and moisture of film increased after the addition of APP into chitosan. It can be said that the barrier properties of the film are improved, as the water vapor permeability and moisture must be as low as possible to extend its shelf life. In addition, the incorporation of APP into chitosan films showed antimicrobial activity against *B. cereus*, *E. coli*, *S. aureus*, and *Salmonella typhimurium* (*S. typhimurium*). de Moraes Crizel *et al.* studied that the influence of the presence of microparticle and flour of olive pomace (as microparticle and flour) in chitosan-based bionanocomposites as food packaging films [135]. Although tensile strength increased by 7.58 MPa and 22.40 MPa after the addition of 10% of olive flour and 10% of microparticles, respectively, the chitosan incorporated with 10% of olive microparticles showed a better mechanical property than with 10% of olive flour. Furthermore, both olive forms improved the antioxidant capability of nanocomposite films. The chitosan films incorporated with both 30% of olive flour and 30% of microparticles indicated that the antioxidant activity against walnut oxidation. It was indicated that the prepared chitosan/olive pomace films showed an improved antioxidant effect on walnuts oxidation, which presented the potential use of olive pomace to extend shelf life in food packaging.

In recent years, chitosan-based composites incorporated with MMT nanoclay are widely considered to enhance the barrier, mechanical, and optical properties of chitosan films [136]. Qui *et al.* investigated that the effect of MMT with different concentration values (1, 3, and 5 wt.%) into chitosan films on mechanical and barrier properties of the chitosan films. The findings revealed that MMT enhanced the water vapor permeability and mechanical properties of the chitosan nanocomposite. Thus, they suggested that MMT could be utilized in food packaging applications [137]. Another study was performed by Pires *et al.* that the effect of two different MMTs (Cloisite®Na^+ and Cloisite®Ca^{+2}) on mechanical, barrier properties, antioxidant, and antimicrobial of chitosan nanocomposite. They indicated that the antioxidant and antimicrobial properties enhanced after chitosan reinforced with MMT, thereby significantly important for the extent of the poultry meat shelf life. The mechanical and barrier properties also improved after introducing MMT. Additionally, they also studied that the incorporated essential oils (rosemary and ginger) into chitosan/MMT nanocomposite to observe the role of essential oils on the chitosan/MMT film. The results showed only enhanced barrier properties after reinforced with essential oil. However, incorporated with essential oils showed no antimicrobial activity for chitosan nanocomposite even if the oxidation process of poultry meat reduced [138]. A similar study to this study was carried out by Souza *et al.* [139]. They developed a chitosan/MMT bionanocomposite film reinforced with ginger essential oil and tested its mechanical and barrier properties. They proved that MMT had presented the potential use to prolong shelf life in food packaging. Many studies have been performed to improve mechanical properties and antibacterial activity related to MMT-supported nanoparticles such as AgNPs and CuO NPs. Lavorgna *et al.* reported chitosan-based nanocomposite incorporated MMT/AgNPs and investigated its antibacterial activity [140]. They displayed that the chitosan/MMT films showed a better improvement after supporting with AgNPs due to the interaction of the chitosan-Ag ions along with chitosan penetration into MMT silicate galleries. The results also presented that the antibacterial activity against *Pseudomonas fluorescens* (*P. fluorescens*) and *Pseudomonas putida (P. putida).* In the study performed by Nouri *et al.* developed, it was indicated that the addition of CuO NPs into chitosan nanocomposite *supporting* reinforced with MMT enhanced antibacterial activity [141]. Their study presented antibacterial activity against *E. coli* and *P. aeruginosa* as gram-negative bacteria *and S. aureus* and *B. cereus* as gram-positive bacteria. Besides, compared with the control film, mechanical and barrier properties were improved.

4.3.3 Other Natural Polymer Nanocomposites as Engineered Food Packaging Materials

As mentioned in the introduction part and shown in Figure 4.3, it is known that there are many natural origin polymers on earth. In addition to mentioned natural polymers in this section, the other natural polymers such as starch, agar, gelatin, collagen, corn zein, and algae could be used in food packaging applications [142–145]. Some studies of gelatin-, starch-, and algae-based nanocomposites will be exemplified in food packaging applications.

4.3.3.1 Gelatin-Based Natural Polymer Nanocomposites in Food Packaging

Gelatin, which is mostly found in animal bones, tendon, and skin, is a protein that is also biodegradable, bifunctionality, abundance, and low cost. Gelation also presents good mechanical and barrier properties [146]. Because of these properties, gelatin has been preferred to form food packaging films [143, 147, 148]. Many researchers reported the gelatin-based nanocomposites to enhance the barrier and mechanical properties and along with the antimicrobial and antioxidant activity of these films [143, 149–151]. Roy and Rhim, for instance, prepared gelatin nanocomposite incorporated with curcumin to analyze its antimicrobial and antioxidant activity [150]. Results clearly showed that the gelatin-based films have strong antimicrobial and antioxidant activity. In a recent study, Liu *et al.* reported the sealable soybean polysaccharide (SSPS)/gelatin blend films to evaluate their mechanical properties for food packaging applications. The findings clearly showed that the addition of gelatin into SSPS films significantly increased the thermal stability, tensile strength, and stretchability while decreased water solubility and permeability of films [152]. Another study was performed by Najwa *et al.* [153]. In this study, they prepared gelatin-based silver (Ag)–kaolin (Kln) composite film to enhance the barrier properties and the antimicrobial properties for application in food packaging. The results showed gelatin reinforced Ag-Kln composite improved barrier properties. Further, this study indicated the antimicrobial activity against *E. coli*, *L. monocytogenes*, *S. aureus*, and *S. typhimurium*. In conclusion, gelatin is one of the reliable materials as food packaging to extend shelf life and ensure food safety.

4.3.3.2 *Starch Natural Polymer Nanocomposites in Food Packaging*

Starch, one of the most used polymers, has been investigated in food packaging due to its biodegradable, non-toxic, low cost, and accessibility properties [154–156]. Starch-based films show poor mechanical properties because of hydrophilicity [154]. Further, these films also exhibit poor thermal properties, wherefore low decomposition temperature and high melting point [157]. In order to avoid these limitations, it is necessary to incorporated starch with nanofillers or other polymers; thereby, mechanical properties have improved to be utilized for food packaging [145, 158, 159]. For instance, Mendes *et al.* reported that the mechanical properties of starch film improved after reinforced lemongrass essential oil [160]. Another study was performed by Diaz-Galindo [161] to the enhancement of mechanical properties of starch film using a grape cane. Besides, improving mechanical properties of starch films, both studies showed antimicrobial activity to *S. aureus* and antifungal activity to *Botrytis cinerea* (*B. cinerea*). On the other hand, many studies have been reported on the antimicrobial activity of starch films incorporated with reduced GO, essential oils, and chitosan NPs as food package material [162–165]. For instance, Shapi'i *et al.* reported that the antimicrobial activity of starch/CNPs nanocomposite films improved. It was also reported that the presence of CNPs significantly important the antimicrobial activity of composites [163]. In a recent study, it was reported that the starch-based films reinforced with phytic acid and chicory root extract to test antibacterial properties of starch films were developed by Jaśkiewicz *et al.* [162]. In conclusion, it can clearly say that starch-based nanocomposite films are promising materials for food packaging.

4.3.3.3 *Algae Natural Polymer Nanocomposites in Food Packaging*

Algae, as a natural polymer, are mostly available all over the earth. They could be used in many different application areas due to their bioactive content, and rich chemical composition. Due to their properties of stabilizing, thickening, and gelling, different products such as carrageenan, agar, and alginate could obtain from it. In the food industry, algae can be used as not only food supplements but also food packaging material [142, 144]. In terms of food packaging, algae show poor moisture property because of high hydrophilicity, which led to high water vapor transmission rate. Thereby, the incorporation of MMT nanoclay, MMT/cellulose NPs, glycerol, sorbitol, and so on, into biopolymer matrix provides to overcome this limitation [166] For instance, Jost *et al.* reported that the barrier properties of algae enhanced after the addition of glycerol. In this study, they also showed that compared to pure alginate, alginate-based nanocomposite with sorbitol did not improve the barrier properties [168]. In addition, the incorporation of antimicrobial and antioxidant agents such as essential oils, nisin, acetic acid, lactic acid, and extracts (pomegranate, ginseng, etc.) provides good antioxidant or antimicrobial activity [166].

4.4 Conclusions

The environment-friendly nanocomposites that are obtained from synthetic process, plant, or animal sources have brought many improvements in food packaging applications via enhanced mechanical and barrier properties, antimicrobial and antibacterial activity, and prolong shelf life. Through decreasing the usage of petroleum-based packaging materials, these novel materials not only find a new way to reducing of environmental pollution concerns but also give a solution economically. Furthermore, these materials are promising to be fulfilled consumers demand for assured safety, high-quality food products, ready-to-eat concept, and extended shelf life.

References

1. EPA, National Overview: Facts and Figures on Materials, Wastes and Recycling, https://www.epa.gov/facts-and-figures-about-materials-waste-and-recycling/national-overview-facts-and-figures-materials (Accessed date 25.02.2022).
2. Sarfraz, J., Gulin-Sarfraz, T., Nilsen-Nygaard, J., Pettersen, M.K., Nanocomposites for Food Packaging Applications: An Overview. *Nanomaterials*, 11, 10, 2021.
3. Singh, R. and Anderson, B.A., The major types of food spoilage: An overview, in: *Understanding and Measuring the Shelf-Life of Food*, R. Steele (Ed.), pp. 3–23, Woodhead Publishing, Cambridge, 2004.
4. Singh, R. and Singh, N., Quality of packaged foods, in: *Innovations in Food Packaging*, J.H. Han (Ed.), pp. 22–24, Academic Press, Cambridge, 2005.
5. Mohanty, F., Swain, S.K., Oprea, A.E., Bionanocomposites for Food Packaging Applications, in: *Nanotechnology Applications in Food*, A.M. Grumezescu (Ed.), pp. 363–379, Academic Press, Cambridge, 2017.
6. Suppakul, P., Miltz, J., Sonneveld, K., Bigger, S.W., Active Packaging Technologies with an Emphasis on Antimicrobial Packaging and its Applications. *J. Food Sci.*, 68, 408, 2003.

7. Brown, H., Williams, J., Kirwan, M., Packaged Product Quality and Shelf Life, in: *Food and Beverage Packaging Technology*, R. Coles and M. Kirwan (Eds.), pp. 59–83, Blackwell Publishing, New Jersey, 2011.
8. Cui, Y., Kumar, S., Rao Kona, B., van Houcke, D., Gas barrier properties of polymer/clay nanocomposites. *RSC Adv.*, 5, 63669, 2015.
9. Enescu, D., Cerqueira, M.A., Fucinos, P., Pastrana, L.M., Recent advances and challenges on applications of nanotechnology in food packaging. A literature review. *Food Chem. Toxicol.*, 134, 110814, 2019.
10. Jamróz, E., Kulawik, P., Kopel, P., The Effect of Nanofillers on the Functional Properties of Biopolymer-Based Films: A Review. *Polymers*, 11, 675, 2019.
11. Müller, K., Bugnicourt, E., Latorre, M., Jorda, M., Echegoyen Sanz, Y., Lagaron, J.M., Miesbauer, O., Bianchin, A., Hankin, S., Bölz, U., Pérez, G., Jesdinszki, M., Lindner, M., Scheuerer, Z., Castelló, S., Schmid, M., Review on the Processing and Properties of Polymer Nanocomposites and Nanocoatings and Their Applications in the Packaging, Automotive and Solar Energy Fields. *Nanomaterials*, 7, 74, 2017.
12. Tan, B. and Thomas, N.L., A review of the water barrier properties of polymer/clay and polymer/graphene nanocomposites. *J. Membr. Sci.*, 514, 595, 2016.
13. Vasile, C., Polymeric Nanocomposites and Nanocoatings for Food Packaging: A Review. *Materials*, 11, 1834, 2018.
14. Abdullah, Z.W. and Dong, Y., *Polyvinyl Alcohol/Halloysite Nanotube Bionanocomposites as Biodegradable Packaging Materials*, pp. 1–50, Springer, Singapore, 2020.
15. Sablani S.S., Polymer nanocomposites for food packaging applications, in: *Food Nanoscience and Nanotechnology*, Hernández-Sánchez H., Gutiérrez-López G. (eds), pp. 205–212, Springer, Cham, 2015.
16. Bordes, P., Pollet, E., Avérous, L., Nano-biocomposites: Biodegradable polyester/nanoclay systems. *Prog. Poly. Sci.*, 34, 125, 2009.
17. Anne-Marie, C. and Halleux, J., Classification of biodegradable polymers, in: *Biodegradable Polymers for Industrial Applications*, R. Smith (Ed.), pp. 3–31, Woodhead Publishing, Cambridge, 2005.
18. Martelli-Tosi, M., Esposto, B.S., Cristina da Silva, N., Tapia-Blácido, D.R., Jafari, S.M., Reinforced nanocomposites for food packaging, in: *Handbook of Food Nanotechnology*, S.M. Jafari (Ed.), pp. 533–574, Academic Press, Cambridge, 2020.
19. Díez-Pascual, A.M., Biodegradable food packaging nanocomposites based on ZnO-reinforced polyhydroxyalkanoates, in: *Food Packaging, A.M. Grumezescu*, pp. 185–221, Academic Press, Cambridge, 2017.
20. Nesic, A.R. and Seslija, S.I., The influence of nanofillers on physical–chemical properties of polysaccharide-based film intended for food packaging, in: *Food Packaging*, A.M. Grumezescu (Ed.), pp. 637–697, Academic Press, Cambridge, 2017.

21. Stoica, M. and Stoica, D., Nanofillers for Food Packaging: Antimicrobial Potential of Metal-Based Nanoparticles. *Cur. Nanotoxic. Prev.*, 1, 44, 2021.
22. Rhim, J.W., Park, H.M., Ha, C.S., Bio-nanocomposites for food packaging applications. *Prog. Poly. Sci.*, 38, 1629, 2013.
23. Singhvi, M., Joshi, D., Adsul, M., Varma, A., Gokhale, D., d-(−)-Lactic acid production from cellobiose and cellulose by Lactobacillus lactis mutant RM2-24. *Green Chem.*, 12, 1106, 2010.
24. Adsul, M.V., Varma, A.J., Gokhale, D.V., Lactic acid production from waste sugarcane bagasse derived cellulose. *Green Chem.*, 9, 58, 2007.
25. Li, G., Zhao, M., Xu, F., Yang, B., Li, X., Meng, X., Teng, L., Sun, F., Li, Y., Synthesis and Biological Application of Polylactic Acid. *Molecules*, 25, 5023, 2020.
26. Deng, K., Chen, H., Zhao, Y., Zhou, Y., Wang, Y., Sun, Y., Evaluation of adaptation of the polylactic acid pattern of maxillary complete dentures fabricated by fused deposition modelling technology: A pilot study. *PloS One*, 13, e0201777, 2018.
27. Liu, Z., Wang, Y., Wu, B., Cui, C., Guo, Y., Yan, C., A critical review of fused deposition modeling 3D printing technology in manufacturing polylactic acid parts. *Int. J. Adv. Manuf. Technol.*, 102, 2877, 2019.
28. Shao, W., He, J., Han, Q., Sang, F., Wang, Q., Chen, L., Cui, S., Ding, B., A biomimetic multilayer nanofiber fabric fabricated by electrospinning and textile technology from polylactic acid and Tussah silk fibroin as a scaffold for bone tissue engineering. *Mat. Sci. Eng. C*, 67, 599, 2016.
29. Ncube, L.K., Ude, A.U., Ogunmuyiwa, E.N., Zulkifli, R., Beas, I.N., Environmental Impact of Food Packaging Materials: A Review of Contemporary Development from Conventional Plastics to Polylactic Acid Based Materials. *Materials*, 13, 4994, 2020.
30. Tang, X.Z., Kumar, P., Alavi, S., Sandeep, K.P., Recent Advances in Biopolymers and Biopolymer-Based Nanocomposites for Food Packaging Materials. *Crit. Rev. Food Sci. Nutr.*, 52, 426, 2012.
31. Bezerra Lima, E.M., Middea, A., Marconcini, J.M., Corrêa, A.C., Fernandes Pereira, J., Vieira Guimarães, A., Firmino de Lima, J., Ramos dos Anjos, M., Miranda de Castro, I., Nunes Oliveira, R., Ramos Moreira, C., de Paiva, M.M., Rangel, F.L.C., Neumann, R., Biodegradable PLA based nanocomposites for packaging applications: The effects of organo-modified bentonite concentration. *J. Appl. Poly. Sci.*, 50907, 1–17, 2021.
32. Ramesh, P., Prasad, B.D., Narayana, K.L., Effect of MMT Clay on Mechanical, Thermal and Barrier Properties of Treated Aloevera Fiber/PLA-Hybrid Biocomposites. *Silicon*, 12, 1751, 2020.
33. Łopusiewicz, Ł., Jędra, F., Mizielińska, M., New Poly(lactic acid) Active Packaging Composite Films Incorporated with Fungal Melanin. *Polymers*, 10, 386, 2018.

34. Kim, H.K., Kim, S.J., Lee, H.S., Choi, J.H., Jeong, C.M., Sung, M.H., Park, S.H., Park, H.J., Mechanical and barrier properties of poly(lactic acid) films coated by nanoclay–ink composition. *J. Appl. Poly. Sci.*, 127, 3823, 2013.
35. Sung, S.H., Chang, Y., Han, J., Development of polylactic acid nanocomposite films reinforced with cellulose nanocrystals derived from coffee silverskin. *Carbohydr. Poly.*, 169, 495, 2017.
36. Di Maio, L., Scarfato, P., Milana, M.R., Feliciani, R., Denaro, M., Padula, G., Incarnato, L., Bionanocomposite Polylactic Acid/Organoclay Films: Functional Properties and Measurement of Total and Lactic Acid Specific Migration. *Packag. Technol. Sci.*, 27, 535, 2014.
37. Şengül, B. and Dilsiz, N., Barrier properties of polylactic acid/layered silicate nanocomposites for food contact applications. *Poly. Sci. Ser. A*, 56, 896, 2014.
38. Vilarinho, F., Andrade, M., Buonocore, G.G., Stanzione, M., Sanches Silva, M.F., A., Monitoring lipid oxidation in a processed meat product packaged with nanocomposite poly(lactic acid) film. *Eur. Poly. J.*, 98, 362, 2018.
39. Othman, S.H., Ling, H.N., Talib, R.A., Naim, M.N., Risyon, N.P., Saifullah, M., PLA/MMT and PLA/Halloysite Bio-Nanocomposite Films: Mechanical, Barrier, and Transparency. *J. Nano Res.*, 59, 77, 2019.
40. Bendahou, D., Bendahou, A., Grohens, Y.H.K., New nanocomposite design from zeolite and poly(lactic acid). *Ind. Crops Prod.*, 72, 107, 2015.
41. Elisabeta, E., Râpă, M., Popa, O., Mustatea, G., Popa, V., Mitelut, A., Popa, M., Polylactic Acid/Cellulose Fibres Based Composites for Food Packaging Applications. *Mater. Plast.*, 54, 673, 2017.
42. Fotie, G., Limbo, S., Piergiovanni, L., Manufacturing of Food Packaging Based on Nanocellulose: Current Advances and Challenges. *Nanomaterials*, 10, 1726, 2020.
43. Herrera, N., Salaberria, A.M., Mathew, A.P., Oksman, K., Plasticized polylactic acid nanocomposite films with cellulose and chitin nanocrystals prepared using extrusion and compression molding with two cooling rates: Effects on mechanical, thermal and optical properties. *Compos. Part A Appl. Sci. Manuf.*, 83, 89, 2016.
44. Herrera, N., Roch, H., Salaberria, A.M., Pino-Orellana, M.A., Labidi, J., Fernandes, S.C.M., Radic, D., Leiva, A., Oksman, K., Functionalized blown films of plasticized polylactic acid/chitin nanocomposite: Preparation and characterization. *Mater. Des.*, 92, 846, 2016.
45. Lizundia, E., Fortunati, E., Dominici, F., Vilas, J.L., León, L.M., Armentano, I., Torre, L., Kenny, J.M., PLLA-grafted cellulose nanocrystals: Role of the CNC content and grafting on the PLA bionanocomposite film properties. *Carbohyd. Poly.*, 142, 105, 2016.
46. Vanitha, R. and Kavitha, C., Development of natural cellulose fiber and its food packaging application. *Mater. Today-Proc.*, 36, 903, 2021.
47. Long, S., Zhong, L., Lin, X., Chang, X., Wu, F., Wu, R., Xie, F., Preparation of formyl cellulose and its enhancement effect on the mechanical and barrier properties of polylactic acid films. *Int. J. Biol. Macromol.*, 172, 82, 2021.

48. Kongkaoroptham, P., Piroonpan, T., Pasanphan, W., Chitosan nanoparticles based on their derivatives as antioxidant and antibacterial additives for active bioplastic packaging. *Carbohyd. Poly.*, 257, 117610, 2021.
49. Thong, C.C., Teo, D.C.L., Ng, C.K., Application of polyvinyl alcohol (PVA) in cement-based composite materials: A review of its engineering properties and microstructure behavior. *Constr. Build Mater.*, 107, 172, 2016.
50. Hoque, M.A., Ahmed, M.R., Rahman, G.T., Rahman, M.T., Islam, M.A., Khan, M.A., Hossain, M.K., Fabrication and comparative study of magnetic Fe and α-Fe2O3 nanoparticles dispersed hybrid polymer (PVA + Chitosan) novel nanocomposite film. *Results Phys.*, 10, 434, 2018.
51. Abu-Thabit, N., Hakeem, A.S., Mezghani, K., Ratemi, E., Elzagheid, M., Umar, Y., Primartomo, A., Al Batty, S., Azad, A.K., Al Anazi, S., Ahmad, A., Preparation of pH-Indicative and Flame-Retardant Nanocomposite Films for Smart Packaging Applications. *Sensors*, 20, 5462, 2020.
52. Bumbudsanpharoke, N. and Ko, S., Nanoclays in Food and Beverage Packaging. *J. Nanomater*, 2019, 1, 2019.
53. Allison, P.G., Moser, R.D., Chandler, M.Q., Caminero-Rodriguez, J.A., Torres-Cancel, K., Rivera, O.G., Goodwin, J.R., Gore, E.R., Weiss, C.A., Mechanical, Thermal, and Microstructural Analysis of Polyvinyl Alcohol/ Montmorillonite Nanocomposites. *J. Nanomater*, 2015, 1, 2015.
54. Turhan, Y., Alp, Z.G., Alkan, M., Doğan, M., Preparation and characterization of poly(vinylalcohol)/modified bentonite nanocomposites. *Micropor. Mesopor. Mater.*, 174, 144, 2013.
55. Sapalidis, A.A., Katsaros, F.K., Steriotis, T.A., Kanellopoulos, N.K., Properties of poly(vinyl alcohol)—Bentonite clay nanocomposite films in relation to polymer–clay interactions. *J. Appl. Poly. Sci.*, 123, 1812, 2012.
56. Strawhecker, K.E. and Manias, E., Structure and Properties of Poly(vinyl alcohol)/Na+ Montmorillonite Nanocomposites. *Chem. Mater.*, 12, 2943, 2000.
57. Abdullah, Z.W. and Dong, Y., Biodegradable and Water Resistant Poly(vinyl) Alcohol (PVA)/Starch (ST)/Glycerol (GL)/Halloysite Nanotube (HNT) Nanocomposite Films for Sustainable Food Packaging. *Front. Mater.*, 6, 58, 2019.
58. Li, H.Z., Chen, S.C., Wang, Y.Z., Preparation and characterization of nanocomposites of polyvinyl alcohol/cellulose nanowhiskers/chitosan. *Compos. Sci. Technol.*, 115, 60, 2015.
59. Sarwar, M.S., Niazi, M.B.K., Jahan, Z., Ahmad, T., Hussain, A., Preparation and characterization of PVA/nanocellulose/Ag nanocomposite films for antimicrobial food packaging. *Carbohyd. Poly.*, 184, 453, 2018.
60. Ben Shalom, T., Nevo, Y., Leibler, D., Shtein, Z., Azerraf, C., Lapidot, S., Shoseyov, O., Cellulose Nanocrystals (CNCs) Induced Crystallization of Polyvinyl Alcohol (PVA) Super Performing Nanocomposite Films. *Macromol. Biosci.*, 19, 1800347, 2019.

61. Wu, Z., Wu, J., Peng, T., Li, Y., Lin, D., Xing, B., Li, C., Yang, Y., Yang, L., Zhang, L., Ma, R., Wu, W., Lv, X., Dai, J., Han, G., Preparation and Application of Starch/Polyvinyl Alcohol/Citric Acid Ternary Blend Antimicrobial Functional Food Packaging Films. *Polymers*, 9, 102, 2017.
62. Yurong, G. and Dapeng, L., Preparation and characterization of corn starch/PVA/glycerol composite films incorporated with ε-polylysine as a novel antimicrobial packaging material. *e-Polymers*, 20, 154, 2020.
63. Liu, B., Xu, H., Zhao, H., Liu, W., Zhao, L., Li, Y., Preparation and characterization of intelligent starch/PVA films for simultaneous colorimetric indication and antimicrobial activity for food packaging applications. *Carbohyd. Poly.*, 157, 842, 2017.
64. Zhang, J., Xu, W.R., Zhang, Y., Li, W., Hu, J., Zheng, F., Wu, Y., Liquefied chitin/polyvinyl alcohol based blend membranes: Preparation and characterization and antibacterial activity. *Carbohyd. Poly.*, 180, 175, 2018.
65. Zhang, J., Xu, W.R., Zhang, Y.C., Han, X.D., Chen, C., Chen, A., In situ generated silica reinforced polyvinyl alcohol/liquefied chitin biodegradable films for food packaging. *Carbohyd. Poly.*, 238, 116182, 2020.
66. Wu, Y., Guan, Y., Gao, H., Zhou, L., Peng, F., Novel high-strength montmorillonite/polyvinyl alcohol composite film enhanced by chitin nanowhiskers. *J. Appl. Poly. Sci.*, 138, e:50344, 2020.
67. Kuswandi, B. and Moradi, M., Improvement of Food Packaging Based on Functional Nanomaterial, in: *Nanotechnology: Applications in Energy, Drug and Food*, S. Siddiquee, G.J.H. Melvin, M.M. Rahman (Eds.), pp. 309–344, Springer, Cham, 2019.
68. Wu, F., Misra, M., Mohanty, A.K., Challenges and New Opportunities on Barrier Performance of Biodegradable Polymers for Sustainable Packaging. *Prog. Poly. Sci.*, 117, 101395, 2021.
69. Narancic, T., Cerrone, F., Beagan, N., O'Connor, K.E., Recent Advances in Bioplastics: Application and Biodegradation. *Polymers*, 12, 920, 2020.
70. Benini, K.C., Montoro, S., Cioffi, M.O., Voorwald, H., Chandrasekharakurup, Polyhydroxyalkanoates and Their Nanobiocomposites With Cellulose Nanocrystals, in: *Design and Applications of Nanostructured Polymer Blends and Nanocomposite Systems*, S. Thomas and R. Shanks (Eds.), pp. 261–285, Elsevier, Amsterdam, 2016.
71. Chiulan, I., Mihaela Panaitescu, D., Nicoleta Frone, A., Teodorescu, M., Andi Nicolae, C., Căşărică, A., Tofan, V., Sălăgeanu, A., Biocompatible polyhydroxyalkanoates/bacterial cellulose composites: Preparation, characterization, and in vitro evaluation. *J.*, 104, 2576, 2016.
72. Manikandan, N.A., Pakshirajan, K., Pugazhenthi, G., Preparation and characterization of environmentally safe and highly biodegradable microbial polyhydroxybutyrate (PHB) based graphene nanocomposites for potential food packaging applications. *Int. J. Biol. Macromol.*, 154, 866, 2020.
73. Zhang, B., Huang, C., Zhao, H., Wang, J., Yin, C., Zhang, L., Zhao, Y., Effects of Cellulose Nanocrystals and Cellulose Nanofibers on the Structure and

Properties of Polyhydroxybutyrate Nanocomposites. *Polymers*, 11, 2063, 2019.
74. Wei, R., Breite, D., Song, C., Gräsing, D., Ploss, T., Hille, P., Schwerdtfeger, R., Matysik, J., Schulze, A., Zimmermann, W., Biocatalytic Degradation Efficiency of Postconsumer Polyethylene Terephthalate Packaging Determined by Their Polymer Microstructures. *Adv. Sci.*, 6, 1900491, 2019.
75. Curtzwiler, G.W., Williams, E.B., Maples, A.L., Davis, N.W., Bahns, T.L., Eliseo De León, J., Vorst, K.L., Ultraviolet protection of recycled polyethylene terephthalate. *J. Appl. Poly. Sci.*, 134, 45181, 2017.
76. Nisticò, R., Polyethylene terephthalate (PET) in the packaging industry. *Poly. Test.*, 90, 106707, 2020.
77. Masmoudi, F., Alix, S., Buet, S., Mehri, A., Bessadok, A., Jaziri, M., Ammar, E., Design and Characterization of a New Food Packaging Material by Recycling Blends Virgin and Recovered polyethylene terephthalate. *Poly. Eng. Sci.*, 60, 250, 2020.
78. Lim, J.W., Lim, W.S., Lee, M.H., Park, H.J., Barrier and structural properties of polyethylene terephthalate film coated with poly(acrylic acid)/montmorillonite nanocomposites. *Packag. Technol. Sci.*, 34, 141, 2021.
79. Meng, X., Qi, P., Sun, J., Li, H., Zhang, S., Liu, X., Gu, X., Fabrication of transparent clay-polymer hybrid coatings on PET film to enhance flame retardancy and oxygen barrier properties. *Prog. Org. Coat.*, 147, 105788, 2020.
80. Kim, J.K., Choi, B., Jin, J., Transparent, water-stable, cellulose nanofiber-based packaging film with a low oxygen permeability. *Carbohyd. Poly.*, 249, 116823, 2020.
81. Kissel, W.J., Han, J., Meyer, J.A., Polypropylene: Structure, Properties, Manufacturing Processes, and Applications, in: *Handbook of Polypropylene and Polypropylene Composites, Revised and Expanded*, H. Karian (Ed.), pp. 10–27, CRC Press, Boca Raton, 2004.
82. Khalaj, M.J., Ahmadi, H., Lesankhosh, R., Khalaj, G., Study of physical and mechanical properties of polypropylene nanocomposites for food packaging application: Nano-clay modified with iron nanoparticles. *Trends Food Sci. Technol.*, 51, 41, 2016.
83. d'Eon, J., Zhang, W., Chen, L., Berry, R.M., Zhao, B., Coating cellulose nanocrystals on polypropylene and its film adhesion and mechanical properties. *Cellulose*, 24, 1877–1888, 2017.
84. Wang, L., Chen, C., Wang, J., Gardner, D.J., Tajvidi, M., Cellulose nanofibrils versus cellulose nanocrystals: Comparison of performance in flexible multilayer films for packaging applications. *Food Packag. Shelf Life*, 23, 100464, 2020.
85. Zhang, W., Zhang, Y., Cao, J., Jiang, W., Improving the performance of edible food packaging films by using nanocellulose as an additive. *Int. J. Biol. Macromol.*, 166, 288, 2021.

86. Gomes, T.S., Visconte, L.L.Y., Pacheco, E.B.A.V., Life Cycle Assessment of Polyethylene Terephthalate Packaging: An Overview. *J. Poly. Environ.*, 27, 533, 2019.
87. Gironi, F. and Piemonte, V., Life cycle assessment of polylactic acid and polyethylene terephthalate bottles for drinking water. *Environ. Prog. Sustain Energy*, 30, 459, 2011.
88. Leejarkpai, T., Mungcharoen, T., Suwanmanee, U., Comparative assessment of global warming impact and eco-efficiency of PS (polystyrene), PET (polyethylene terephthalate) and PLA (polylactic acid) boxes. *J. Clean. Prod.*, 125, 95, 2016.
89. Lorite, G.S., Rocha, J.M., Miilumäki, N., Saavalainen, P., Selkälä, T., Morales-Cid, G., Gonçalves, M.P., Pongrácz, E., Rocha, C.M.R., Toth, G., Evaluation of physicochemical/microbial properties and life cycle assessment (LCA) of PLA-based nanocomposite active packaging. *LWT*, 75, 305, 2017.
90. Lyu, J.S., Lee, J.-S., Han, J., Development of a biodegradable polycaprolactone film incorporated with an antimicrobial agent via an extrusion process. *Sci. Rep.*, 9, 20236, 2019.
91. Cesur, S., Köroğlu, C., Yalçın, H.T., Antimicrobial and biodegradable food packaging applications of polycaprolactone/organo nanoclay/chitosan polymeric composite films. *J. Vinyl Addit. Techn.*, 24, 376, 2018.
92. Radhakrishnan, S., Nagarajan, S., Belaid, H., Farha, C., Iatsunskyi, I., Coy, E., Soussan, L., Huon, V., Bares, J., Belkacemi, K., Teyssier, C., Balme, S., Miele, P., Cornu, D., Kalkura, N., Cavaillès, V., Bechelany, M., Fabrication of 3D printed antimicrobial polycaprolactone scaffolds for tissue engineering applications. *Mater. Sci. Eng. C*, 118, 111525, 2021.
93. Prasad, A. and Kandasubramanian, B., Fused deposition processing polycaprolactone of composites for biomedical applications. *Poly-Plast. Tech. Mat.*, 58, 1365, 2019.
94. Endres, H.J., Siebert-Raths, A., Matyjaszewski, K., Moller, M., Performance Profile of Biopolymers Compared to Conventional Plastics, in: *Polymer Science: A Comprehensive Reference*, pp. 317–353, Elsevier, Amsterdam, 2012.
95. Bliley, J.M. and Marra, K.G., Polymeric Biomaterials as Tissue Scaffolds, in: *Stem Cell Biology and Tissue Engineering in Dental Sciences*, A. Vishwakarma, P. Sharpe, S. Shi, M. Ramalingam (Eds.), pp. 149–161, Academic Press, Boston, 2015.
96. Uzunlu, S. and Niranjan, K., Laboratory antimicrobial activity of cinnamaldehyde and pomegranate-based polycaprolactone films. *J. Appl. Poly. Sci.*, 134, 45347, 2017.
97. Correa, J.P., Molina, V., Sanchez, M., Kainz, C., Eisenberg, P., Massani, M.B., Improving ham shelf life with a polyhydroxybutyrate/polycaprolactone biodegradable film activated with nisin. *Food Packag. Shelf Life*, 11, 31, 2017.
98. Pereira de Abreu, D.A., Cruz, J.M., Paseiro Losada, P., Active and Intelligent Packaging for the Food Industry. *Food Rev. Int.*, 28, 146, 2012.

99. Dey, D., Dharini, V., Periyar Selvam, S., Rotimi Sadiku, E., Mahesh Kumar, M., Jayaramudu, J., Nath Gupta, U., Physical, antifungal, and biodegradable properties of cellulose nanocrystals and chitosan nanoparticles for food packaging application. *Mater. Today: Proc.*, 38, 860, 2021.
100. Xu, Y., Liu, X., Jiang, Q., Yu, D., Xu, Y., Wang, B., Xia, W., Development and properties of bacterial cellulose, curcumin, and chitosan composite biodegradable films for active packaging materials. *Carbohyd. Poly.*, 260, 117778, 2021.
101. Pandey, V.K., Upadhyay, S.N., Niranjan, K., Mishra, P.K., Antimicrobial biodegradable chitosan-based composite Nano-layers for food packaging. *Int. J. Biol. Macromol*, 157, 212, 2020.
102. Liu, Y., Ahmed, S., Sameen, D.E., Wang, Y., Lu, R., Dai, J., Li, S., Qin, W., A review of cellulose and its derivatives in biopolymer-based for food packaging application. *Trends Food Sci. Technol.*, 112, 532–546, 2021.
103. Rånby, B., Natural cellulose fibers and membranes: Biosynthesis, in: *Encyclopedia of Materials: Science and Technology*, Buschow *et al.* (eds), pp. 5938–5943, Elsevier, Amsterdam, 2001.
104. Lim, D.B.K. and Gong, H., Highly stretchable and transparent films based on cellulose. *Carbohyd. Poly.*, 201, 446, 2018.
105. Shokri, J. and Adibkia, K., Application of cellulose and cellulose derivatives in pharmaceutical industries. *Cellulose - Medical, Pharm. Electronic Appl.*, 47, 47–66, 2013.
106. Jorfi, M. and Foster, E.J., Recent advances in nanocellulose for biomedical applications. *J. Appl. Poly. Sci.*, 132, 41719, 2015.
107. Yilmaz, T. and Kus, M., Advanced applications of green materials in bioelectronics applications, in: *Applications of Advanced Green Materials*, S. Ahmed (Ed.), pp. 631-661, Woodhead Publishing, Cambrigde, 2021.
108. González, A., Gastelú, G., Barrera, G.N., Ribotta, P.D., Álvarez Igarzabal, C.I., Preparation and characterization of soy protein films reinforced with cellulose nanofibers obtained from soybean by-products. *Food Hydrocoll.*, 89, 758, 2019.
109. Irimia-Vladu, M., Głowacki, E.D., Voss, G., Bauer, S., Sariciftci, N.S., Green and biodegradable electronics. *Mater. Today*, 15, 340, 2012.
110. Irimia-Vladu, M., "Green" electronics: biodegradable and biocompatible materials and devices for sustainable future. *Chem. Soc Rev.*, 43, 588, 2014.
111. Rudnik, E., *Compostable Polymer Materials: Definitions, Structures, and Methods of Preparation*, pp. 10–36, Elsevier, Amsterdam, 2008.
112. Rudin, A. and Choi, P., Biopolymers, in: *The Elements of Polymer Science & Engineering* (3rd Edition), pp. 521–535, Academic Press, Cambridge, 2013.
113. Huang, J., Ma, X., Yang, G., Alain, D., Introduction to Nanocellulose, in: *Nanocellulose: From Fundamentals to Advanced Materials*, pp. 1–20, Wiley-VCH Verlag GmbH & Co. KGaA, Weinheim, 2019.
114. Yu, H.Y., Zhang, H., Song, M.L., Zhou, Y., Yao, J., Ni, Q.Q., Nanospheres, From Cellulose, Nanorods to Nanofibers: Various Aspect Ratio Induced

Nucleation/Reinforcing Effects on Polylactic Acid for Robust-Barrier Food Packaging. *ACS Appl. Mater. Interfaces*, 9, 43920, 2017.
115. Yu, Z., Wang, W., Kong, F., Lin, M., Mustapha, A., Cellulose nanofibril/silver nanoparticle composite as an active food packaging system and its toxicity to human colon cells. *Int. J. Biol. Macromol.*, 129, 887, 2019.
116. Gan, I. and Chow, W.S., Antimicrobial poly(lactic acid)/cellulose bionanocomposite for food packaging application: A review. *Food Packag. Shelf Life*, 17, 150, 2018.
117. Khosravi, A., Fereidoon, A., Khorasani, M.M., Naderi, G., Ganjali, M.R., Zarrintaj, P., Saeb, M.R., Gutiérrez, T.J., Soft and hard sections from cellulose-reinforced poly(lactic acid)-based food packaging films: A critical review. *Food Packag. Shelf Life*, 23, 100429, 2020.
118. Fortunati, E., Armentano, I., Zhou, Q., Iannoni, A., Saino, E., Visai, L., Berglund, L.A., Kenny, J.M., Multifunctional bionanocomposite films of poly(lactic acid), cellulose nanocrystals and silver nanoparticles. *Carbohyd Poly.*, 87, 1596, 2012.
119. Luzi, F., Fortunati, E., Jiménez, A., Puglia, D., Chiralt, A., Torre, L., P.L.A., Nanocomposites Reinforced with Cellulose Nanocrystals from Posidonia oceanica and ZnO Nanoparticles for Packaging Application. *J. Renew. Mater.*, 5, 103, 2017.
120. Fortunati, E., Rinaldi, S., Peltzer, M., Bloise, N., Visai, L., Armentano, I., Jiménez, A., Latterini, L., Kenny, J.M., Nano-biocomposite films with modified cellulose nanocrystals and synthesized silver nanoparticles. *Carbohyd. Poly.*, 101, 1122, 2014.
121. Pantani, R., Gorrasi, G., Vigliotta, G., Murariu, M., Dubois, P., PLA-ZnO nanocomposite films: Water vapor barrier properties and specific end-use characteristics. *Eur. Poly. J.*, 49, 3471, 2013.
122. Shalumon, K.T., Anulekha, K.H., Nair Sodium alginate/poly(vinyl alcohol)/nano ZnO composite nanofibers for antibacterial wound dressings. *Int. J. Biol. Macromol.*, 49, 247, 2011.
123. Mujtaba, M., Morsi, R.E., Kerch, G., Elsabee, M.Z., Kaya, M., Labidi, J., Khawar, K.M., Current advancements in chitosan-based film production for food technology; A review. *Int. J. Biol. Macromol.*, 121, 889, 2019.
124. Wang, H., Qian, J., Ding, F., Emerging Chitosan-Based Films for Food Packaging Applications. *J. Agric. Food Chem.*, 66, 395, 2018.
125. Cazón, P. and Vázquez, M., Applications of Chitosan as Food Packaging Materials, in: *Sustainable Agriculture Reviews*, vol. 36, G. Crini and E. Lichtfouse (Eds.), pp. 81–123, Springer, Cham, 2019.
126. Haghighi, H., Licciardello, F., Fava, P., Siesler, H.W., Pulvirenti, A., Recent advances on chitosan-based films for sustainable food packaging applications. *Food Packag. Shelf Life*, 26, 100551, 2020.
127. Thapa, B. and Narain, R., Mechanism, current challenges and new approaches for non-viral gene delivery, in: *Polymers and Nanomaterials for Gene Therapy*, R. Narain (Ed.), pp. 1–27, Woodhead Publishing, Cambridge, 2016.

128. Ali, A. and Ahmed, S., A review on chitosan and its nanocomposites in drug delivery. *Int. J. Biol. Macromol*, 109, 273, 2018.
129. Zhang, X., Xiao, G., Wang, Y., Zhao, Y., Su, H., Tan, T., Preparation of chitosan-TiO2 composite film with efficient antimicrobial activities under visible light for food packaging applications. *Carbohyd. Polym.*, 169, 101, 2017.
130. Qin, Y., Liu, Y., Yuan, L., Yong, H., Liu, J., Preparation and characterization of antioxidant, antimicrobial and pH-sensitive films based on chitosan, silver nanoparticles and purple corn extract. *Food Hydrocoll*, 96, 102, 2019.
131. Kumar, S., Mukherjee, A., Dutta, J., Chitosan based nanocomposite films and coatings: Emerging antimicrobial food packaging alternatives. *Trends Food Sci. Tech.*, 97, 196, 2020.
132. Dizaj, S.M., Lotfipour, F., Barzegar-Jalali, M., Zarrintan, M.H., Adibkia, K., Antimicrobial activity of the metals and metal oxide nanoparticles. *Mater.*, 44, 278, 2014.
133. Almasi, H., Jafarzadeh, P., Mehryar, L., Fabrication of novel nanohybrids by impregnation of CuO nanoparticles into bacterial cellulose and chitosan nanofibers: Characterization, antimicrobial and release properties. *Carbohyd. Polym.*, 186, 273, 2018.
134. Riaz, A., Lei, S., Akhtar, H.M.S., Wan, P., Cheni, D., Jabbar, S., Abid, M., Hashim, M.M., Zeng, X., Preparation and characterization of chitosan-based antimicrobial active food packaging film incorporated with apple peel polyphenols. *Int. J. Biol. Macromol.*, 114, 547, 2018.
135. de Moraes Crizel, T., de Oliveira Rios, A., Alves, V.D., Bandarra, N., Moldão-Martins, M., Hickmann Flôres, S., Active food packaging prepared with chitosan and olive pomace. *Food Hydrocoll*, 74, 139, 2018.
136. da Silva, S.B., de Souza, D., Lacerda, L.D., Food Applications of Chitosan and its Derivatives, in: *Chitin and Chitosan: Properties and Applications*, A.M.V.D. Broek and C.G. Boeriu (Eds.), pp. 315–347, John Wiley & Sons Ltd., New Jersey, 2020.
137. Qin, Y.Y., Zhang, Z.H., Li, L., Yuan, M.L., Fan, J., Zhao, T.R., Physio-mechanical properties of an active chitosan film incorporated with montmorillonite and natural antioxidants extracted from pomegranate rind. *J. Food Sci. Technol.*, 52, 1471, 2015.
138. Pires, J.R.A., de Souza, V.G.L., Fernando, A.L., Chitosan/montmorillonite bionanocomposites incorporated with rosemary and ginger essential oil as packaging for fresh poultry meat. *Food Packag. Shelf Life*, 17, 142, 2018.
139. Souza, V.G.L., Pires, J.R.A., Vieira, É.T., Coelhoso, I.M., Duarte, M.P., Fernando, A.L., Activity of chitosan-montmorillonite bionanocomposites incorporated with rosemary essential oil: From in vitro assays to application in fresh poultry meat. *Food Hydrocoll.*, 89, 241, 2019.
140. Lavorgna, M., Attianese, I., Buonocore, G.G., Conte, A., Del Nobile, M.A., Tescione, F., Amendola, E., MMT-supported Ag nanoparticles for chitosan nanocomposites: Structural properties and antibacterial activity. *Carbohyd. Polym.*, 102, 385, 2014.

141. Nouri, A., Yaraki, M.T., Ghorbanpour, M., Gupta, V.K., Enhanced Antibacterial effect of chitosan film using Montmorillonite/CuO nanocomposite. *Int. J. Biol. Macromol.*, 109, 1219, 2018.
142. Ścieszka, S. and Klewicka, E., Algae in food: a general review. *Crit. Rev. Food Sci. Nutr.*, 59, 3538, 2019.
143. Al-Tayyar, N.A., Youssef, A.M., Al-hindi, R., Antimicrobial food packaging based on sustainable Bio-based materials for reducing foodborne Pathogens: A review. *Food Chem.*, 310, 125915, 2020.
144. Kontominas, M.G., Use of Alginates as Food Packaging Materials. *Foods*, 9, 1440, 2020.
145. Zhong, Y., Godwin, P., Jin, Y., Xiao, H., Biodegradable polymers and green-based antimicrobial packaging materials: A mini-review. *Adv. Ind. Eng. Polym. Res.*, 3, 27, 2020.
146. Amjadi, S., Hamishehkar, H., Ghorbani, M., A novel smart PEGylated gelatin nanoparticle for co-delivery of doxorubicin and betanin: A strategy for enhancing the therapeutic efficacy of chemotherapy. *Mater. Sci. Eng. C*, 97, 833, 2019.
147. Lacroix, M. and Vu, K.D., Edible Coating and Film Materials: Proteins, in: *Innovations in Food Packaging*, J.H. Han (Ed.), pp. 277–304, Academic Press, San Diego, 2014.
148. Chentir, I., Kchaou, H., Hamdi, M., Jridi, M., Li, S., Doumandji, A., Nasri, M., Biofunctional gelatin-based films incorporated with food grade phycocyanin extracted from the Saharian cyanobacterium Arthrospira sp. *Food Hydrocoll.*, 89, 715, 2019.
149. Nur Hanani, Z.A., Roos, Y.H., Kerry, J.P., Use and application of gelatin as potential biodegradable packaging materials for food products. *Int. J. Biol. Macromol.*, 71, 94, 2014.
150. Roy, S. and Rhim, J.W., Preparation of antimicrobial and antioxidant gelatin/curcumin composite films for active food packaging application. *Colloids Surf. B Biointerfaces*, 188, 110761, 2020.
151. Riahi, Z., Priyadarshi, R., Rhim, J.W., Bagheri, R., Gelatin-based functional films integrated with grapefruit seed extract and TiO2 for active food packaging applications. *Food Hydrocoll.*, 112, 106314, 2021.
152. Liu, C., Huang, J., Zheng, X., Liu, S., Lu, K., Tang, K., Liu, J., Heat sealable soluble soybean polysaccharide/gelatin blend edible films for food packaging applications. *Food Packag. Shelf Life*, 24, 100485, 2020.
153. Nur Amila Najwa, I.S., Mat Yusoff, M., Nur Hanani, Z.A., Potential of Silver-Kaolin in Gelatin Composite Films as Active Food Packaging Materials. *Food Packag. Shelf Life*, 26, 100564, 2020.
154. Hassan, B., Chatha, S.A.S., Hussain, A.I., Zia, K.M., Akhtar, N., Recent advances on polysaccharides, lipids and protein based edible films and coatings: A review. *Int. J. Biol. Macromol.*, 109, 1095, 2018.
155. Mose, B. and Maranga, S.M., A review on starch based nanocomposites for bioplastic materials. *J. Mater. Sci. Eng. B*, 1, 239, 2011.

156. Thakur, R., Pristijono, P., Scarlett, C.J., Bowyer, M., Singh, S.P., Vuong, Q.V., Starch-based films: Major factors affecting their properties. *Int. J. Biol. Macromol.*, 132, 1079, 2019.
157. Wang, X.L., Yang, K.K., Wang, Y.Z., Properties of Starch Blends with Biodegradable Polymers. *J. Macromol. Sci. C*, 43, 385, 2003.
158. Wei, D., Wang, H., Xiao, H., Zheng, A., Yang, Y., Morphology and mechanical properties of poly(butylene adipate-co-terephthalate)/potato starch blends in the presence of synthesized reactive compatibilizer or modified poly(butylene adipate-co-terephthalate). *Carbohyd. Polym.*, 123, 275, 2015.
159. Imre, B. and Pukánszky, B., Compatibilization in bio-based and biodegradable polymer blends. *Eur. Polym. J.*, 49, 1215, 2013.
160. Mendes, J.F., Norcino, L.B., Martins, H.H.A., Manrich, A., Otoni, C.G., Carvalho, E.E.N., Piccoli, R.H., Oliveira, J.E., Pinheiro, A.C.M., Mattoso, L.H.C., Correlating emulsion characteristics with the properties of active starch films loaded with lemongrass essential oil. *Food Hydrocoll.*, 100, 105428, 2020.
161. Díaz-Galindo, E.P., Nesic, A., Cabrera-Barjas, G., Mardones, C., von Baer, D., Bautista-Baños, S., Dublan Garcia, O., Physical-Chemical Evaluation of Active Food Packaging Material Based on Thermoplastic Starch Loaded with Grape cane Extract. *Molecules*, 25, 1306, 2020.
162. Jaśkiewicz, A., Budryn, G., Nowak, A., Efenberger-Szmechtyk, M., Novel Biodegradable Starch Film for Food Packaging with Antimicrobial Chicory Root Extract and Phytic Acid as a Cross-Linking Agent. *Foods*, 9, 1696, 2020.
163. Shapi'i, R.A., Othman, S.H., Nordin, N., Kadir Basha, R., Nazli Naim, M., Antimicrobial properties of starch films incorporated with chitosan nanoparticles: *In vitro* and *in vivo* evaluation. *Carbohyd. Polym.*, 230, 115602, 2020.
164. Narayanan, K.B., Park, G.T., Han, S.S., Antibacterial properties of starch-reduced graphene oxide–polyiodide nanocomposite. *Food Chem.*, 342, 128385, 2021.
165. Iamareerat, B., Singh, M., Sadiq, M.B., Anal, A.K., Reinforced cassava starch based edible film incorporated with essential oil and sodium bentonite nanoclay as food packaging material. *J. Food Sci. Technol.*, 55, 1953, 2018.
166. Senturk Parreidt, T., Müller, K., Schmid, M., Alginate-Based Edible Films and Coatings for Food Packaging Applications. *Foods*, 7, 170, 2018.
167. Annu, A.A. and Ahmed, S., Eco-friendly natural extract loaded antioxidative chitosan/polyvinyl alcohol based active films for food packaging. *Heliyon*, e06550, 2021.
168. Jost, V., Kobsik, K., Schmid, M., Noller, K., Influence of plasticiser on the barrier, mechanical and grease resistance properties of alginate cast films. *Carbohyd. Polym.*, 110, 309, 2014.

5

Novel Nanostructured Inclusions in Biopolymers to Form Advanced Materials for Packaging in the Food Industry

Bratin Sengupta

Department of Chemical and Biological Engineering and RENEW Institute, University at Buffalo, Buffalo, NY, USA

Abstract

Food packing is undergoing a mammoth change in terms of materials used for packaging and food coatings. One major reason for the change is the switch to environmentally friendly biodegradable polymer materials. To this end, the most available biodegradable materials that are researched are of inferior mechanical and barrier properties. However, these properties are greatly improved when these biopolymers are amalgamated with reinforcing fillers, provided these have at least one nanoscale dimension to improve synergy among them and the polymer matrix. With the advent of nanotechnology over the past few decades, interesting strategies have been developed to engineer these nanocomposites to enhance interaction with the matrix and not only to improve mechanical and barrier properties but also to incorporate favorable attributes such as antimicrobial and antioxidant. Tracking these recent developments of nanoengineered materials, particularly the inorganic counterpart and their composites for use in food packaging applications, this chapter reviews the current developments in nanocomposite materials and explores prospects of development that would provide a possible direction for future research.

***Keywords*:** Nanocomposites, food packaging, mechanical properties, barrier properties

Email: bratinse@buffalo.edu

Annu, Tanima Bhattacharya and Shakeel Ahmed (eds.) Nanotechnology in Intelligent Food Packaging, (113–134) © 2022 Scrivener Publishing LLC

5.1 Introduction

Considering the growing population and the subsequent demand for fresh produce, it is critical to have a steady supply chain coupled with excellent storage of food [1, 2]. With the advent of globalization, there has been a steep rise in the trade of exotic food throughout the world. To meet the requirements of this growing food market, there is a need for proper packing material that preserves the food from the time of its production to consumption through its entire supply chain. The food packaging industry is expected to grow steadily by about 30% from 2018 to USD 378.5 billion in 2022 to meet the global demands and reduce food waste [3, 4]. Over 95% of the materials used for food packaging presently are polymers derived from petroleum sources and most of these are non-biodegradable and would remain in the environment for hundreds of years before they can decompose [5, 6]. The problem with these food packaging materials is so large that these contribute to almost 30% of the total plastic waste on earth with more than 50% of them being discarded after single use [7, 8]. These accumulated plastics are seldom recycled (~10%) due to various complexities, and they are either incinerated or end up in landfills [3]. Thermal incineration is the only viable option to fully break down these polymers, but these generate furans, polychlorinated biphenyls which are detrimental to the environment [9]. Synthetic polymers are of very critical concern as these not only causes landfill problems but also often ends up in oceans as micro-plastics and causes a tremendous problem for the ecosystem [10, 11]. Figure 5.1 shows the polymer used, its fate after use for 65 years between 1950 and 2015 [12]. About 55% of all the synthetic plastic produced during this time is discarded which is a cause of major environmental pollution. It is known that the plastic generated from the packaging industry is the most notorious due to its very short life span and is the one that is discarded the most. Thus, there is an active search for a bio-degradable alternative to this recalcitrant plastic packaging problem.

Biopolymers serve as an alternative to the menace of non-degradable packing materials but their market volume is still very low (~1%) [13]. Although, recently, the cost of production of biopolymers has come down considerably, it still fails to capture the global market. The common biopolymers that are mainly used for packaging materials are collagen, casein, polylactic acid (PLA), alginate, etc. [14]. The failure to be an attractive alternative lies in the fact of their lacking mechanical strength [15]. Other issues facing the application of biopolymers are its poor gas and moisture

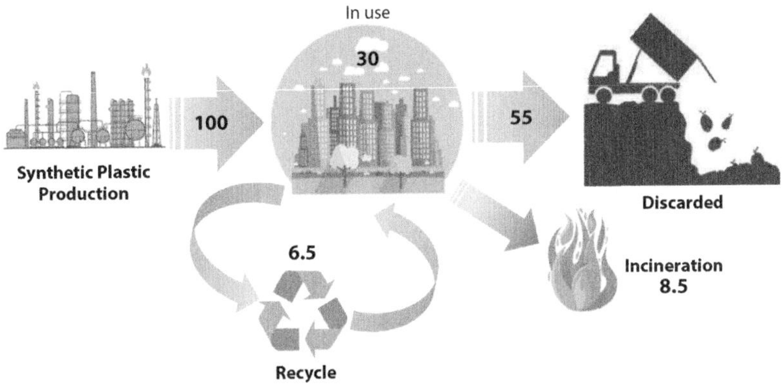

Figure 5.1 Fate of all plastic produced between 1950 to 2015. A total of 8300 million metric ton plastic was produced the numbers represents the percentage of plastic that went into each stream.

barrier properties, brittleness, poor processing condition, and, finally, cost of mass production [16].

Nanotechnology-based mitigation of these problems seems to be a promising way forward. Of US$ 3 trillion research grants funding discoveries, about 50% of them are to develop products targeted toward solving food packaging problems [17]. A major portion of this research is to make biodegradable polymers more sustainable and solve their limitations to make them an alternative to synthetic polymers.

In this chapter, the advancements in research toward making biocomposites polymers a true alternative to petroleum-derived synthetic polymers have been reviewed. It has been observed that including nanostructured materials into the polymer matrix has greatly improved the properties of these biopolymers and has shown promise to be used as a food packaging material. In this chapter, we will mainly focus on literature reports that use inorganic nanostructured inclusions to enhance the packaging property of biopolymers.

5.2 Biopolymers and Biodegradability

Biopolymers are polymers that are obtained from natural resources [18]. Whereas biodegradable polymers are the one which gets degraded eventually into carbon dioxide (CO2) and water (H2O) as shown in Figure 5.2a [19]. Although many of the biopolymers available in the market are biodegradable (capable of undergoing natural enzymatic degradation

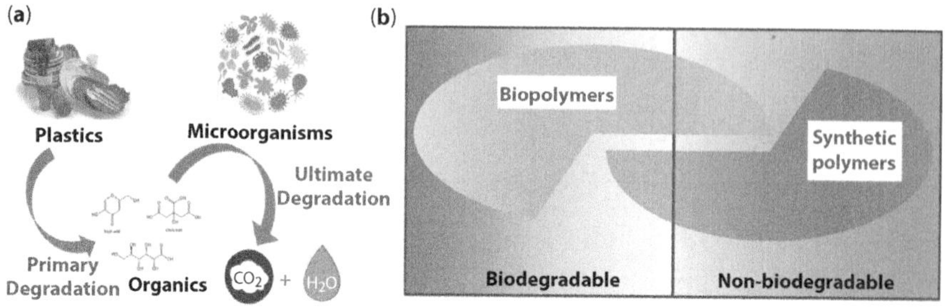

Figure 5.2 (a) Pathways of degradation of polymers into its organic monomers or constituents then eventually into water and carbon dioxide via microorganism activity. (b) Classification of polymer materials for food packaging based on either biodegradability or its source of synthesis.

into CO_2, CH_4, H_2O, inorganic components, or biomass within a specified time), some of them are not. For example, bio-PE (polyethylene) and bio-PET (polyethylene terephthalate), although synthesized from biological sources, are not biodegradable in nature [20]. Similarly, there are synthetic polymers like poly-ε caprolactone and polybutylene succinate which are biodegradable because of the presence of an ester bond that is easily degradable in nature by enzymatic reactions initiated by microorganisms. The three main routes of synthesis of biopolymers are 1) obtain them directly from bio-resources (proteins and carbohydrates), 2) chemical synthesis either from biomass or petrochemical sources, and 3) via microbial fermentation [21]. Biopolymers which are synthesized via their natural route are carbon neutral even if they do not ultimately degrade into H_2O and CO_2 and are accounted as environmentally friendly as they are carbon neutral. This is because, when we discard a bio-based polymer or sent it for incineration, the CO_2 generated can be utilized for photosynthesis by biological resources from which we derive bio-polymers, thus making the net carbon generation for the whole life cycle of the process zero [20]. Thus, in summary, there are two major classifications for polymer materials used for packaging—one based on their origin of development and the other based on their fate after use (Figure 5.2b). From the figure, it is evident that we need packaging material that is biodegradable or is made of bio-based polymers for a more eco-friendly solution. Since the majority of the biodegradable polymers are bioplastics, it is important to consider their future as a potential food packaging material and assess the challenges faced toward much broader use of these in everyday food packaging. The most common biopolymers or bioplastics that have been investigated in the early 1990s are PLA and microbial polyesters [22, 23].

PLA was synthesized from lactic acid obtained by fermentation of glucose sources while microbial polyesters were synthesized by microorganisms using glucose and plant oil as feedstock [24]. Although the biopolymer/biodegradable plastic industry is picking up pace, reaching 2.11 million tonnes production in 2019 (1.17 tonnes biodegradable), among 20 classes of biopolymers, only three (PLA, starch, and polybutylene) make up 95% of the total. PLA and starch are bio-based, while polybutylene (polybutylene succinate, PBS; polybutylene adipate terephthalate, PBAT)–based biodegradable polymers are made from fossil fuels but bio-based alternatives of their synthesis are also available [25]. Starch blends, PLA, and PBS/PBAT are majorly manufactured by Novamont (Italy), NatureWorks (USA), and BASF (Germany), respectively [25]. Apart from these major biodegradable polymers, polyhydroxyalkanoates (PHAs), polyvinyl alcohol (PVA), polypropylene carbonate (PPC), and polycaprolactone (PCL) have some presence in the market. However, these require more research for economic production and superior properties to capture the market.

For these polymers to be suitable for food packaging use, they need to have certain properties like gas barrier, mechanical strength, and moderate stretchability among many others (Figure 5.3) [18]. However, most of them suffer from a lack of these properties. One way of solving this problem would be to blend them with a commercial synthetic polymer. However, the challenge with having non-biodegradable micro-plastics when the

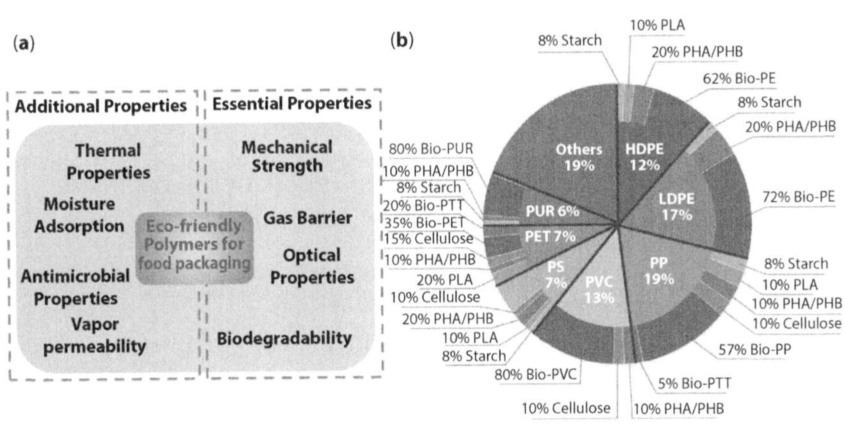

Figure 5.3 (a) Desirable properties of eco-friendly food packaging materials. (b) Possible replacement of commonly used synthetic polymers by blend of biopolymers (PHB, polyhydroxybutyrate; PTT, polytrimethylene terephthalate; PVC, polyvinyl chloride; PUR, polyurethane). Adopted with permission from [3] (Copyright 2020 American Chemical Society).

biodegradable part of the blend degrades is so grand that it is not an ecologically viable option. The proper blend of biopolymers in a certain ratio sometimes gives a composite that often has similar properties as some of the commercially available polymers used in the food packaging industry (Figure 5.3b). However, many of these are not sustainable because one or the other biopolymer in the blend is difficult or uneconomical to manufacture. One viable option that gained success recently is the use of filler composites within the polymeric blend that increases the desirable properties of the polymers [13]. A larger, micron-size filler component faces challenges of incorporation into the polymer blend because of weak interaction at the interface, hence causing defects. However, when these filler reinforcements are nanostructured to have at least one dimension in the nanometer range, they show excellent miscibility and forms favorable end products with enhanced properties.

5.3 Improvement of Biopolymers Using Nanostructured Materials

Among the various polymer inclusions that have been researched upon for improving the property of the polymer to make it suitable for the food packaging industry, the inorganic and non-natural–based nanostructured inclusions seem to provide the maximum property enhancements. The various nanocomposites that have been studied include silver [nanoparticles (AgNPs) or nanoclusters (AgNCs)], zinc oxide nanoparticles (ZnNPs), graphene, carbon nanotubes (CNT), titanium oxide, nanoclays, and layered double hydroxides [26–30].

5.3.1 Carbon-Based Nanostructured Inclusions: Graphene, Graphene Oxide, and Carbon Nanotubes

The carbon-based nanomaterials family includes some of the most commonly used nanostructured materials—graphene, GO, reduced GO (r-GO), and CNT (single-walled or multi-walled). Graphene is ideally a single atom thick material with densely packed sp2 carbon atoms. Its derivatives, GO, are formed by the chemical oxidation of graphene or directly from graphite. GO has ample oxidized groups (hydroxyl and carboxylic groups) at the edges and the defect sites and epoxy groups at the planer sites. The reduction of GO results in the formation of r-GO where most oxidized groups are eliminated via the reducing agent. All graphene,

GO, and r-GO are considered as two-dimensional materials with one of its dimensions approaching one atom thickness. CNTs, on the other hand, are fibrous in morphology with a diameter ranging around 10 nm for single-wall CNT and ~35 nm for multiwall CNTs. The length of these often are larger than 100s of microns and they are often considered as one-dimensional materials. CNTs have been traditionally included in mixed matrix membranes to improve the mechanical strength and selectivity of membranes [31]. Recently, Pattanshetti and coworkers included multi-walled CNT (MWCNT, diameter of 30–45 nm) into gelatin melt to incorporate water and oil resistance to their food packaging material [32]. The inclusion of MWCNT also increased the antimicrobial activity of the packaging. While speculation of the harmful effect of CNT on food material remains, they coated the inner walls with garlic nanoparticles to minimize the contact of the MWCNT/gelatin composite with the food. Kim *et al.* recently used a mixture of CNT and GO (graphene oxide) in a PLA matrix to show improved mechanical, optical, and oxygen barrier property [33]. A schematic of their fabrication of CNTGO-PLA nanocomposite is given in Figure 5.4a. Chao and coworkers found increased mechanical properties, reduced water absorption, and better barrier properties by incorporating GO into the PVA matrix [34]. Recently, GO-AgNPs were incorporated into PVA via a facile one-pot synthesis (Figure 5.4b) that shows great barrier properties as well as microbial properties due to the addition of silver [35].

Generally, it has been noticed that incorporating GO into the polymer melt is easier compared to graphene because of the functional groups in GO allowing better dispersion in polymer. Similarly, CNTs when are

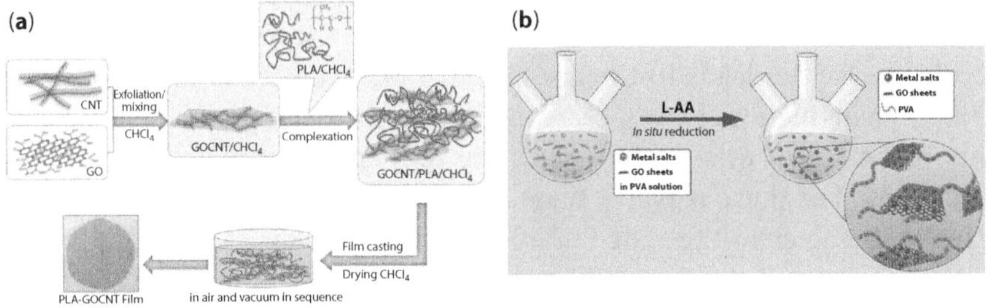

Figure 5.4 Fabrication of carbon nanomaterial based packaging material. (a) The prepared CNT-GO dispersions via sonication were incorporated into the PLA matrix via solution casting method (b) One pot synthesis of nanocomposites improves the possibility of large scale production. The PVA/AgNPs-GO nanocomposite is prepared by reduction (using vitamin C) of Ag ions in the same pot containing GO and PVA matrix. Reprinted with permission from [33] (Copyright 2020, with permission from Elsevier) and [35] (Copyright 2019, with permission from Elsevier).

functionalized with, say, poly-dopamine, it results in better dispersion in the matrix [36]. These strategies can prevent aggregation of the nanomaterials in the matrix resulting in a defect-free final film. CNTs with proper modifications were dispersed in a starch matrix shown improved film properties with higher mechanical strength, oxygen barrier, and increased crystallinity which are crucial for packaging materials [37]. Graphene-based materials like GO or reduced GO also impart antimicrobial properties to the food packaging just by themselves [38]. Although the reasons for the antimicrobial property of graphene and its derivatives are still under speculation, there is three major hypothesis for its special properties: (a) physical damage to the cell membrane due to it two-dimensional geometry; (b) it surrounds the microbial cell, and because of their excellent barrier properties, it cuts off nutrient flow into the cell; and (c) by carrying out a redox reaction with the organic matters, it generates oxygen which eventually harms the microbial cell [39–41]. The antimicrobial activity seems to be strongest in GO, followed by r-GO, and lowest in graphene [42]. Overall increment of mechanical strength, barrier properties, and introduction of antimicrobial properties by introducing carbon-based nanomaterials (CNT, GO, r-GO, graphene) can be achieved. These make these nanocomposites a promising material for food packaging applications.

5.3.2 Nanostructured Metal Oxide Inclusions: Silver Nanoparticles, Zinc Oxide, and Titanium Oxide

Silver has traditionally shown high antimicrobial properties against both gram-positive and gram-negative bacteria, fungi, and certain viruses [43]. Similar antimicrobial and catalytic properties are observed with zinc and titanium oxide (ZnO and TiO_2). Although the incorporation of antimicrobial activities leads to the use of metal or metal oxide inclusions into biopolymer blends, it was seen that these nanocomposites often show improvement of other food packaging properties compared to the neat polymers. For these reasons, these metals are nanostructured into particles of diameter in the range of 10–100 nm. Shankar *et al.* recently prepared a composite of PLA with the incorporation of lignin and AgNPs that not only showed excellent barrier against UV and antimicrobial properties but higher degradation temperature and tensile strength of 45.4 MPa compared to 38.6 MPa for the pristine PLA [44]. Poly(3-hydroxybutyrate-co-3-hydroxy valerate) or PHBV was *in situ* doped with Silver nanocomposites (0.04% w/w) to decrease oxygen permeability by 56% while maintaining optical and mechanical properties [45]. Furthermore, this new nanocomposite showed good microbial resistance against the two most common pathogens present in food—*Listeria monocytogenes* and *Salmonella enterica* for over months [45]. ZnNPs have been doped into the PLA matrix at 1%, 3%, and 5% to obtain a homogeneous ZnNP-doped PLA film. The O_2 and CO_2 permeability decreased by 18% and 17% by doping PLA by only 1% ZnNPs. Moreover, temperature-independent antimicrobial activity was also observed [46]. ZnNPs were doped into chitosan films by Priyadarshi, and Negi showed a 77% increase in tensile modulus and one order magnitude decrease of moisture permeability through the ZnNPs/Chitosan nanocomposite films [47]. Titanium oxide nanoparticles (TiNPs) have been added to chitosan to improve its antimicrobial property for the packaging of red grapes [48]. A general idea of the preparation of metal nanoparticle or metal oxide nanoparticle doped biopolymer films is given in Figure 5.5 [49].

AgNPs and AgNCs impart antimicrobial property via charge or production of reactive oxygen groups (O^{2-} or OH radical). The AgNPs attach themselves to the cell membrane of the target microbe and create an electron-dense region at the attachment site which causes intercellular matrix shrinking and finally rapture of the cell membrane [50]. The Ag^+ ions generated from the AgNPs interact with the intercellular contents, for example, binds to the active sites of the cell proteins and interrupts the normal metabolic pathways [51]. AgNCs, on the other hand, disrupt microbial function by generating reactive oxygen groups via the generation of Ag^+

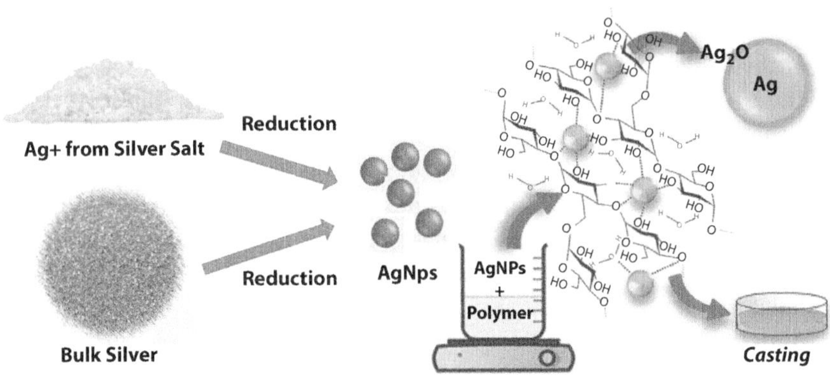

Figure 5.5 General preparation of AgNPs composite material. The AgNPs are made from either bottom up or top down approach via reduction. The AgNPs are then doped in the polymer melt and casted to make thin films. Similar approaches can be adopted for other metal oxide NPs-polymer nanocomposites Adopted with permission from [49] (Copyright 2019, with permission from Elsevier).

ions which damage the cell membrane [52]. ZnNPs induce resistance to microbes via two distinctive methods—(a) direct electrostatic interaction of positive ZnNPs with the negatively charged phospholipids and polysaccharides of the cell membranes causes charge-based damage, and (b) via the production of reactive oxygen groups at higher ZnNPs concentration which interacts with the cell membrane and intercellular organelles to damage it [53, 54]. The antimicrobial properties of TiNPs are from reactive oxygen groups generated by the nanoparticles [55].

One interesting thing to note for the incorporation of metal or metal oxide nanoparticles into polymer blends is the amount to which we can add the nanoparticles. It is usually seen that by increasing the number of nanoparticles in the polymer blend, the desirable properties of the nanocomposite generally increase but beyond a certain point, the nanoparticles cease to be well dispersed in the polymer and start aggregating. This causes drastic failure of the quality of packaging material. Seldom defective films are formed where there is an aggregation of nanoparticles. In summary, apart from imparting antimicrobial characteristics, the inclusion of metal or metal oxide nanoparticles also improves mechanical properties, decreases gas permeance, hence improving barrier properties, and thus shows good promise in being a nanocomposite material for the food packaging industry.

5.3.3 Other Nanostructured Inclusions: Nanoclays, Layered Double Hydroxides, Silicon Dioxide

Nanoclays are aluminum silicates that are two-dimensional sheets obtained as stacked clusters due to van der Waals's force in nature. They have ~1 nm thickness with several micrometers in other dimensions. These are low-cost materials and have been proven to impart good mechanical and barrier properties when added to polymer blends. This is due to their high surface area (~800 $m^2\ g^{-1}$) and large aspect ratio (~50–1,000) [56]. This makes the transport path for gas and moisture more tortuous within the polymer blend slowing down its movement and hence providing more effective barrier properties.

Abdollahi *et al.* observed a 40%–45% decrease in water vapor permeance by adding 3% of montmorillonite (MMT) in chitosan and it also increased the mechanical strength [57]. MMT/AgNps nanoparticles were used in the PVA matrix to create a nanocomposite for food packaging which showed low water adsorption capacity, more mechanical strength, and low visible light transmission [58].

A class of inorganic material, layered double hydroxide (LDHs), having the same structure as brucite is often blended with polymers to generate composites for the food packaging industry. There are three main types of nanocomposites with LDH and polymers—(a) the LDHs can be intercalated with the polymers where the polymer moves in and occupy the space between the layered sheets; (b) the LDHs is exfoliated into since layers and are introduced in the blend, or (c) the LDHs remain in their cluster form and are blended into the polymer melt [60]. Wu *et al.* incorporated Zn-Al LDHs into Glycerol plasticized Starch to increase tensile strength and reduce water vapor permeation rates (desirable characters of packaging materials) but at the cost of lower thermal stability [59]. Three-dimensional silver-based coordination polymers were prepared by self-assembly and were incorporated into acrylated epoxidized soybean oil biodegradable films (Figure 5.6a) to generate an antimicrobial coating to preserve food [61]. These films showed over 99.9% bacterial eradication when tested against four bacteria. Recently, Li and coworkers incorporated a nanoparticle prepared from a low-cost, biodegradable prolamin protein (Zein) into a matrix made of cellulose nanofibrils, derived from bacteria (Figure 5.6b) [62].

Incorporation of these protein nanoparticles showed improved mechanical strength (61% and 46% increase in Young's modulus and tensile strength as compared to pristine bacterial cellulose). They speculate that interactions due to hydrogen bonding between the zein nanoparticles and the matrix caused improvement of these mechanical properties [62].

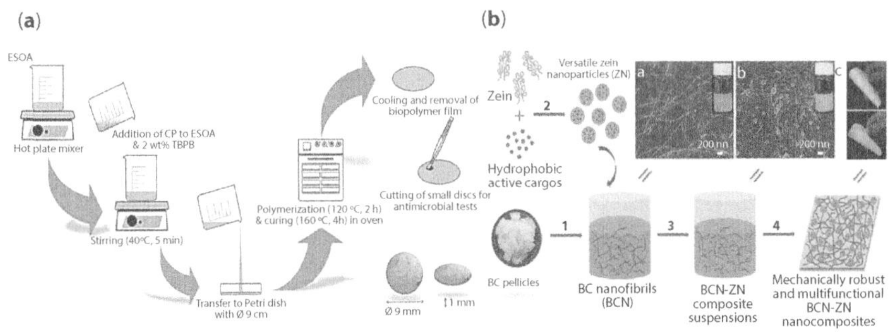

Figure 5.6 (a) Fabrication of Silver coordination polymer incorporated soya bean oil based biodegradable polymer films that shows excellent antimicrobial properties. (b) Fabrication of Zein (protein) nanoparticle based bacterial cellulose composites that shows improved mechanical properties (inset: SEM image and optical image of the pristine bacterial cellulose film and Zein Nanoparticle incorporated bacterial cellulose films). Reprinted with permission from [61] (Copyright 2021 American Chemical Society) and [62] (Copyright 2020 American Chemical Society).

These results show that low-cost inclusions like nanoclays, LDHs, and low-cost proteins in the biopolymer melts are effective in imparting desirable characteristics in the material which makes it suitable for use in the food packaging industry. Table 5.1 summarizes the different inorganic and

Table 5.1 Summary of different nanostructured inclusions in the biopolymer matrix.

Class	Inclusion	Matrix	Property enhancement	Reference
Nanostructured carbon materials	MWCNT	Gelatin	Decreased water and oil swelling, antimicrobial properties	[32]
	GO/CNT	PLA	67% reduction of oxygen transmission rate, 74% higher tensile strength, improved UV-vis protection	[33]
	GO	PVA	212% increase in tensile strength, 105% moisture adsorption reduction, improved barrier properties	[34]
	GO-AgNPs	PVA	4°C increase of glass transition temperature, 33% increase in elastic modulus, 42% reduction in moisture absorption, and improved antimicrobial properties	[35]
	CNT	Starch	Increase in film crystallinity, 100% increase in tensile strength, decreased oxygen permeability	[37]

(Continued)

Table 5.1 Summary of different nanostructured inclusions in the biopolymer matrix. (*Continued*)

Class	Inclusion	Matrix	Property enhancement	Reference
Metal and Metal oxide nanostructured materials	AgNPs/ Lignin	PLA	17.8% increase in tensile strength, better UV barrier, higher degradation temperature	[44]
	AgNPs	PHBV	Additional antimicrobial properties, 56% reduction of O_2 permeability, no change in mechanical and optical properties	[45]
	ZnNPs	PLA	17 and 18% decrease in CO2 and O2 permeance; antimicrobial properties	[46]
	ZnNPs	Chitosan	77% increase in the tensile module, one order decrease in vapor permeance, antimicrobial properties	[47]
	TiNPs	Chitosan	90% increase in tensile strength, improved wettability, decreased light transmission, and antimicrobial properties	[48]

(*Continued*)

Table 5.1 Summary of different nanostructured inclusions in the biopolymer matrix. (*Continued*)

Class	Inclusion	Matrix	Property enhancement	Reference
Nanoclay	MMT	Chitosan	~45% decrease in water vapor permeance, 20% increase in tensile strength, promising antimicrobial property	[57]
	MMT/Ag	PVA	56% increase in tensile strength, water adsorption capacity decreased by 75%, imparts lower light transmittance, antimicrobial property	[58]
Layered Double Hydrides	Zn-Al LDHs	Starch	Tensile strength doubled, 88% decrease in water vapor permeability, lower thermal stability	[59]

carbon-based nanostructured materials that have been studied in the literature for inclusion in the biopolymer matrix and their corresponding property improvements.

5.4 Concerns of Application of Nanostructured Inclusions in Biopolymers

The two major concern is with the use of these nanocomposite materials for food packaging: (a) what if the inorganic nanostructured inclusion of the polymer melt gets displaced from the polymer matrix and end up in the food itself and (b) what kind of environmental threats these nano-engineered materials have when the whole nanocomposite degrades in nature after their use as food packaging material. The dislocation of nanostructures from their polymer matrix depends on the condition of its use—the temperature, the humidity, the pH of the environment, and even the chemical composition of the food material in contact with the packaging material [63]. The size of the nanostructured inclusion is also crucial as for smaller particles, the diffusion is proportional to the inverse of the size of the particles [64]. The release of nanostructured inclusions is seen to increase with the increase in operating conditions. Also, lower pH conditions promote the release of nanostructures from the matrix [65, 66]. The study of the metabolism of nanostructures is essential not only for they can contaminate the food but also end up in the environment during degradation. Few studies show that in the environment of the mammalian gastrointestinal tract, the AgNPs get oxidized and agglomerates into larger particles [67]. Nanostructured materials from the dissociation of composites pose a grand threat to the environment. Nanostructured materials released into the soil eventually end up in water bodies and are taken in by plants [68, 69]. It is detrimental for plants to intake these nanomaterials, for example, carbon-based nanomaterials impede nutrient absorption and TiO_2 reduces photosynthesis and ion adsorption in plants [70]. Then, again, certain nanostructures promote plant growth [71]. This uncertainty indicates the need for more research in this area of toxicity due to nanostructured materials. For nanocomposites, apart from determining the property change due to nanostructure inclusion, a detailed study on leaching of these nanomaterials from the nanocomposites is very essential. Interestingly, research has been conducted recently to biologically synthesize the inorganic nanoparticle not only to make them more biocompatible but also eco-friendly by not producing hazardous waste materials during their production [72]. AgNPs were synthesized using aqueous extract of oil cake of *Madhuca latifolia* L. species which can be incorporated into biopolymers to be used as food packaging [73]. Recent approaches of production of AgNPs using extracts of *Salvadora persica* as the reducing agents to might be useful to replace toxic chemicals to produce these nanoparticles [74]. Such innovations might show a possibility to make biopolymers with inorganic inclusions less toxic and much more feasible option.

5.5 Conclusions

The application of nanostructured materials into polymer blends enhances the material property of food packaging material to a great extent. These inclusions solve many problems which hinder the practical use of biopolymers for extensive food packaging. The simple preparation of these nanocomposites is very attractive and the wide array of nanostructures and polymer matrix has allowed researchers to mix and match to find a nanocomposite with excellent packaging property and this makes research lucrative. Large-scale manufacturing of these nanocomposites will improve markets for biopolymers and, in turn, will help in the reduction of CO_2 as these are carbon neutral and inch toward solving the grand challenge of micro-plastic pollution. Before the commercialization of these nanocomposite materials, all the safety concerns must be addressed. There is a need for more scientific data concerning the dislocation of nanomaterials from the polymer matrix into the food and toxic effects of these materials on the ecosystem to assess the future of nanostructured materials and their large-scale usability.

Acknowledgment

The author acknowledges the University at Buffalo, The State University of New York, and Rensselaer Polytechnic Institute, New York for their support.

References

1. Sengupta, B., Priyadarshinee, R., Roy, A., Banerjee, A., Malaviya, A., Singha, S., Mandal, T., Kumar, A., Toward sustainable and eco-friendly production of coffee: Abatement of wastewater and evaluation of its potential valorization. *Clean Technol. Environ. Policy*, 22, 995–1014, 2020.
2. Kumar, A., Roy, A., Priyadarshinee, R., Sengupta, B., Malaviya, A., Dasguptamandal, D., Mandal, T., Economic and sustainable management of wastes from rice industry: Combating the potential threats. *Environ. Sci. Pollut. Res.*, 24, 34, 26279–26296, 2017.
3. Zhao, X. and Cornish, K., and Vodovotz, Y., Narrowing the gap for bioplastic use in food packaging: An update. *Environ. Sci. Technol.*, 54, 8, 4712–4732, 2020.
4. Kumar, A., Sengupta, B., Dasgupta, D., Mandal, T., Datta, S., Recovery of value added products from rice husk ash to explore an economic way for recycle and reuse of agricultural waste. *Rev. Environ. Sci. Biotechnol.*, 15, 1, 47–65, 2016.
5. Kumar, A., Priyadarshinee, R., Singha, S., Sengupta, B., Roy, A., Dasgupta, D., Mandal, T., Biodegradation of alkali lignin by Bacillus flexus RMWW II: analyzing performance for abatement of rice mill wastewater. *Water Sci. Technol.*, 80, 9, 1623–1632, 2019.

6. Arrieta, M.P., Peponi, L., López, D., López, J., Kenny, J.M., An overview of nanoparticles role in the improvement of barrier properties of bioplastics for food packaging applications, in: *Food Packaging*, pp. 391–424, Elsevier, London, U.K., 2017.
7. Luijsterburg, B. and Goossens, H., Assessment of plastic packaging waste: Material origin, methods, properties. *Resour. Conserv. Recycl.*, 85, 88–97, 2014.
8. Hopewell, J., Dvorak, R., Kosior, E., Plastics recycling: Challenges and opportunities. *Philos. Trans. R. Soc. B Biol. Sci.*, 364, 1526, 2115–2126, 2009.
9. Verma, R., Vinoda, K.S., Papireddy, M., Gowda, A.N.S., Toxic pollutants from plastic waste-a review. *Proc. Environ. Sci.*, 35, 701–708, 2016.
10. Kumar, A., Jash, A., Priyadarshinee, R., Sengupta, B., Dasguptamandal, D., Halder, G., Mandal, T., Removal of catechol from aqueous solutions by adsorption using low cost activated carbon prepared from Eichhornia crassipes. *Desalin. Water Treat.*, 73, 389–398, 2017.
11. Gong, J. and Xie, P., Research progress in sources, analytical methods, eco-environmental effects, and control measures of microplastics. *Chemosphere*, Article 126790, 2020.
12. Geyer, R., Jambeck, J.R., Law, K.L., Production, use, and fate of all plastics ever made. *Sci. Adv.*, 3, 7, e1700782, 2017.
13. Ezati, P. and Rhim, J.-W., Fabrication of Quercetin-loaded biopolymer films as functional packaging materials. *ACS Appl. Polym. Mater.*, 3, 2131–2137, 2021.
14. Turan, D., Gunes, G., Kilic, A., Perspectives of bio-nanocomposites for food packaging applications, in: *Bionanocomposites for Packaging Applications*, M. Jawaid and S. Swain (Eds.), pp. 1–32, Springer, Cham, Switzerland, 2018.
15. Oberlintner, A., Bajić, M., Kalčíková, G., Likozar, B., Novak, U., Biodegradability study of active chitosan biopolymer films enriched with Quercus polyphenol extract in different soil types. *Environ. Technol. Innov.*, 21, 101318, 2021.
16. Hu, B., Biopolymer-based lightweight materials for packaging applications, in: *Lightweight materials from biopolymers and biofibers*, pp. 239–255, ACS Publications, Washington D.C., U.S.A., 2014.
17. Vasile, C., Polymeric nanocomposites and nanocoatings for food packaging: A review. *Mater. (Basel).*, 11, 10, 1834, 2018.
18. Rhim, J.-W., Park, H.-M., Ha, C.-S., Bio-nanocomposites for food packaging applications. *Prog. Polym. Sci.*, 38, 10–11, 1629–1652, 2013.
19. Sawada, H., ISO standard activities in standardization of biodegradability of plastics—Development of test methods and definitions. *Polym. Degrad. Stab.*, 59, 1–3, 365–370, 1998.
20. Iwata, T., Biodegradable and bio-based polymers: Future prospects of eco-friendly plastics. *Angew. Chemie Int. Ed.*, 54, 11, 3210–3215, 2015.
21. Robertson, G.L., *Food packaging: principles and practice*, CRC Press, Florida, U.S.A., 2016.

22. Jiménez, A., Peltzer, M., Ruseckaite, R., *Poly (lactic acid) science and technology: Processing, properties, additives and applications*, Royal Society of Chemistry, UK, 2014.
23. Laycock, B., Halley, P., Pratt, S., Werker, A., Lant, P., The chemomechanical properties of microbial polyhydroxyalkanoates. *Prog. Polym. Sci.*, 38, 3–4, 536–583, 2013.
24. Lenz, R.W. and Marchessault, R.H., Bacterial polyesters: Biosynthesis, biodegradable plastics and biotechnology. *Biomacromolecules*, 6, 1, 1–8, 2005.
25. Jia, M.Z., Biodegradable plastics: breaking down the facts, Green Peace International, 2020. https://www.greenpeace.org/static/.planet4-eastasia-stateless/84075f56-biodegradable-plastics-report.pdf.
26. Barik, S., Khandual, A., Behera, L., Badamali, S.K., Luximon, A., Nano-Mg–Al-layered double hydroxide application to cotton for enhancing mechanical, UV protection and flame retardancy at low cytotoxicity level. *Cellulose*, 24, 2, 1107–1120, 2017.
27. Huang, J.-Y., Limqueco, J., Chieng, Y.Y., Li, X., Zhou, W., Performance evaluation of a novel food packaging material based on clay/polyvinyl alcohol nanocomposite. *Innov. Food Sci. Emerg. Technol.*, 43, 216–222, 2017.
28. Carbone, M., Donia, D.T., Sabbatella, G., Antiochia, R., Silver nanoparticles in polymeric matrices for fresh food packaging. *J. King Saud Univ.*, 28, 4, 273–279, 2016.
29. Lim, H.N., Huang, N.M., Loo, C.H., Facile preparation of graphene-based chitosan films: Enhanced thermal, mechanical and antibacterial properties. *J. Non. Cryst. Solids*, 358, 3, 525–530, 2012.
30. Peter, A., Tegla, D., Giurgiulescu, L., Cozmuta, A.M., Nicula, C., Cozmuta, L.M., Vagelas, I., Development of Ag/TiO2-SiO2-Coated food packaging film and its rile in preservation of green lettuce during storage. *Carpathian J. Food Sci. Technol.*, 7, 4, 88–96, 2015.
31. Yu, L., Lin, F., Xiao, W., Luo, D., Xi, J., CNT@ polydopamine embedded mixed matrix membranes for high-rate and long-life vanadium flow batteries. *J. Memb. Sci.*, 549, 411–419, 2018.
32. Pattanshetti, A., Pradeep, N., Chaitra, V., Uma, V., Synthesis of multi-walled carbon nanotubes (MWCNTs) from plastic waste & analysis of garlic coated gelatin/MWCNTs nanocomposite films as food packaging material. *SN Appl. Sci.*, 2, 4, 1–7, 2020.
33. Kim, Y., Kim, J.S., Lee, S.-Y., Mahajan, R.L., Kim, Y.-T., Exploration of hybrid nanocarbon composite with polylactic acid for packaging applications. *Int. J. Biol. Macromol.*, 144, 135–142, 2020.
34. Cao, Y.-C., Wei, W., Liu, J., You, Q., Liu, F., Lan, Q., Zhang, C., Liu, C., Zhao, J., The Preparation of graphene reinforced poly(vinyl alcohol) antibacterial nanocomposite thin film. *Int. J. Polym. Sci.*, 2015, 407043, 2015.
35. Cobos, M., De-La-Pinta, I., Quindós, G., Fernández, M.J., Fernández, M.D., One-step eco-friendly synthesized silver-graphene oxide/poly (vinyl alcohol) antibacterial nanocomposites. *Carbon, N. Y.*, 150, 101–116, 2019.

36. Faraguna, F., Pötschke, P., Pionteck, J., Preparation of polystyrene nanocomposites with functionalized carbon nanotubes by melt and solution mixing: Investigation of dispersion, melt rheology, electrical and thermal properties. *Polym. (Guildf).*, 132, 325–341, 2017.
37. Liu, S., Li, X., Chen, L., Li, L., Li, B., Zhu, J., Understanding physicochemical properties changes from multi-scale structures of starch/CNT nanocomposite films. *Int. J. Biol. Macromol.*, 104, 1330–1337, 2017.
38. Zou, X., Zhang, L., Wang, Z., Luo, Y., Mechanisms of the antimicrobial activities of graphene materials. *J. Am. Chem. Soc.*, 138, 7, 2064–2077, 2016.
39. Karahan, H.E., Wang, Y., Li, W., Liu, F., Wang, L., Sui, X., Riaz, M.A., Chen, Y., Antimicrobial graphene materials: the interplay of complex materials characteristics and competing mechanisms. *Biomater. Sci.*, 6, 4, 766–773, 2018.
40. Lukowiak, A., Kedziora, A., Strek, W., Antimicrobial graphene family materials: Progress, advances, hopes and fears. *Adv. Colloid Interface Sci.*, 236, 101–112, 2016.
41. Lu, X., Feng, X., Zhang, X., Chukwu, M.N., Osuji, C.O., Elimelech, M., Fabrication of a desalination membrane with enhanced microbial resistance through vertical alignment of graphene oxide. *Environ. Sci. Technol. Lett.*, 5, 10, 614–620, 2018.
42. Liu, S., Zeng, T.H., Hofmann, M., Burcombe, E., Wei, J., Jiang, R., Kong, J., Chen, Y., Antibacterial activity of graphite, graphite oxide, graphene oxide, and reduced graphene oxide: Membrane and oxidative stress. *ACS Nano*, 5, 9, 6971–6980, 2011.
43. Hulikere, M.M. and Joshi, C.G., Characterization, antioxidant and antimicrobial activity of silver nanoparticles synthesized using marine endophytic fungus-Cladosporium cladosporioides. *Process Biochem.*, 82, 199–204, 2019.
44. Shankar, S. and Rhim, J.-W., and Won, K., Preparation of poly (lactide)/lignin/silver nanoparticles composite films with UV light barrier and antibacterial properties. *Int. J. Biol. Macromol.*, 107, 1724–1731, 2018.
45. Castro-Mayorga, J.L., Fabra, M.J., Lagaron, J.M., Stabilized nanosilver based antimicrobial poly (3-hydroxybutyrate-co-3-hydroxyvalerate) nanocomposites of interest in active food packaging. *Innov. Food Sci. Emerg. Technol.*, 33, 524–533, 2016.
46. Marra, A., Silvestre, C., Duraccio, D., Cimmino, S., Polylactic acid/zinc oxide biocomposite films for food packaging application. *Int. J. Biol. Macromol.*, 88, 254–262, 2016.
47. Priyadarshi, R. and Negi, Y.S., Effect of varying filler concentration on zinc oxide nanoparticle embedded chitosan films as potential food packaging material. *J. Polym. Environ.*, 25, 4, 1087–1098, 2017.
48. Zhang, X., Xiao, G., Wang, Y., Zhao, Y., Su, H., Tan, T., Preparation of chitosan-TiO2 composite film with efficient antimicrobial activities under visible light for food packaging applications. *Carbohydr. Polym.*, 169, 101–107, 2017.

49. Ortega, F., García, M.A., Arce, V.B., Nanocomposite films with silver nanoparticles synthesized *in situ*: Effect of corn starch content. *Food Hydrocoll.*, 97, 105200, 2019.
50. Dakal, T.C., Kumar, A., Majumdar, R.S., Yadav, V., Mechanistic basis of antimicrobial actions of silver nanoparticles. *Front. Microbiol.*, 7, 1831, 2016.
51. Rai, M.K., Deshmukh, S.D., Ingle, A.P., Gade, A.K., Silver nanoparticles: The powerful nanoweapon against multidrug-resistant bacteria. *J. Appl. Microbiol.*, 112, 5, 841–852, 2012.
52. Javani, S., Lorca, R., Latorre, A., Flors, C., Cortajarena, A.L., Somoza, A., Antibacterial activity of DNA-stabilized silver nanoclusters tuned by oligonucleotide sequence. *ACS Appl. Mater. Interfaces*, 8, 16, 10147–10154, 2016.
53. Sarwar, S., Chakraborti, S., Bera, S., Sheikh, I.A., Hoque, K.M., Chakrabarti, P., The antimicrobial activity of ZnO nanoparticles against Vibrio cholerae: Variation in response depends on biotype. *Nanomedicine Nanotechnology, Biol. Med.*, 12, 6, 1499–1509, 2016.
54. Ng, Y.H., Leung, Y.H., Liu, F.Z., Ng, A.M.C., Gao, M.H., Chan, C.M.N., Djurišić, A.B., Leung, F.C.C., Chan, W.K., Antibacterial activity of ZnO nanoparticles under ambient illumination—The effect of nanoparticle properties. *Thin Solid Films*, 542, 368–372, 2013.
55. Parham, S., Wicaksono, D.H.B., Bagherbaigi, S., Lee, S.L., Nur, H., Antimicrobial treatment of different metal oxide nanoparticles: A critical review. *J. Chin. Chem. Soc*, 63, 4, 385–393, 2016.
56. Majeed, K., Jawaid, M., Hassan, A., Bakar, A.A., Khalil, H.P.S.A., Salema, A.A., Inuwa, I., Potential materials for food packaging from nanoclay/natural fibres filled hybrid composites. *Mater. Des.*, 46, 391–410, 2013.
57. Abdollahi, M., Rezaei, M., Farzi, G., A novel active bionanocomposite film incorporating rosemary essential oil and nanoclay into chitosan. *J. Food Eng.*, 111, 2, 343–350, 2012.
58. Mathew, S., Snigdha, S., Mathew, J., Radhakrishnan, E.K., Biodegradable and active nanocomposite pouches reinforced with silver nanoparticles for improved packaging of chicken sausages. *Food Packag. Shelf Life*, 19, 155–166, 2019.
59. Wu, D., Chang, P.R., Ma, X., Preparation and properties of layered double hydroxide-carboxymethylcellulose sodium/glycerol plasticized starch nanocomposites. *Carbohydr. Polym.*, 86, 2, 877–882, 2011.
60. Tronto, J., Bordonal, A.C., Naal, Z., Valim, J.B., Conducting polymers/layered double hydroxides intercalated nanocomposites, in Y. Mastai (ed.), *Materials Science - Advanced Topics*, pp. 3–30, IntechOpen, London, 2013.
61. Fernandes, T.A., Costa, I.F.M., Jorge, P., Sousa, A.C., André, V., Cerca, N., Kirillov, A.M., Silver (I) Coordination Polymers Immobilized into Biopolymer Films for Antimicrobial Applications. *ACS Appl. Mater. Interfaces, 13*, 12836–12844, 2021, 2021.

62. Li, Q., Gao, R., Wang, L., Xu, M., Yuan, Y., Ma, L., Wan, Z., Yang, X., Nanocomposites of bacterial cellulose nanofibrils and zein nanoparticles for food packaging. *ACS Appl. Nano Mater.*, 3, 3, 2899–2910, 2020.
63. Pillai, K.V., Gray, P.J., Tien, C.-C., Bleher, R., Sung, L.-P., Duncan, T.V., Environmental release of core–shell semiconductor nanocrystals from free-standing polymer nanocomposite films. *Environ. Sci. Nano*, 3, 3, 657–669, 2016.
64. Weiner, R.G., Sharma, A., Xu, H., Gray, P.J., Duncan, T.V., Assessment of mass transfer from poly(ethylene) nanocomposites containing noble-metal nanoparticles: A systematic study of embedded particle stability. *ACS Appl. Nano Mater.*, 1, 9, 5188–5196, 2018.
65. Soleimani, J., Ghanbarzadeh, B., Dehghannya, J., Baheri Islami, S., Sorouraddin, S.M., Comparative numerical study of titanium and silver nano-particles migration from nano-composite of polystyrene into simulants on experimental data basis. *Int. J. Food Eng.*, 13, 12, 0170091, 2017.
66. Hannon, J.C., Kerry, J.P., Cruz-Romero, M., Azlin-Hasim, S., Morris, M., Cummins, E., Kinetic desorption models for the release of nanosilver from an experimental nanosilver coating on polystyrene food packaging. *Innov. Food Sci. Emerg. Technol.*, 44, 149–158, 2017.
67. Axson, J.L., Stark, D.I., Bondy, A.L., Capracotta, S.S., Maynard, A.D., Philbert, M.A., Bergin, I.L., Ault, A.P., Rapid kinetics of size and pH-dependent dissolution and aggregation of silver nanoparticles in simulated gastric fluid. *J. Phys. Chem. C*, 119, 35, 20632–20641, 2015.
68. Kumar, A., Singha, S., Sengupta, B., Dasgupta, D., Datta, S., Mandal, T., Intensive insight into the enhanced utilization of rice husk ash: Abatement of rice mill wastewater and recovery of silica as a value added product. *Ecol. Eng.*, 91, 270–281, 2016.
69. Kumar, A., Sengupta, B., Kannaujiya, M.C., Priyadarshinee, R., Singha, S., Dasguptamandal, D., Mandal, T., Treatment of coke oven wastewater using ozone with hydrogen peroxide and activated carbon. *Desalin. Water Treat.*, 69, 352–365, 2017.
70. Singh, J. and Lee, B.-K., Effects of Nano-TiO2 particles on bioaccumulation of 133Cs from the contaminated soil by soybean (Glycine max). *Process Saf. Environ. Prot.*, 116, 301–311, 2018.
71. Zhu, Y., Xu, F., Liu, Q., Chen, M., Liu, X., Wang, Y., Sun, Y., Zhang, L., Nanomaterials and plants: Positive effects, toxicity and the remediation of metal and metalloid pollution in soil. *Sci. Total Environ.*, 662, 414–421, 2019.
72. Eltarahony, M., Abu-Serie, M., Hamad, H., Zaki, S., Abd-El-Haleem, D., Unveiling the role of novel biogenic functionalized CuFe hybrid nanocomposites in boosting anticancer, antimicrobial and biosorption activities. *Sci. Rep.*, 11, 1, 1–22, 2021.
73. Biswal, A.K. and Misra, P.K., Biosynthesis and characterization of silver nanoparticles for prospective application in food packaging and biomedical fields. *Mater. Chem. Phys.*, 250, 123014, 2020.

74. Arshad, H., Sami, M.A., Sadaf, S., Hassan, U., Salvadora persica mediated synthesis of silver nanoparticles and their antimicrobial efficacy. *Sci. Rep.*, 11, 1, 1–11, 2021.

6

Natural Biopolymeric Nanotechnology-Based Food Packaging Materials with Antimicrobial Properties

Hitesh Chopra[1], Pooja Mittal[2], Rupesh K. Gautam[3*] and Mohammad Amjad Kamal[4,5,6,7]

[1]*Chitkara College of Pharmacy, Chitkara University, Rajpura Punjab, India*
[2]*School of Pharmaceutical Sciences, RIMT University, Mandi Gobindgarh Punjab, India*
[3]*MM School of Pharmacy, Maharishi Markandeshwar University, Sadopur, Ambala, Haryana, India*
[4]*Institutes for Systems Genetics, Frontiers Science Center for Disease-Related Molecular Network, West China Hospital, Sichuan University, Chengdu, China*
[5]*King Fahd Medical Research Center, King Abdulaziz University, Jeddah, Saudi Arabia*
[6]*Department of Pharmacy, Faculty of Allied Health Sciences, Daffodil International University, Dhaka, Bangladesh*
[7]*Enzymoics, Novel Global Community Educational Foundation, Peterlee Place, Hebersham NSW, Australia*

Abstract
The environmental concerns raise issues like non-biodegradable nature of petro-chem–based food packaging materials and consumer demands to have better packaging materials and raise the need to develop biodegradable materials for various applications. The natural biopolymers can fulfill the requirements of consumers. There has been a growing interest of world toward bio-polymeric materials, due to exhaustive nature of fossils. Although natural polymers possess shortcomings like poor mechanical strength and reproducibility related issues, the use of nanotechnology in their production can abolish all the issues. The bio-polymeric nanocomposites have shown excellent mechanical and water retention properties and decreased gaseous exchange as well as intelligent

**Corresponding author*: drrupeshgautam@gmail.com

Annu, Tanima Bhattacharya and Shakeel Ahmed (eds.) Nanotechnology in Intelligent Food Packaging, (135–156) © 2022 Scrivener Publishing LLC

systems with nanosensors properties help in tracking the condition of packed food. Moreover, the active moieties can also be incorporated to it to impart desired functional properties such as anti-microbial antioxidant, biological, and biocatalyst properties. This chapter showcases deep review of various natural biopolymers which can be used for nanotechnology with anti-microbial properties for food packing applications.

***Keywords*:** Nanotechnology, nanocomposites, biopolymers, food packing, natural polymers, anti-microbial, hydrogels, starch

6.1 Introduction

A vast variety of food materials, biodegradable as well as nonrenewable materials, are being produced daily, which are being utilized as food packaging materials. Modifications are also done to improve their properties. The waste materials present in our environment can cause greater hazards to the human body and degrade or harm the ecosystem also. However, creation of biopolymeric materials for the food packaging by using these waste products can be an excellent idea to save the human being as well as climate. Nanotechnology can impart a great impact to the food and packing materials. Nanotechnology-based polymers, polysaccharides, and proteins can impart good role in area of food packaging by providing the safety to food, enhancing the shelf life along with the antibacterial properties. The antimicrobial and antioxidant activities can be added by the various packaging materials and also by the various other components of the natural origin. Generally, the trend in the packaging had been evolved from the inert as well as polluted plastic waste, which was processed to impart the antimicrobial properties biodegradable and edible type of packaging for the storage of food materials. By altering the public perception and industry focus from the conventional packing materials to biocompatible, biodegradable, and antimicrobial materials from the waste will enhance the shelf life of the products as well. This idea will be able to reduce the agricultural waste as well as plastic pollution from the environment. Antimicrobial properties can also be given by various natural polysaccharides like chitosan, proteins, cellulose derivatives, and starch from vegetable and animal origin and some biopolymers like polylactic and polyvinyl alcohol can also be used (individually or in blend with other components) as antimicrobial compounds. These natural antimicrobial compounds are found in plants such as clove, cinnamon, thyme, ginger, and organo, among others, and have the potential to be used as antimicrobial packaging materials. Pediocin, nisin, and other bacteriocins, for example, can be isolated from the waste products of bacteria and fungi. Various synthetic compounds can also be used for other food industries such as bread, vegetables, and pastry packaging. In this chapter, we explained the major recent findings in biodegradable and antimicrobial packaging, which can be made more efficient by the use of nanotechnology-based principles [1].

6.2 Natural Antimicrobials Used in Packaging of Food Products

Antimicrobials intercalated with food packing materials can serve a promising technology that impregnates antimicrobials compounds into the packaging materials and delivers them over a longer period of time, imparting their antimicrobial properties to the food packed inside and extending their shelf life. They have different effects on the different types of pathogens depending upon their physiology. Identification of the microbes can supposed to be very helpful for choosing the antimicrobial compound for a particular packing of the products. Microbes can be identified by the nature of their cell wall as they are of basic two types: Gram +ve and Gram −ve. Similarly, on the basis of oxygen requirements: aerobes and anaerobes; on the basis of growth stages: spores and vegetative cells; optimal growth temperature: mesopills, thermophills, psychotropics, etc. Some antimicrobials act on the inhibition of cell wall growth or other acts by inhibiting essentials metabolic pathways of microbes. For example, lysozyme causes the destruction of cell wall of microbes without interfering the metabolic pathways, while EDTA (ethylene diamine tetra acetic acid) and lactoferrin act like the coupling agents for charged polymers. Therefore, antimicrobials perform two duties: microcidal and microstatics. The most used antimicrobials compounds which are being utilized in food packaging are mentioned in Table 6.1 [2, 3].

Table 6.1 Potential antimicrobials for the food packaging.

Class	Antimicrobial agents and biopolymers	Applications
Plant extracts [8, 39]	Allyl-isothiocyanate, brassicaerucic acid oil, grape seed extract, cinnamaldehyde, eugenol, hop beta acid, rosemary oil, oregano oil, basil oil, other essential oils.	Active against *E. coli*, *S. aureus*, *S. typhimurium*, etc. LDPE (low density polyethylene films) films are active against *E. coli* for the packing of cheddar cheese etc.
Polysaccharides and derivatives [4, 9, 10]	Starch, cellulose, and pullulan.	LDPE films of various agents and nisin-coated polymeric films can be prepared which act on *L. monocytogenes*. HPMC (hydroxyl propyl methyl cellulose) films for the packing of tofu.
Proteins and enzymes [12]	Corn zein, glucose-oxidase, lactopeoxidase nisin, milk proteins, protein isolate, peanut-protein whey-protein isolates, pediocin, wheat-gluten, collagen/gelatin lysozyme, subtilin, lacticin.	Cellulose and gelatin films have excellent gas barrier and mechanical properties. Nisin is thought to be added in almost all the agents to prepare films because of its strong bactericidal effects. Nisin treated films and alginate and nisin coated films can be made which acts on bacteria's, yeast and molds.
Derived from lipids [16, 32]	Rice bran wax, beeswax, sugar cane wax, bay berry wax, carnauba wax,	They are particularly used to prepare the water resistant coating over the packaging materials. The coat will act as water gas barrier and increase the shelf life of the products.
Chelating agents [52]	EDTA	When we combine zein with disodium-ethylene-diamine-tetra-acetic acid (Na_2EDTA) and lysozyme, it will destroy the microbes.

6.3 Types of Various Biopolymers with Antimicrobial Activities

6.3.1 Starch as Nanocomposites for Food Packaging

Starch is the most common polysaccharide that is used in packaging of food materials. Jayakumar *et al.* formulated starch-PVA–based composite film by incorporation of nanospheres of ZnO and phytochemicals such as nutmeg oil and Jamun extract [1]. It showed anti-microbial activity against *Salmonella typhimurium*. The anti-microbial action of film can be accounted due to presence of nutmeg oil and ZnO NPs. Similar results has also been proved by various other researchers. The film showed pH sensitive behavior as its color changed due to Jamun extract presence. At low pH, film showed dark violet color changes to cherry red and showed brownish color in basic pH and light green at neutral pH.

Meira *et al.* developed starch/halloysite/nisinnanocomposite films that showed anti-microbial properties [2]. The presence of halloysite causes alteration in the physical properties, while other characteristic parameters such as Young's modulus and tensile strength decreased with addition of nisin. The anti-microbial property was evaluated against the *Listeria monocytogenes*, Clostridium perfringens, and *Staphylococcus aureus* and showed promising results when inoculated on Minas Frescal cheese.

Issa *et al.* formulated packaging material film as antibacterial film which was made up of sweet potato starch (PS) [3]. The study was conducted for over 8 days on baby spinach leaves. The film showed no sensory effect on the product. The presence of higher amount of Thyme oil is lead to more desirable odor, color, and overall acceptability. The film was able to control the growth of *E. coli* and *S. typhimurium*.

Nafchi *et al.* developed films composed of nanorod-based zinc oxide derived from bago starch and bovine gelatin [4]. Characterization of bionanocomposite films was performed. It could be correlated with the concentration of ZnO nanorods. The presence of water plays significant role as plasticizer. As the concentration of water decreases, the flexibility and E of films decreased and inversely TS and YM increases. The nanocomposite films of starch, montmorillonite, carvacol, and thymol were prepared using extrusion method [5]. In comparison to neat thermoplastic starch used, the mechanical and thermal properties were increased many folds and showed controlled delivery of essential oils used. The films showed antimicrobial effect on the *Botrytis cinerea*–inoculated strawberries.

Researchers developed films based on thermoplastic starch and layered silicates, using Carvacrol as model drug [6]. It was found that the concentration of clay directly affects the TM and thermal resistance of film. Higher concentration of clay caused formation of intercalation, due to complexes formation and hinders the release of Carvacrol, thus increasing the half-life of drug.

Xin *et al.* formulated zein/PS-based film of curcumin matrixed in chitosan nanoparticles [7]. Organoleptically, the film was smooth and round. The preservative efficacy of *Schizothorax prenati* fillets was evaluated, and films were able to delay the physicochemical changes in the fillets and increased shelf life by 15 days and showed antimicrobial activity.

6.3.2 Cellulose as Nanocomposites for Food Packaging

Dairi and colleagues used triethyl acetate, cellulose acetate, silver nanoparticles, gelatin-modified montmorillonite nanofiller (AgM), and Thymol (Th) to create nanofilms. Curcuma longa tuberous extract was used to make silver nanoparticles in a biological way. The resulting silver particles had a spherical appearance and a face centric cubic crystalline structure. The addition of clay raises the glass transition temperature of cellulose acetate, but thymol decreases at the same time. However, the presence of clay improved thermal stability. The films demonstrated both antimicrobial (against *E. coli*) and antioxidant activity (as determined by the DPPH test) [8].

Marrez *et al.* formulated nano composite films of cellulose acetate and nano silver that showed anti-bacterial activity against the foodborne pathogens [9]. The silver nanoparticles of size 5 to 18 nm were synthesized using the acids such as pyrogallic acid and gallic acid, quercitin, and rutin with MIC in range of 16.7 to 46.7 $\mu g\,ml^{-1}$. The antibacterial activity was checked against the *S. aureus*, *B. cereus*, *S. typhi*, *E. coli*, *K. pneumoniae*, and *Pseudomonase* spp. Despite of presence of quercitin, there was no cytotoxic effect on the larvae.

Saravanakumar and his group worked on the development of polymeric films with antibacterial properties, using the sodium alginate (as plasticizer), copper oxide nanoparticles embedded in cellulose nanowhisker [10]. The cellulose nanowhiskers improved the barrier properties and limited moisture penetration to the food. While due to presence of CuO NPs, the microbial contamination was minimized. The films thus prepared showed antibacterial activity against the *S. aureus*, *E. coli*, *Salmonella* sp., *C. albicans*, and *Trichoderma* spp.

Lazić *et al.* evaluated silver nanoparticles coated with dextran coating on antimicrobial films of cellulose nanofibrils [11]. The dextran not only was able to act as dispersing media for silver nanoparticles but also acts as moisture sealant also to reduce oxygen permeability and enhanced the shelf life of the food against the bacterial growth. It also reduced the oxygen transmission rate and hydrophilicity also it inhibited *E. coli* after five repeated cycle up to 99.9%. up on exposure of 24 hours. The presence of dextran into film matrix caused denser and tightly packed film structures.

Gedarawatte *et al.* worked on the controlling the spoilage of meat using bacterial cellulose nanocrytals entrapping the nisin and effective against the lactic acid [12]. The formation of nanocrystals occurred through hydrolysis and complexation reaction. The control nanocrystals showed zeta potential of −43 mv, while drug loaded showed ≥−30 mV. The bacterial cellulose-based film was able to shown the response against the two gram-positive vacuum-packaged beef spoilage bacteria, *L. rhamnosus* LBM1 and *L. mesenteroides* LBM2, and showed promising results for its use as antimicrobial film.

Similarly, the Pirsa worked on the active and intelligence-based bacterial cellulose films coupled with polypyrrole–zinc oxide nanocomposite [13]. The films thus formed were used in the packaging of chicken thigh, and effect of storage time and temperature was evaluated based on central composite design. It was found that film decreased the growth of microbial load and controlled the variation on pH. It was estimated to increase the rheological properties and shelf life of chicken thigh, thus acting as antimicrobial and intelligent film.

6.3.3 Protein as Nanocomposites for Food Packaging

Mohammadi *et al.* formulated whey protein-based biodegradable film which was incorporated with nanofiber (CSNF) [14]. Due to presence of CSNF, the water solubility and permeability both decreased. The CSNF were finely distributed throughout, while after addition of CiEO, the TS decreased. The film also showed UV light barrier properties along with antibacterial properties, which may be helpful in extending shelf life of perishable food products.

Yu *et al.* developed multifunctional soy protein-based films, using cellulose nanofibril, lactic acid, and *Cedrusdeodara* pine needle extract (PNE) [15]. The TS of film enhanced due to presence of cellulose nanofibril and presence of PNE enhanced the light barrier property and antioxidant property along with antimicrobial property against the food borne pathogen, such as *Escherichia coli* O157:H7, *Staphylococcus aureus*, *Salmonella* typhimurium, and *Listeria* monocytogenes.

Kaimi *et al.* worked on probiotic nanocomposite film of cellulose nanofiber, polydextrose incorporated whey protein isolate having Lactobacillus plantarum as probiotic [16]. The cellulose nanofiber, whey protein, and polydextrose presence lowered the water vapor permeability, and mechanical properties were enhanced. The film showed inhibition against the *P. aeruginosa*, *S. aureus*, and *E. coli* bacteria.

Ghadetaj *et al.* formulated whey protein isolate–based films containing emulsions and nanoemulsions essential oil of *Grammosciadium pterocarpum* Bioss [17]. The FE-SEM results provided evidence that the presence of nanoemulsion enhanced the integrity of film as well as the mechanical properties but decreased the water vapor transmission rate. The films also showed antimicrobial and antioxidant potential.

Arfat *et al.* formulated the nanocomposite films of fish protein isolate (FPI) and fish skin gelatin (FSG) by utilizing the essential oil from basil leaf along with Zn oxide nanoparticles [18]. With increase in the level of basil leaf essential oil, the TS decreased and E increased. However, as the ZnO NPs were added, the TS increased and E decreased. The films showed antibacterial action against the food borne pathogenic bacteria and helped in extending the shelf life of stored product.

Xiao *et al.* developed films using the cellulose nanocrystals, soy protein isolates, and zinc nanoparticles [19]. The TS, water vapor permeability, and water resistance properties showed comparable differences due to presence of cellulose nanocrystals. The hydrogen bonding between moieties was found to be responsible for compact and homogeneity in film structure.

The researchers developed chitosan, potato protein, linseed oil, and ZnO NP-based biopolymeric films for meat storage [20]. The ZnO NPs caused improvement in transparency and TS, while linseed provided good elastic properties. The films showed excellent moisture barrier capability. The raw meat was stored at 4°C and was acceptable until the 7 days of storage and reduced chances of pH variation and bacterial counts.

6.3.4 Chitosan as Nanocomposites for Food Packaging

Pandey and his co-workers formulated silver nanoparticles based on chitosan-medicated green synthesis and blended with the PVA to form the fibrous composite nanolayers using electrospinning technique [21]. The nanocomposite confirmed the antimicrobial activity against the both gram positive (*Listeria monocytogenes*) and negative species (*Escherichia coli*).

When meat was stored in the packaging material, it showed extension of shelf life of material to one week.

Cruz *et al.* compared antimicrobial potential of low and medium molecular weighed chitosan and organic acids with commercial mixtures for preservation of meat products as Sulac-01® and Articoat DLP-02® [22]. The low and medium weighed chitosan exhibited highest antimicrobial activity.

Lin *et al.* developed chitosan/okra powder/nano-silicon aerogel composite films [23]. The films were shown to have good mechanical and barrier properties. The films had anti-bacterial action against the both positive and negative bacteria that promote its use as food packaging material.

The group of researchers molded blends of poly (ε-caprolactone) (PCL) with chitosan and grapefruit seed extracts (GFSE) [24]. The effect of variation of GFSE/chitosan layer was evaluated. However, the presence of GFSE made films more brittle and films were effective against the *E. coli* and *P. aeruginosa* for about 120 hours. The packaging application was evaluated against salmon for growth of *E. coli* and was found effective.

Researchers developed food packaging based on Mahua oil using polyurethane and chitosan added in various concentration along with ZnO nanoparticles [25] Due to presence of ZnO, the antibacterial properties were enhanced. The film was found to be biodegradable in nature as, after 28 days, the 86% of film eroded and extended the shelf life of carrot for about 9 days.

Niu *et al.* developed films of cellulose nanofiber, treated with rosin and polylactic acid as reinforcement filler, and used chitosan for coating purpose [26]. The FTIR results confirmed the esterification of cellulose nanofibril with rosin and greatly affected the mechanical properties of film. The film showed inhibitory action against the *E. coli* and *B. subtilis*.

Mohammadi *et al.* developed nanofiber with antibacterial activity by electrospinning gelatin, chitosan, and 3-phenyllactic acid (PLA) [27]. The presence of PLA improved the microstructure of fibers. At lower pH, the formation of hydrogen bonding takes place between chitosan and PLA and, in turn, decreased the crystallinity. The fiber was able to reduce the growth of *Salmonella enteric*, Enteritidis, and *Staphylococcus aureus* in 30 min.

Researchers investigated the antimicrobial films made of chitosan and ZnO NPs [28]. The ZnO NPs formed were added to chitosan, chitosan/glycerol, and chitosan/PVA. Out of them, films with PVA showed higher strain break and barrier properties. The films were able to restrict the growth of *Brevibacterium lactofermentum* and *Escherichia coli*.

Wang *et al.* fabricated nanocomposite films of carboxymethyl chitosan and nano MgO [29]. There exists strong dense layer of MgO on the CMCS matrix. The films showed better thermal stability, UV barrier properties, as well as water insolubility and showed prominent results in food packaging applications especially in case of water rich foods. The films showed anti-microbial property against *Listeria monocytogenes* and *Shewanella baltica.*

6.3.5 Silica as Nanocomposites for Food Packaging

Silica, along with carbon nanoparticles, has been used widely in chemical polymer industry. Silica nanoparticles have enormously large surface area and smaller particle size, making it prominent candidate for nanoscale application. The nanoscale silica can be fabricated by various methods as reported by researchers [30–32].

Biswas *et al.* fabricated antimicrobial thin film of Ag-assisted bio-based silica-carbon nanoparticles, using 3D printing technique [33]. The silica-carbon hybrid was obtained via pyrolysis of rice husk. The antimicrobial activity was evaluated on the inoculum of *Salmonella* Enteritidis. On increasing the concentration of Ag, the more bacteriostatic effect was observed. The silver release test was conducted in chicken thigh, and evaluated via AAS. The films were stored at 4°C for 96 hours. It was found that the films were able to retard release of Ag to the product making extension in shelf life of product.

Bi *et al.* prepared silicon dioxide nanoparticles nanofilms (nano-SiO_2) by using D-α-tocopheryl polyethylene glycol 1000 succinate (TPGS) into chitosan (CS) films [34]. The strength of film was due to compact inner microstructure and hydrogen bonding. The films formed were found to have lower moisture content, water vapor rate, and oxygen permeability and were able to increase the stability of soybean oil.

Wu *et al.* prepared chitosan films incorporated with mesoporous silica nanoparticles and chitosan by solvent vesting method [35]. The films exhibited pH-responsive nature and showed inhibitory action against the *Staphylococcus aureus* and *Escherichia coli.*

The researchers developed cinnamon essential oil-based mesoporous silica nanoparticles, for protecting and preserving the white mushrooms (*Agaricusbisporus*) from CNRMA 03.0371 strain (*Mucorcircinelloides*) and FJ09 species (*Mucor sp.*) [36]. The films were prepared using the casting method which resisted the growth of microbes in white mushrooms.

The nanosilica coating along with chitosan has been applied on the Loquat fruit for testing its chilling injury symptoms [37]. The fruit was stored at 5°C for 40 days. The decrease in total soluble solids and titratable acidity was also evaluated. Increase in levels of glucose and fructose has also been found along with the other antioxidants.

Venkatesan *et al.* developed thin transparent films of SiO_2, and solvent casting technique was used [38]. Due to presence of silicon dioxide, the tensile strength increased along with increase in contact angle and decreased oxygen permeability and showed antimicrobial activity against the *S. aureus* and *E. coli*.

Melendez *et al.* prepared poly(3-hydroxybutyrate-co-3-hydroxyvalerate) (PHBV), eugenol, and silica nanoparticles films, possessing antimicrobial activity [39]. The films were prepared using electrospinning technique and annealing later. The films were presented with high mechanical strength and barrier toward water vapor and limonene. The antimicrobial activity was found against *S. aureus* and *E. coli*.

6.3.6 Carbon Nanoubes as Nanocomposites for Food Packaging

The carbon nanotubes are the huge cylinders of large molecular weight with hexagonal arrangement of hybridized carbon atoms. It may have wall of single and multiple layers.

The carbon nanotubes have been found to load cinnamaldehyde, polylactic acid, and poly(ε-caprolactone) [40]. The films were prepared using the solvent evaporation method. The film was found to possess UV barrier properties along with antimicrobial action against the *Staphylococcus aureus* and *Escherichia coli*. The UV barricade features were mainly due to the presence of aromatic ring in the cinnamaldehyde. The film was able to release the cinnamaldehyde entrapped for 21 days. It has also been found to increase the shelf life of shredded cooked meat, for about 40days [41]. The carbon nanotubes along with the allyl isothiocyanate as film packaging material was found effective against the *Salmonella choleraesuis* on shredded chicken. The presence of allyl isothiocyanate has decreased the TS of film and increase the E. Due to presence of allyl isothiocyanate the chances of microbial contamination decreased and reduced the changes in color.

Researchers prepared nanocomposite films of carbon nanotubes, gallic acid, and carboxymethyl konjac glucomannan and gelatin as matrix [42]. The amide linkage and hydrogen bonding was responsible for overall strength of film and further confirmed by FTIR study. The films exhibited strong activity against the *Staphylococcus aureus* and *Escherichia coli* and was found to possess shelf life–extending property.

The graphene oxide along with multiwalled carbon nanotubes were treated with furcellaran films (FUR) and Silver nanoparticles [43].

Liu *et al.* developed electrospunned fibers of polylactic acid, carbon nanotubes, and chitosan to prepare the packaging film for the packing of strawberry [44]. The chitosan played a major role, as the concentration of chitosan increased, the mechanical properties and swelling ratio also increased first and degraded further after 7% wt. Similar observation was found in antimicrobial property against gram positive and negative strain.

6.4 Recent Patents in the Field of Nanocomposite Food Packaging Applications

Various patents have been filed and granted for the inventions related to the antimicrobial nanotechnology-based food packaging. These section overviews about the recent patents granted/published.

The US patent no. US9332751B2 discloses the use of halloysite nanotubes as antimicrobial food packaging material [45]. Carvacrol and cinnamaldehydewere used as antimicrobial agent as they are well evidenced for their antimicrobial activity against *Campylobacter jejuni*, *Escherichia coli*, *Listeria monocytogenes*, and *Salmonella enterica*. The benefit of using halloysite nanotubes lies in the as they have empty cavities that present greater area for them to get entrapped and release the agents for longer interval of time at controlled rate. They also provide advantage such as easy loading of active components via sonication and vacuum filling process. The nanotubes has tortuous path that does not let the gas molecules to pass through it and enter the food medium. The hydrophilic surface of nanotubes provides high aspect ratio and makes them effective feelings for polymeric nanocomposites, thus preventing gas permeation and acts as barrier for food packaging.

As per the US patent no. US20100216908A1, the inventors disclosed method of preparing antimicrobial polymer nanocomposite using the organoclay [46]. The process included the mixing the antimicrobial agent with the clay to form the organoclay and later dispersing the organoclay with the polymer matrix. The nanocomposite thus formed has advantage of not leaching out the antimicrobial compound. Earlier the method of producing antimicrobial polymer included the addition of inorganic/organic compounds to polymer matrix or chemically bonding of the biocidal moieties to polymer matrices. However, due to poor compatibility between biocide and polymer, there occurs decrease in mechanical properties and other properties required. Moreover, it also leaches biocides easily leading to effect for lower time and environmental effects too. However, the method as described in invention is less prominent to leaching issue.

The Chinese patent no. CN105907105A describes use of zein fibers along with nano zinc for their use as food packaging material [47]. Zein exists as natural polymer that has good biocompatibility and is biodegradable in nature and on degradation produces amino acids and polypeptides. The nano zinc oxide produces antibacterial action here.

The patent no. CN102816440B describes the method of preparing food packaging material using the isolated soy protein and nano silver antibacterial composites [48]. The material is prepared by mixing the isolated protein to silver nitrate, *in situ* reduction of silver into silver nano particles followed by preparation on composite membrane structure. Here, the soy protein acts as base material along with reducing agent and stabilizer for nano silver. The formulation has antibacterial effects. The isolated soy protein is used because it is green in nature and utilizes reducibility of tyrosine in soy protein isolate is good in alkaline conditions. Moreover, due to presence of various groups such as amino, carboxyl presents good affinity toward the silver.

The World Patent WO2014153210A1 discloses use of edible films formed from plants or their parts [49]. The coatings described here utilize cellulose nanomaterial and nano calcium carbonate. The coatings thus formed were able to sustain against the moisture loss and UV damage to the fresh fruits both pre- and post-harvest. The antimicrobial agents such as thyme oil, clove oil, oregano, lemongrass, marjoram, cinnamon, coriander, or combinations thereof can be selected as additive agents.

CN104829874A describes method of preparation of gelatin and ZnO nanocomposite film. The gelatin used here is derived from Tuna head [50]. The use of Tuna head helps in reusing the processing left overs. The ZnO nano solution along with gelatin is mixed and forms a film forming solution along with the antioxidant and plasticizer on a polyvinyl chloride crystal plate followed by scraping, drying, and peeling of film. The film thus formed has capacity to inhibit the lipid peroxides and pathogenic microorganisms from growth and to prolong the shelf life of food and enhance the food quality.

The Chinese patent no. CN104610657A discloses the method of preparation of antimicrobial nanocellulose-based bottle packaging for food [51]. The bottle packaging is mainly for milk products. The bottle utilizes polylysine or volatile oil as fungi stat, xylogen or nanocellulose as photoresist, paraffin as lubricant, sodium chloride as inorganic salt, and poly(oxyethylene glycol) 400 or Liquid Macrogol as auxiliary components.

6.5 Types of Structures

Hydrogels are three dimensional, cross-linked networks of polymers, capable of imbibing large amount of water or biological fluids. The extent of cross-linking between polymer chains helps in controlling the pore size, amount of swelling, and water carrying capacity, leading to enhancement in its bioavailability. Hydrogels, a "smart drug delivery system", has shown an outstanding feature in entrapping the drug within its entangled polymeric network without dissolving the polymer. They are considered as smart gel because of their quick property to receive the stimuli and to act according it [48–50]. Different kinds of stimuli that have been investigated in varying the tone of drug delivery are represented in Figure 6.1.

Hydrogels are multifaceted; they are commonly used in clinical application in wide range including tissue regeneration, regenerative medicine biomedical application anti-tumor activity osteoarthritis therapy, antibacterial activity revascularization, etc. Presently, hydrogel has explored itself as a reliable drug delivery tool to load the nanoparticle or drug due to its ability to form a sustainable or controlled depot for a prolonged period within the targeted site, thus ideal for the local delivery.

Main subject of this review is to address a short introduction to hydrogel, its passage from ancient to modern age, formation and functioning, classification, application, and future aspects.

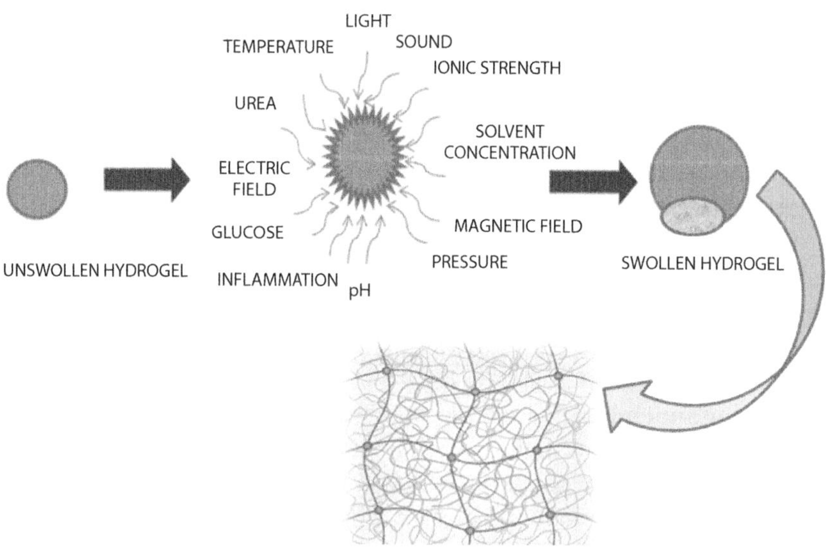

Figure 6.1 Stimuli-responsive hydrogels.

The terms gel and hydrogels sound interchangeable but there lies a difference between them.

Gels are semisolid systems consisting of dispersions of small or large molecules in an aqueous liquid vehicle rendered jelly-like by the addition of a gelling agent. Hydrogels are colloidal gel in which water is a continuous phase, which absorbs large amount of water cross-linking facilitates insolubility in water and provides required mechanical strength and physical integrity [51].

6.5.1 Hydrogels Preparation

According to definition, hydrogels are polymeric network. These hydrophilic polymers can absorb water in different fraction ranging from several thousand time of their own molecular weight as well as density of the water-loving group present on the polymer. Their water absorbing property is mainly due to the presence of hydrophilic group like OH, -CONH–, -$CONH_2$–, and –SO_3H present on the backbone of the polymer. To form stable linkage hydrogels, polymers allow binding, either noncovalent linkage like hydrogen bond, ionic, or hydrophobic bonding and physical entanglements, or by covalent cross-linkages [47]. All these methods will adequately permit the polymer chain to remain in contact with solvent molecule and swell fully in solvated state. A general presentation for the preparation of hydrogels is shown in Figure 6.2.

6.5.2 Applications of Hydrogels in Packaging of Food

Hydrogels can be potentially used for the packaging of food items because of their helpful nature by virtue. Their applications in food technology can be considered by virtue of their swelling properties on absorption of water. Likewise, absorbent pads and water/moisture removal systems have gained a lot of interest because of their potential to reduce the microbial content, and they can act as antimicrobials. Absorption of fluid by the packaging during the storage time is essential as it will terminate the growth of microbes in the food. Hydrogels have vast properties and they acts by absorption of solvents/fluids, which is the reason behind their use as food packaging systems. Moreover, the absorbent materials should be such that it should maintain the good presentation of the food and should provide this sensory attribute to the food at lower cost. It should be able to maintain the integrity of the food system as well as should be able to enhance the shelf life of the products. Considering the all advantageous and the

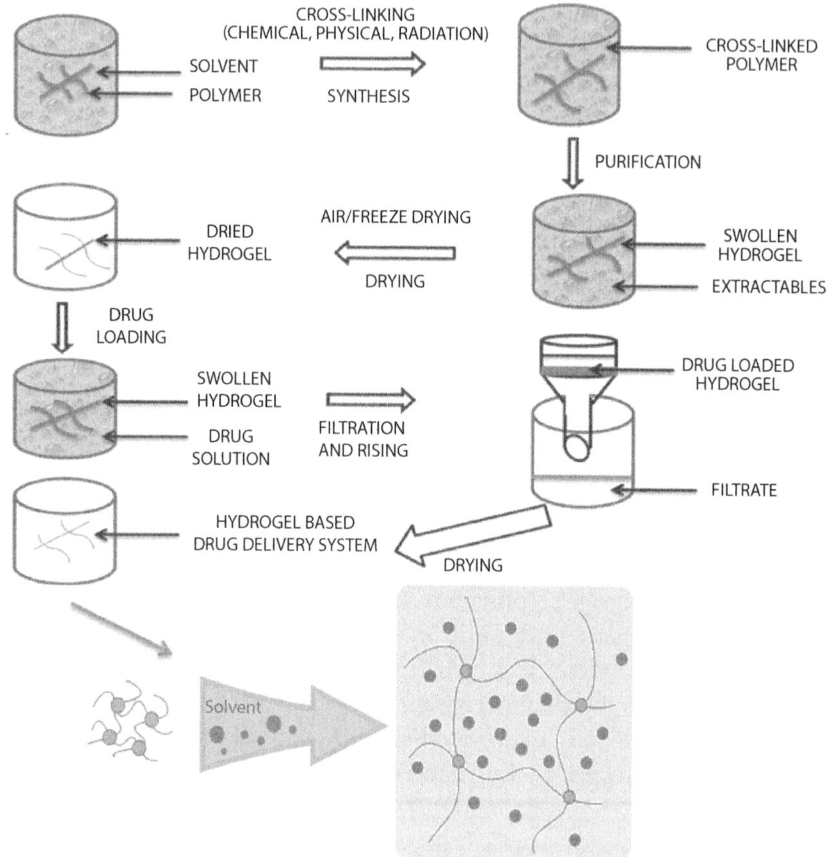

Figure 6.2 Various steps for the formulation of hydrogels.

requirements, the hydrogels can be the best fitted systems for the packaging of the food systems [46–49].

6.6 Conclusion and Future Prospective

From the last decade, the applications of biodegradable materials are highly demandable as they are the eco-friendly and bio-decomposable. Nanotechnology is providing the promising facts in this regard as utilization of this technique could enhance the food protection and can decrease the wastage and spoliation of the foods. There are many other strategies that can be used to combat the food loss or food spoilage. However, antimicrobial packing is gaining much of researcher's interest and antimicrobial packing along with the use of nanotechnology is highly in demand. Many natural and abundant polymers can be used for serving the purpose, but they cannot be used in their crude stage. All the natural and synthetic polymers can be used with some modifications like blending or cross-linking with other polymers so that the requirements of the degradable industry can be fulfilled.

In future, the integration of nanotechnology along with these food grade materials or the food packaging polymers can prove to be more beneficial. They can increase the mechanical strength and elasticity of the food packaging materials. Some antimicrobial compounds suffer from the high loss rate because of their internal volatilization rate, and further research is necessary to improve the durability and efficiency of these materials.

References

1. Jayakumar, A., Heera, K.V., Sumi, T.S., Joseph, M., Mathew, S., Praveen, G., Nair, I.C., Radhakrishnan, E.K., Starch-PVA composite films with zinc-oxide nanoparticles and phytochemicals as intelligent pH sensing wraps for food packaging application. *Int. J. Biol. Macromol.*, 136, 395–403, 2019.
2. Meira, S.M., Zehetmeyer, G., Scheibel., J.M., Werner., J.O., Brandelli., A., Starch-halloysitenanocomposites containing nisin: Characterization and inhibition of Listeria monocytogenes in soft cheese, in: *LWT-Food Science and Technology*, vol. 68, pp. 226–234, 2016.
3. Issa, A., Ibrahim, S.A., Tahergorabi, R., Impact of sweet potato starch-based nanocomposite films activated with thyme essential oil on the shelf-life of baby spinach leaves. *Foods*, 6, 43, 2017.
4. Nafchi, A.M., Nassiri, R., Sheibani, S., Ariffin, F., Karim, A.A., Preparation and characterization of bionanocomposite films filled with nanorod-rich zinc oxide. *Carbohydr. Polym.*, 96, 233–239, 2013.
5. Campos-Requena, V.H., Rivas, B.L., Pérez, M.A., Figueroa, C.R., Figueroa, N.E., Sanfuentes, E.A., Thermoplastic starch/clay nanocomposites loaded with essential oil constituents as packaging for strawberries– *In vivo* antimicrobial synergy over Botrytis cinerea. *Postharvest Biol. Technol.*, 129, 29–36, 2017.
6. Campos-Requena, V.H., Rivas, B.L., Pérez, M.A., Garrido-Miranda, K.A., Pereira, E.D., Release of essential oil constituent from thermoplastic starch/layered silicate bionanocomposite film as a potential active packaging material. *Eur. Polym. J.*, 109, 64–71, 2018.
7. Xin, S., Xiao, L., Dong, X., Li, X., Wang, Y., Hu, X., Sameen, D.E., Qin, W., Zhu, B., Preparation of chitosan/curcumin nanoparticles based zein and

potato starch composite films for Schizothoraxprenati fillet preservation. *Int. J. Biol. Macromol.*, 164, 211–221, 2020.
8. Dairi, N., Ferfera-Harrar, H., Ramos, M., Garrigós, M.C., Cellulose acetate/AgNPs-organoclay and/or thymol nano-biocomposite films with combined antimicrobial/antioxidant properties for active food packaging use. *Int. J. Biol. Macromol.*, 121, 508–523, 2019.
9. Marrez, D.A., Abdelhamid, A.E., Darwesh, O.M., Eco-friendly cellulose acetate green synthesized silver nano-composite as antibacterial packaging system for food safety. *Food Packag. Shelf Life*, 20, 100302, 2019.
10. Saravanakumar, K., Sathiyaseelan, A., Mariadoss, A.V.A., Xiaowen, H., Wang, M.H., Physical and bioactivities of biopolymeric films incorporated with cellulose, sodium alginate and copper oxide nanoparticles for food packaging application. *Int. J. Biol. Macromol.*, 153, 207–214, 2020.
11. Lazić, V., Vivod, V., Peršin, Z., Stoiljković, M., Ratnayake, I.S., Ahrenkiel, P.S., Nedeljković, J.M., Kokol, V., Dextran-coated silver nanoparticles for improved barrier and controlled antimicrobial properties of nanocellulose films used in food packaging. *Food Packag. Shelf Life*, 26, 100575, 2020.
12. Gedarawatte, S.T., Ravensdale, J.T., Al-Salami, H., Dykes, G.A., Coorey, R., Antimicrobial efficacy of nisin-loaded bacterial cellulose nanocrystals against selected meat spoilage lactic acid bacteria. *Carbohydr. Polym.*, 251, 117096, 2020.
13. Pirsa, S. and Shamusi, T., Intelligent and active packaging of chicken thigh meat by conducting nano structure cellulose-polypyrrole-ZnO film. *Mater. Sci. Eng. C*, 102, 798–809, 2019.
14. Mohammadi, M., Mirabzadeh, S., Shahvalizadeh, R., Hamishehkar, H., Development of novel active packaging films based on whey protein isolate incorporated with chitosan nanofiber and nano-formulated cinnamon oil. *Int. J. Biol. Macromol.*, 149, 11–20, 2020.
15. Yu, Z., Dhital, R., Wang, W., Sun, L., Zeng, W., Mustapha, A., Lin, M., Development of multifunctional nanocomposites containing cellulose nanofibrils and soy proteins as food packaging materials. *Food Packag. Shelf Life.*, 21, 100366, 2019.
16. Karimi, N., Alizadeh, A., Almasi, H., Hanifian, S., Preparation and characterization of whey protein isolate/polydextrose-based nanocomposite film incorporated with cellulose nanofiber and L. plantarum: A new probiotic active packaging system, in: *LWT*, vol. 121, p. 108978, 2020.
17. Ghadetaj, A., Almasi, H., Mehryar, L., Development and characterization of whey protein isolate active films containing nanoemulsions of GrammosciadiumptrocarpumBioss. essential oil. *Food Packag. Shelf Life*, 16, 31–40, 2018.
18. Arfat, Y.A., Benjakul, S., Prodpran, T., Sumpavapol, P., Songtipya, P., Properties and antimicrobial activity of fish protein isolate/fish skin gelatin film containing basil leaf essential oil and zinc oxide nanoparticles. *Food Hydrocoll.*, 41, 265–273, 2014.

19. Xiao, Y., Liu, Y., Kang, S., Wang, K., Xu, H., Development and evaluation of soy protein isolate-based antibacterial nanocomposite films containing cellulose nanocrystals and zinc oxide nanoparticles. *Food Hydrocoll.*, 106, 105898, 2020.
20. Wang, C., Chang, T., Dong, S., Zhang, D., Ma, C., Chen, S., Li, H., Biopolymer films based on chitosan/potato protein/linseed oil/ZnO NPs to maintain the storage quality of raw meat. *Food Chem.*, 332, 127375, 2020.
21. Pandey, V.K., Upadhyay, S.N., Niranjan, K., Mishra, P.K., Antimicrobial biodegradable chitosan-based composite Nano-layers for food packaging. *Int. J. Biol. Macromol.*, 157, 212–219, 2020.
22. Cruz-Romero, M.C., Murphy, T., Morris, M., Cummins, E., Kerry, J.P., Antimicrobial activity of chitosan, organic acids and nano-sized solubilisates for potential use in smart antimicrobially-active packaging for potential food applications. *Food Control*, 34, 393–397, 2013.
23. Lin, D., Zheng, Y., Huang, Y., Ni, L., Zhao, J., Huang, C., Chen, X., Chen, X., Wu, Z., Wu, D., Chen, H., Investigation of the structural, physical properties, antioxidant, and antimicrobial activity of chitosan-nano-silicon aerogel composite edible films incorporated with okara powder. *Carbohydr. Polym.*, 250, 116842, 2020.
24. Wang, K., Lim, P.N., Tong, S.Y., San Thian, E., Development of grapefruit seed extract-loaded poly (ε-caprolactone)/chitosan films for antimicrobial food packaging. *Food Packag. Shelf Life*, 22, 100396, 2019.
25. Indumathi, M.P. and Rajarajeswari, G.R., Mahua oil-based polyurethane/chitosan/nano ZnO composite films for biodegradable food packaging applications. *Int. J. Biol. Macromol.*, 124, 163–174, 2019.
26. Niu, X., Liu, Y., Song, Y., Han, J., Pan, H., Rosin modified cellulose nanofiber as a reinforcing and co-antimicrobial agents in polylactic acid/chitosan composite film for food packaging. *Carbohydr. Polym.*, 183, 102–109, 2018.
27. Mohammadi, M., Mirabzadeh, S., Shahvalizadeh, R., Hamishehkar, H., Development of novel active packaging films based on whey protein isolate incorporated with chitosan nanofiber and nano-formulated cinnamon oil. *Int. J. Biol. Macromol.*, 149, 11–20, 2020.
28. Boura-Theodoridou, O., Giannakas, A., Katapodis, P., Stamatis, H., Ladavos, A., Barkoula, N.M., Performance of ZnO/chitosan nanocomposite films for antimicrobial packaging applications as a function of NaOH treatment and glycerol/PVOH blending. *Food Packag. Shelf Life*, 23, 100456, 2020.
29. Wang, Y., Cen, C., Chen, J., Fu, L., MgO/carboxymethyl chitosan nanocomposite improves thermal stability, waterproof and antibacterial performance for food packaging. *Carbohydr. Polym.*, 236, 116078, 2020.
30. Li, J., Shirai, T., Fuji, M., Rapid carbothermal synthesis of nanostructured silicon carbide particles and whiskers from rice husk by microwave heating method. *Adv. Powder Technol.*, 24, 838–843, 2013.

31. Ma, X., Zhou, B., Gao, W., Qu, Y., Wang, L., Wang, Z., Zhu, Y., A recyclable method for production of pure silica from rice hull ash. *Powder Technol.*, 217, 497–501, 2012.
32. Shen, Y., Zhao, P., Shao, Q., Porous silica and carbon derived materials from rice husk pyrolysis char. *Microporous Mesoporous Mater.*, 188, 46–76, 2014.
33. Biswas, M.C., Tiimob, B.J., Abdela, W., Jeelani, S., Rangari, V.K., Nano silica-carbon-silver ternary hybrid induced antimicrobial composite films for food packaging application. *Food Packag. Shelf Life*, 19, 104–113, 2019.
34. Bi, F., Zhang, X., Liu, J., Yong, H., Gao, L., Liu, J., Development of antioxidant and antimicrobial packaging films based on chitosan, D-α-tocopheryl polyethylene glycol 1000 succinate and silicon dioxide nanoparticles. *Food Packag. Shelf Life*, 24, 100503, 2020.
35. Wu, C., Zhu, Y., Wu, T., Wang, L., Yuan, Y., Chen, J., Hu, Y., Enhanced functional properties of biopolymer film incorporated with curcurmin-loaded mesoporous silica nanoparticles for food packaging. *Food Chem.*, 288, 139–145, 2019.
36. Zhang, R., Cheng, M., Wang, X., Wang, J., Bioactive mesoporous nano-silica/potato starch films against molds commonly found in post-harvest white mushrooms. *Food Hydrocoll.*, 95, 517–525, 2019.
37. Song, H., Yuan, W., Jin, P., Wang, W., Wang, X., Yang, L., Zhang, Y., Effects of chitosan/nano-silica on postharvest quality and antioxidant capacity of loquat fruit during cold storage. *Postharvest Biol. Technol.*, 119, 41–48, 2016.
38. Venkatesan, R. and Rajeswari, N., Preparation, mechanical and antimicrobial properties of SiO2/poly (butylene adipate-co-terephthalate) films for active food packagin. *Silicon*, 11, 2233–2239, 2019.
39. Melendez-Rodriguez, B., Figueroa-Lopez, K.J., Bernardos, A., Martínez-Máñez, R., Cabedo, L., Torres-Giner, S., M Lagaron, J., Electrospun antimicrobial films of poly (3-hydroxybutyrate-co-3-hydroxyvalerate) containing eugenol essential oil encapsulated in mesoporous silica nanoparticles. *Nanomater.*, 9, 227, 2019.
40. Cui, R., Jiang, K., Yuan, M., Cao, J., Li, L., Tang, Z., Qin, Y., Antimicrobial film based on polylactic acid and carbon nanotube for controlled cinnamaldehyde release. *J. Mater. Res. Technol.*, 9, 10130–10138, 2020.
41. Dias, M.V., Nilda de Fátima, F.S., Borges, S.V., de Sousa, M.M., Nunes, C.A., de Oliveira, I.R.N., Medeiros, E.A.A., Use of allylisothiocyanate and carbon nanotubes in an antimicrobial film to package shredded, cooked chicken meat. *Food Chem.*, 141, 3, 3160–3166, 2013.
42. Wang, L., Lin, L., Pang, J., A novel glucomannan incorporated functionalized carbon nanotube films: synthesis, characterization and antimicrobial activity. *Carbohydr. Polym.*, 116619, 1–9, 2020.
43. Jamróz, E., Khachatryan, G., Kopel, P., Juszczak, L., Kawecka, A., Krzyściak, P., Kucharek, M., Bębenek, Z., Zimowska, M., Furcellaran nanocomposite films: The effect of nanofillers on the structural, thermal, mechanical and

antimicrobial properties of biopolymer films. *Carbohydr. Polym.*, 116244, 1–12, 2020.
44. Liu, Y., Wang, S., Lan, W., Qin, W., Fabrication of polylactic acid/carbon nanotubes/chitosan composite fibers by electrospinning for strawberry preservation. *Int. J. Biol. Macromol.*, 121, 1329–1336, 2019.
45. H. Unal, S. Unal, Y.Z. Menceloglu, F.C. Cebeci, Food packaging material with antibacterial, ethylene scavenging and barrier properties. U.S. Patent 9,332,751, assigned to Sabanci Universitesi, 2016.
46. F. Gao and R. Nigmatullin, Antimicrobial polymer nanocomposites. U.S. Patent Application 12/663,929, Nottingham Trent University, 2010.
47. W. Huajie, C. Ying, C. Meng, C. Tingting, M. Liwei, P. Wei, Zein antimicrobial packaging material coated with nano-zinc oxide and preparation method. Chinese Patent Application CN105907105A, 2016.
48. C. Xin, Z. Shan, Y. Jinrong, F. Xiang, S. Zhengzhong, Isolated soy protein-nano silver antibacterial composite membrane and preparation method thereof. Chinese Patent Application CN102816440B, 2015.
49. Y. Zhao, J. Simonsen, G. Cavender, J. Jung, L.H. Fuchigami, Nano-cellulose edible coatings and uses thereof. World Patent Application WO2014153210A1, 2014.
50. C. Changfeng, W. Bin, S. Kunlai, Application of gelatin-ZnO nano composite film. Chinese Patent Application CN104829874A, 2015.
51. D. Bingjian, L. Shuangshuang, L. Menglin, L. Youwei, Antimicrobial plastic bottle for food. Chinese Patent ApplicationCN104610657A, 2015.
52. Unalan, L.U. and Korel, F., Active packaging for ground beef patties by edible zein films incorporated with partially purified lysozyme and Na2EDTA. *Int. J. Food Sci. Technol.*, 46, 6, 1289–1295, 2011.

7

Nanotechnology in Food Packaging and Its Regulatory Aspects

Pooja Mittal[1], Anjali Saharan[2], Ramit Kapoor[3], Kashish Wilson[4] and Rupesh K. Gautam[2*]

[1]School of Pharmaceutical Sciences, RIMT University, Mandi Gobindgarh, Punjab, India
[2]MM School of Pharmacy, Maharishi Markandeshwar University, Sadopur, Ambala, Haryana, India
[3]Amity Institute of Pharmacy, Amity University, Noida, India
[4]MM College of Pharmacy, Maharishi Markandeshwar University, Mullana, Ambala, Haryana, India

Abstract
Nanomaterials have attracted the interest of all industries because of their astonishing properties. They are the site of interest for food industries as well. Nanomaterials are also known as the shining stars because of their extraordinary features such as nanosize range, antimicrobial nature, and mechanical and optical catalytic properties. They have been frequently used in the packaging of foods. Metal nanoparticles such as silver and gold nanoparticles impart important characteristics in the field of food packaging, while other metallic nanoparticles are rarely used in food packaging. The antimicrobial properties provided by silver nanoparticles are the major reason behind their frequent use in the food industry because of their antimicrobial property toward food which helps to improve the shelf life of the packed foods along with freshness. However, safety concerns are originating as some chances of migration of nanoparticles to the food have been reported. This is still the topic of debate that how much extent of migration is safe so the government has set up some limits and regulations concerning the safety issues of the food industries. Since nanoparticles can have a serious impact on human health, so government and stakeholders must take some serious steps for the release of rules and regulations about their limits, which can be tolerated

**Corresponding author*: drrupeshgautam@gmail.com

Annu, Tanima Bhattacharya and Shakeel Ahmed (eds.) Nanotechnology in Intelligent Food Packaging, (157–174) © 2022 Scrivener Publishing LLC

by the human body without causing any injurious effects. This chapter aims to review the role of nanomaterials in the packaging of foods, their migration to the foodstuff, limitations, management of the migrations, and the present status of the regulations across the country. It will further help the researchers to perform a nanomaterial risk assessment and management studies.

Keywords: Nanomaterials, nanoparticles, silver nanoparticles, antimicrobials, nanotechnology

7.1 Introduction

Nanomaterials are the most important feature of nanoscience. From the last decade, nanotechnology that involves nanostructures and nanomaterials became a part of interdisciplinary sciences and has grown explosively worldwide. It can revolutionize the processes through which the various products are created, and it can also change the nature and range of functionalities. It will have a significant commercial impact for sure [1–3].

Nanoscale materials are those materials whose size is lesser than 100 nm (approximately). Due to the emergence of many unique properties like optical, magnetic, and electrical, nanomaterials have emerged with a great impact in medicines, electronics, and other fields [3, 4].

7.1.1 Sources of Nanomaterials

Some nanomaterials are found naturally but only a few are of particular interest. These are referred to as engineered nanomaterials, which are intended for use in a variety of commercial products and processes. Sunscreens, cosmetics, sporting goods, stain-resistant clothes, tyre, and numerous other items are being utilized in the various diagnosis of diseases and delivery of the drugs.

The engineered nanoparticles are widely are extremely small in size with numerous benefits such as increased relative surface area and new quantum effects, which connect to their unique properties [3]. They possess a higher surface area to volume ratio than any other particulate forms are intended for their chemical reactivity [5, 6].

Nanoparticles structures were studied for a decade and are in the global market also. Textiles that are stain-resistant and wrinkle-free, cosmetics, sunscreens, electronics, paints, and varnishes are just a few of the commercial products available today [2].

7.1.2 History

Many everyday consumer goods, ranging from non-dairy coffee creamer to car tyres, optical fibers, and catalyst supports contain nanosized amorphous silica particles. For the formation of nanocrystals, the inert gas evaporation method was used to generate the nanoparticles [1–3].

The recent discovery of maya blue paint is the outcome of nanotechnology wherein nanostructured hybrid materials are used. The actual sources of the acid shield are not known but the authentic samples revealed that the materials consisted of sharp needle clay crystals [31].

As shown by the fabrication of synthetic samples, the beautiful blue color tone can only be obtained if the combination of nanoparticles along with superlattice is present [6–8].

In both inorganic and organic, nanophase engineering is utilized to control electric, optical, catalysis, magnetic, and electrical properties in the day-to-day increasing assets. Clusters of nanophase are generally made by the separation of huge clusters and then fusing them to make a large cluster or suspending them in a compressed liquid-like nanophase silicon which is having different properties than regular silicon which can be further used as a semiconductor in different devices [8–10].

7.1.3 Classification of Nanomaterials

Nanomaterials are extremely small, with a minimum dimension of 100 nm. Different dimensions such as single-dimensional double dimension and triple dimension can also be there. Nanoparticles can possess any shape such as spherical, tube shape, or any irregular shape and can exist as a single, fused, or aggregated form. Popular nanomaterials are nanotubes, quantum dots, dendrimers, and fullerenes [32].

Nanomaterials are materials with a grain size of fewer than 50 nm or dimensionality of less than 50 nm. According to the research by Richard W. Siegel, different nanomaterials can be synthesized by various dimensions of modulations [11–15].

7.1.4 Examples of Various Nanomaterials

The various nanomaterials are given in Figure 7.1.

Nano semiconductors are well known for their non-linear optical properties. In semiconductors, quantum detention consequences in Q-particles can cause a change in specific properties, such as infrared optoelectronic devices comprised of nanostructured materials that

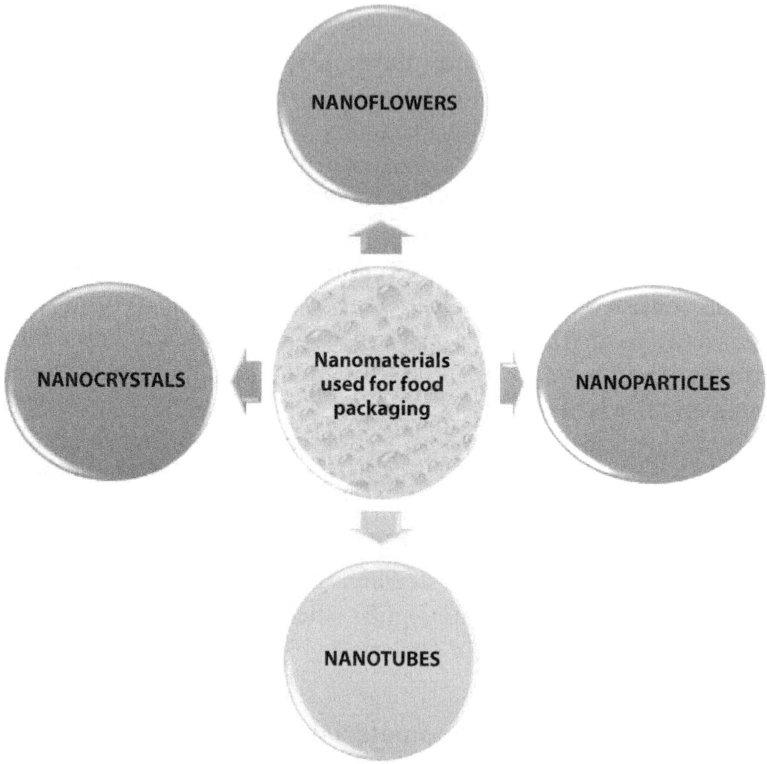

Figure 7.1 Nanomaterials used for packaging of food.

are widely used in solar cells nowadays. Nanosized metallic powders have been used to create porous coats, gas-tight materials, and dense components. Their ductility and cold welding capabilities render them ideal for metal bonding and hence for the electronic industry [16–18]. Nanostructured materials have a distinct impact on catalytic properties also. They have shown major advantages in chemical transformation processes and the electrocatalysis process. Chiral modifiers when used on the surface of metal nanoparticles, enantioselective catalysis can be achieved [19, 20, 33]. Metallic thin films of nanoparticles are becoming more popular nowadays as they can produce gas sensors that improve selectivity and sensitivity such as carbon dioxide methane and aromatic hydrocarbons. Manganese oxide–based nanomaterials are utilized to form rechargeable batteries for various vehicles [34].

7.2 Properties of Nanomaterials

Nanomaterials are substances made from nanoparticles, they possess a dramatic shift in their properties. Due to slight variation in quantum size effect from molecular to nanoscales, the electronic properties of the materials got changed. The surface area to volume ratio also gets changed due to change in the thermal, electrical, and mechanical properties of the molecules [2, 3]. It can also happen that the insulator materials can also start behaving like conductors at the nanoscale.

The physicochemical properties of the nanoparticles can also be determined by the particles' size and shape. Chemical composition, surface area, crystal lattice structure, stability, aggregation, surface energy, etc. An increase in surface area to volume ratio leads to more reactive surfaces. The pharmacological activity of the nanomaterials is heavily impacted by the decrease in size [32]. On interaction with water or other media, nanomaterials rearrange their crystal structure.

The surface coating affects the electrical, physicochemical, and mechanical properties of the materials. Free radicals are generated when there is an interaction of the surface of these materials with oxygen, ozone, or transition materials [10, 11, 21].

The interaction between particles at the nanoscale level is caused by either van der Waals forces or by strong covalent bonds. Polyelectrolytes can be also used to alter the surface properties of nanomaterials [12, 13, 22].

7.3 Nanomaterials for Food Packaging

The role of nanotechnology in the terms of food packaging is growing at an alarming rate with numerous benefits, and few side effects that might leak into the food compositions and have a negative influence on human health. By considering this concern, more advancements are raised which have highly perceptible packaging using nanomaterials which also helps to extend the shelf life of food materials. Nanomaterials act by making the coatings of metal thick to inhibit the permeation of air and other gases inside the food products to prevent their spoilage. This can be achieved using various nanomaterials such as polyolefins, polyesters, and polyamides [14–16].

7.3.1 Antimicrobial Food Packaging

To inhibit microbial growth, silver nanoparticles are used to make antibacterial coating using zinc oxide and magnesium oxides will act as a filter which helps in the prevention of food spoilage and expanding their shelf life [17, 18, 23, 24]. Some of the commonly used examples of these nanomaterials are such as zinc oxide nanoparticles coated silicate, montmorillonite (MMT), silver nanoparticles, and titanium dioxide nanoparticles. These nanomaterials formed a barrier against the outer environmental gases and prevent them from degradation. They can also act as oxygen scavengers and antimicrobials and have more tolerance toward temperature [18–20, 25].

7.3.2 Silver Nanoparticles

They possess antimicrobial properties and helps in the protection of food materials. They are also having antiviral properties and anti-fungal properties. They act by adherence to the surface of the microbes and lead to the destruction of the lipopolysaccharides by the formation of a pit inside their cell wall. The addition of the silver nanoparticles onto the polymers also enhances the shelf life of the food materials [2, 3, 26–28].

7.3.3 Nanoclay

It was used in combination with the polymers and enhances the physical strength of the packaging and acts as a barrier for gas and moisture, for example, MMT.

7.3.4 Titanium Nanoparticles

Titanium nanomaterials are metallic oxides with high thermostability. Titanium is inert and acts by damaging the microbial cell walls by generating the reactive oxygen species of hydrogen radicals [4–7, 29, 30].

7.3.5 Nanostarch

Starch is natural, biodegradable, and non-toxic polysaccharide which was mainly used in food, paper, pharmaceuticals, and rubber packaging due to its low cost and eco-friendly nature. It was mainly found in the two forms such as single line amylose or branching form of amylopectin. Nanostarch can be prepared only after the separation of amorphous and crystalline complexes. Various techniques such as enzyme hydrolysis, acid hydrolysis, amylose extraction, and precipitation techniques can be utilized for the formation of these complexes [8–10, 21].

7.3.6 Carbon Nanotubes

Carbon nanotubes when converted into polymers such as polyvinyl alcohol, propylene, or nylon can be used for packaging food materials because of their antimicrobial properties. Mainly, carbon nanotubes are of two types such as single atom thick nanotubes and the second is with various concentric nanotubes. Both tubes were found to possess higher tensile strength when added to the polymer matrix [11–13, 22].

7.3.7 Nanosilica

They are mainly used for the coating of hydrophobic materials. They act by making non-adhesive coatings that can turn food into free-flowing materials [9, 10].

7.3.8 Copper Nanoparticles

Among these, copper oxide nanoparticles are mainly used. Antimicrobial activity is achieved by inhibiting the growth of bacteria, viruses, and fungi. Because of their large surface area, they can interrelate with the cytoplasm,

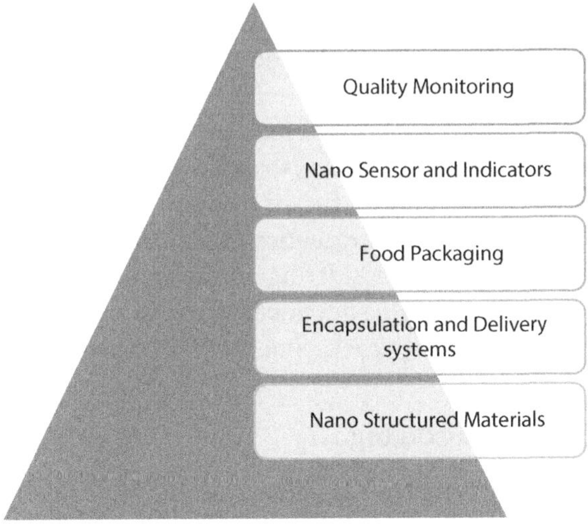

Figure 7.2 Roles of nanomaterials in the food industry.

resulting in an increased antimicrobial impact. These are primarily used in chemical engineering, food and biomedical field, and food packaging materials [11, 21].

The various functions performed by nanomaterials in the nutraceutical industry are mentioned in Figure 7.2.

7.4 Drawbacks of Existing Packaging Materials

The existing packaging methods have a negative influence on the environment due to their intact nature of non-biodegradability and recycle caused environmental pollutions on a wider scale. As a result of all this, the packaging industry consumes only a few parts of the packaging material. The main issue raised is if the shelf life of the product can be increased by preventing it from external factors such as sunlight, humidity, changes in weather conditions, etc. To overcome all issues of stability, shelf life, and contamination of the food products, nanotechnology was used in the field of packaging [20].

7.5 Proactive Packaging

Packaging is defined as techniques that were used for components and containers with the prime objective to protect, carry, identify, and retail any product. It is also known to provide essential information and instructions of products to consumers. Nanotechnology has amplified different mechanisms to serve functional benefits such as by providing real-time information on the quality of products to the consumer along with protection from internal and external factors. Nanomaterials in packaging define the shelf life of food and help consumers to avoid food of expired date. Expiration dates act as vague guesses to the consumers to define the air on the side of caution which encourages consumers to discard the spoil aged food. Improved barriers enhance the life of the nutraceuticals by exerting their antifungal and antibacterial effects. So, by considering all the abovementioned aspects as there are two types of packaging developed such as smart or intelligent packaging and active or protective packaging [9–11, 21] (Figure 7.3).

7.5.1 Protective Packaging

Protective packaging is defined as packaging with the main motive to preserve the food by constructively interacting with the atmosphere to expand their shelf life or accomplish with few characteristics. The prime role of this

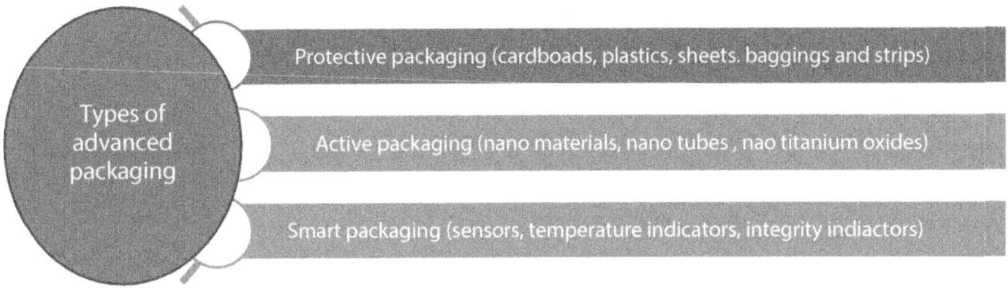

Figure 7.3 Various types of food packaging.

packaging is the protection and preservation of the product from any harm or destruction caused by shipping or warehousing. The protective packaging is made of hard-core things such as cardboard, plastics, and metals. The types of packaging are variable according to the product demands for various purposes [9–11].

7.5.2 Active Packaging

It includes the role of some additives which are used to maintain the quality of the product with the increase in their shelf life. In this type of packaging, nanomaterials interact with the food or surrounding materials directly or indirectly to provide superiority to the quality of food. For example, silver nanoparticles and coating are known to have anti-microbial properties when used in touch or in presence of the other materials such as oxygen. Also, nano magnesium, copper oxides, nano titanium dioxide, and carbon nanotubes are used for antimicrobial food packaging in the upcoming era [7, 29, 30].

7.6 Mechanism of Packaging

The packaging acts as a barrier between the protective cover and food commodity to ensure not the direct contact of the food from the external environment with the prime objective to prevent it from degradation. The diffusion of the molecules is directly across the membranes in a single way, leading to contamination on a rapid scale. While in the case of smart and intelligent packaging, the flow of molecules across the membrane follows the tortuous pathway which ensures not the direct and easy diffusion of the air with the food and increased the shelf life of the commodity by following a polymer matrix pathway [28], in which nanoelectronics were used and have the potential to control the environment inside the packaging and alert consumers about the degradation [4–7, 28]. Functions performed by the packaging materials are enlisted in Figure 7.4.

Innovation and advancements in the material sciences have given positive responses in the future terms of the science with positive impacts on the safety and efficiency of the products. As a concern of this smart packaging became the need of the hour to overcome various alarming issues. The new packaging was coined by special characters of words as intelligent and smart packaging. It is defined as a system that was used to examine the state of packed food to gain information related to the quality of the packaged food for a particular period of storage and transport [26–28]. For the detection of pathogens and microbial contamination, biochemical changes in food nanomaterials were used to prevent degradation and spoilage of the food. Nanoparticles act as reactive units in packaging structures that

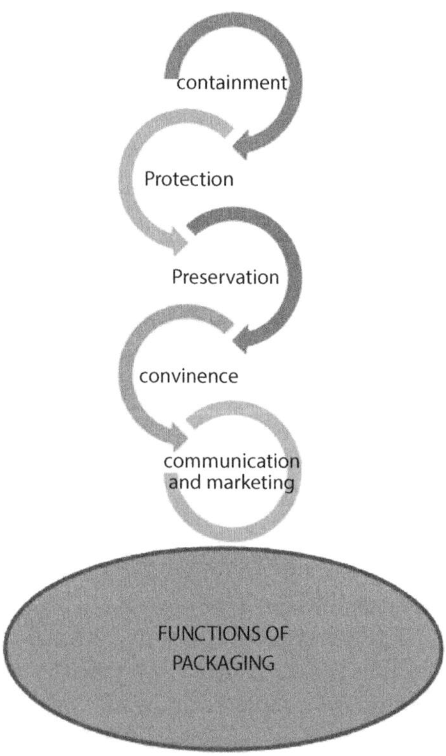

Figure 7.4 Functions of packaging materials.

are used to provide information regarding the situation of the storage products. In divergence to active components, intelligent components do not have the objective to liberate their components in the food. The current studies and discoveries related to nanotechnology used for detection are such as spoilage indicators, oxygen pointers, products recognitions, and traceability. On the large scale, mainly, three types were used in bulks, which include radio frequency identification systems also well known as active tags, smart sensors, time-temperature indicators, and integrity indicators [2, 3, 26–28].

7.6.1 Radio Frequency Identification Systems (Active Tags)

They are also known as active tags which are electronic information-based systems used mainly to utilize the radio frequency which was used to relocate the information from a label stick to a map which was further used to recognize the entity mechanically. This method is just a preceding of the physical tracking method of determination. Some of the factors such as strength, reading range, and capacity to withstand extreme temperature and pressure make them unique. They can sense up to 100 meters along with various gaps simultaneously. Nanotechnology is also facilitating sensor packaging to integrate cheap, nano-enabled radio frequency identification tags that are very tiny, elastic, and can be published on slender labels [6, 7, 29, 30].

7.6.2 Smart Sensors

They are defined as equipment that is used to identify the physical magnitude of material and translate it into legible larger signals. Signals are sensed frequently to adjust the interior atmosphere of food objects and properties. The main parameters to be observed were of are oxygen content, temperature, and pathogens which are an indicator of accurate alarming and shelf life of the products. Some of the common examples of these are gold nanoparticles integrated enzymes used for microbial identification; gas sensors are associated with the condition of the food products. Also, nano barcodes and tagging were used to determine the contamination. Moreover, the packages incorporated with nanosensors provide information regarding enzyme yields during the breakdown of food molecules, hence making them insecure for human utilization [11, 21].

7.6.3 Time-Temperature Indicators (TTIs)

They both are defined as a rate-limiting step that was used to establish the kinetics of physical, chemical, and microbial spoilage of food. As it was reported that the time was used to determine the information related to the threshold temperature. Three major types of TTIs are present in the market: critical temperature, partial history, and full history indicators. Overall, TTIs are small labels that retain the calculation of time-temperature of a fragile product from fabrication to consumption [12].

7.6.4 Freshness Indicators

They are mainly used to ensure the quality of the food products in context to microbial expansions or chemical alterations in the quality of food that can be easily monitored from the freshness indicators. A visual response was monitored from the reactions that occur between the microbial escalations metabolites. As a result of this, it was used as an indicator for package that gives the knowledge about the microbial quality of food, hence alerting the consumer about its freshness. The researchers of Sejong University developed carbon dioxide indicators made up of aqueous solutions of chitosan or whey protein isolate. The presence of this gas was identified by the alteration in intelligibility by the pH-dependent whey. The major drawback of a freshness indicator is that it gives the color change indications as contaminations even if the product is free from any sensory or quality decline. The existence of some target metabolites is not essentially a sign of deprived quality [7, 29].

7.6.5 Integrity Indicators

These are used to ensure the integrity of the package along with the fabrication and dispersal chain a leak indicator is used. By using this, visual oxygen levels in the makers' food with less oxygen content were monitored and color changes were observed with changes in the concentration of the visual oxygen along with redox dyes. The major problem in this indicator is that the apparatus should be extremely responsive and the leftover oxygen in the container is vulnerable to indicators. Oxygen arriving via a leakage may also be utilized by the natural microbes available in the food [7, 29]. The various types and their components are enlisted in Figure 7.5.

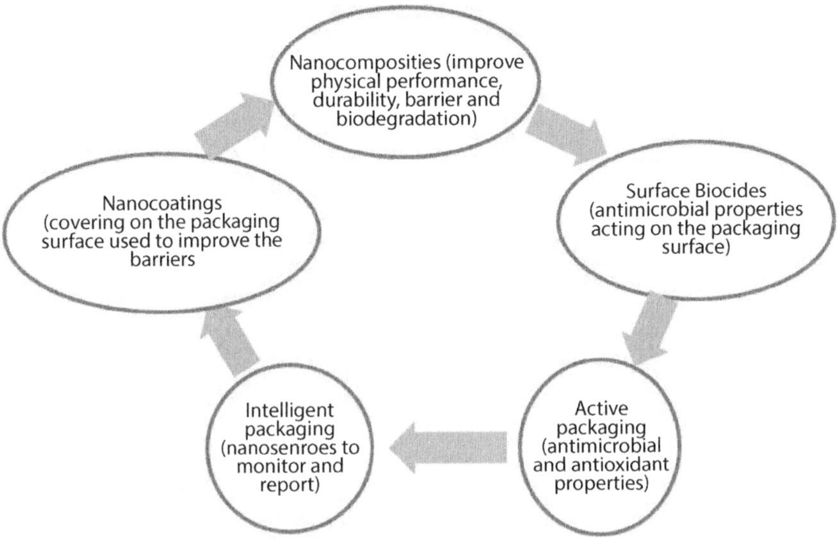

Figure 7.5 Various types and components of packaging materials.

7.7 Smart Packaging

Intelligent packing and active packing are two concepts that are different from each other and are connected in the way of providing a new way to food packaging. The intelligent packaging does not act directly on the packed food but it monitors the conditions of the packed food while active packaging works on the surroundings of the food to enhance the shelf life. Therefore, intelligent packaging is the component that senses and provides the information while active packaging is the component that takes the action. When we combine both the components, we will get "smart packaging". Smart packaging can sense the microbial and pathogenic development inside the food or production of any spoil gases inside the pack can also be sensitive. Therefore, it acts as a safety device or safety tracing device. Currently, Nestle, British Airways, and Monoprix Supermarket are using chemical sensors, where the color changes in the foods are easily detectable [4–10].

While talking about the manufacturing method for these intelligent and smart packages, we cannot neglect the conventional methods to produce packaging materials as most of the newer and novel packages are

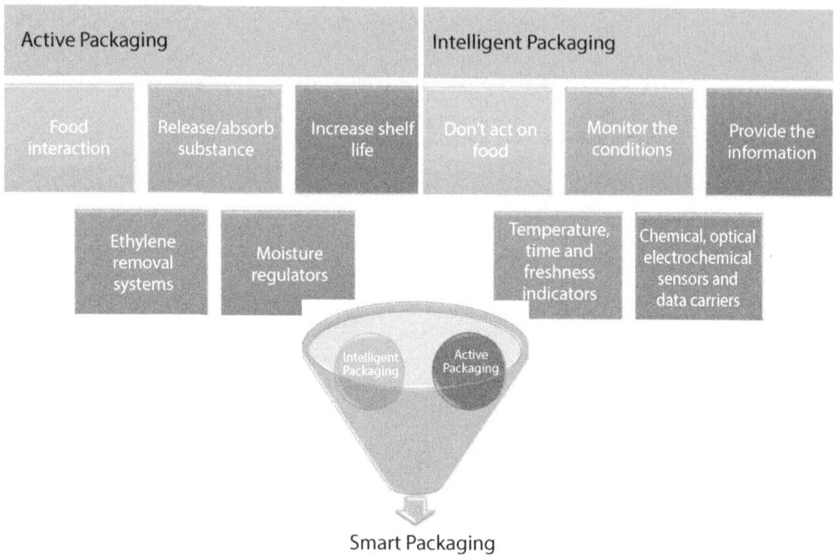

Figure 7.6 Comparative view of active and intelligent packaging systems.

also manufactured by the conventional methods. The two techniques can be there to insert the active components: either the active component is inserted directly into the polymer matrix/ films or the sachets/pads can be added to the package. Moreover, the intelligent part is inserted into the primary or secondary packet. The intelligent and smart packaging ways are popular worldwide especially in the USA and Japan, where the products packed in smart packaging materials are commercialized widely. However, in Europe, there is a lag in the commercialization of these products. The reason behind the lag can be the lenient drug and food regulations. To date, the only act regulating the use of intelligent and smart packaging is Regulation 450/2009 which has established the specific requirements for the materials that are coming in direct contact with the foods [4–10, 28–30]. A diagram of active and intelligent packaging in detail is shown in Figure 7.6.

7.8 Public Concerns and Regulations for Nanomaterials

In Europe, all types of packaging materials that are being used are in direct contact with the foods. They should follow the legislations about food contact materials. It also includes the evaluation of overall mitigation limits (OMLs) and specification limits (SMLs). Accordingly, Legislation 1935/2004/EC states that any material that is coming in direct contact with the food should be manufactured under Good Manufacturing Practices under normal conditions. The contents of the packaging materials should not move toward the food materials and deteriorate them. However, regulation (EU) 2016/1416 is of particular interest in the case of active and intelligent packaging as it states the labeling requirements of the same that the labeling of the intelligent materials should not give false information about the conditions of the foods [2, 8, 10]. Some more regulations particularly for the materials are available such as regulation EU no 10/2011 and regulation no EU/2016/1416 are of special interest in the case of packaging. Regulation 450/2009/EC is specific for the active and intelligent packaging materials [1, 2, 8, 10, 13].

7.9 Conclusion and Future Prospective

The applicability of nanotechnology branch into the foods and nutraceuticals for the improvement of quality, stability, shelf life, efficacy, and safety is the underdevelopment stages, and soon, we will be commercializing the products. Although, nanomaterials possess a wide range of applications in the packaging of foods, the safety of foods and the future of nanotechnology in nutraceuticals remains uncertain still. Moreover, the packaging involving nanotechnology and nanotechnology-based food products enhances the relation between both consumers and the producers; however, the facts about the safety enhancement and shelf-life enhancement are still under research. Besides safety issues, many queries raised regarding the ethical practices to involve nanotechnology in food packaging needs to be addressed yet. The public's reaction would be influenced by how companies, governments, and non-governmental organizations approach the problem of nanotechnology in food packaging. As a result, before nanomaterials are used in food touch or packaging, they must be monitored.

References

1. Amenta, V., Aschberger, K., Arena, M., Bouwmeester, H., Moniz, F.B., Brandhoff, P. *et al.*, Regulatory aspects of nanotechnology in the Agri/feed/food sector in EU and non-EU countries. *Regul. Toxicol. Pharmacol.*, 73, 1, 463–76, 2015.

2. Dimitrijevic, M., Karabasil, N., Boskovic, M., Teodorovic, V., Vasilev, D., Djordjevic, V. *et al.*, Safety aspects of nanotechnology applications in food packaging. *Proc. Food Sci.*, 5, 57–60, 2015.
3. Duncan, T.V., Applications of nanotechnology in food packaging and food safety: Barrier materials, antimicrobials, and sensors. *J. Colloid Interface Sci.*, 363, 1, 1–24, 2011.
4. Garcia, C.V., Shin, G.H., Kim, J.T., Metal oxide-based nanocomposites in food packaging: Applications, migration, and regulations. *Trends Food Sci. Technol.*, 82, 21–31, 2018.
5. Grumezescu, A., *Novel approaches of nanotechnology in food*, Academic Press, London, UK, 2016.
6. Hernández-Muñoz, P., Cerisuelo, J.P., Domínguez, I., López-Carballo, G., Catalá, R., Gavara, R., Nanotechnology in food packaging, in: *Nanomaterials for Food Applications*, pp. 205–32, Elsevier, MDPI Switzerland, 2019.
7. Imran, M., Revol-Junelles, A.-M., Martyn, A., Tehrany, E.A., Jacquot, M., Linder, M. *et al.*, Active food packaging evolution: Transformation from micro-to nanotechnology. *Crit. Rev. Food Sci. Nutr.*, 50, 9, 799–821, 2010.
8. Ludlow, K., The readiness of Australian food regulation for the use of nanotechnology in food and food packaging. *Univ. Tasman. Law Rev.*, 26, 177, 2007.
9. Majid, I., Nayik, G.A., Dar, S.M., Nanda, V., Novel food packaging technologies: Innovations and future prospective. *J. Saudi Soc. Agric. Sci.*, 17, 4, 454–62, 2018.
10. Mihindukulasuriya, S. and Lim, L.-T., Nanotechnology development in food packaging: A review. *Trends Food Sci. Technol.*, 40, 2, 149–67, 2014.
11. Mustafa, F. and Andreescu, S., Nanotechnology-based approaches for food sensing and packaging applications. *RSC Adv.*, 10, 33, 19309–36, 2020.
12. Rai, V.R. and Bai, J.A., *Nanotechnology applications in the food industry*, CRC Press, Boca Raton, 2018.
13. Raj, B. and Matche, R., Safety and regulatory aspects of plastics as food packaging materials. *Woodhead Publishing Series in Food Science, Technology and Nutrition*, pp. 335–357, 2012.
14. Ramkumar, C., Vishwanatha, A., Saini, R., Regulatory aspects of nanotechnology for food industry, in: *Nanotechnology Applications in Dairy Science: Packaging, Processing, and Preservation*, vol. 169, 2019, 9780429425370–7.
15. Rossi, M., Passeri, D., Sinibaldi, A., Angjellari, M., Tamburri, E., Sorbo, A. *et al.*, Nanotechnology for food packaging and food quality assessment. *Adv. Food Nutr. Res.*, 82, 149–204, 2017.
16. Sarwal, A., Singh, G., Singh, K., Garg, S., Recent interventions for nanotechnology based drug products: Insights into the regulatory aspects. *Curr. Pharm. Des.*, 24, 43, 5219–28, 2018.
17. Singh, T., Shukla, S., Kumar, P., Wahla, V., Bajpai, V.K., Rather, I.A., Application of nanotechnology in food science: Perception and overview. *Front. Microbiol.*, 8, 1501, 2017.

18. Sozer, N. and Kokini, J.L., Nanotechnology and its applications in the food sector. *Trends Biotechnol.*, 27, 2, 82–9, 2009.
19. Wesley, S.J., Raja, P., Raj, A.A., Tiroutchelvamae, D., Review on-nanotechnology applications in food packaging and safety. *Int. J. Eng. Res.*, 3, 11, 645–51, 2014.
20. Yam, K.L. and Lee, D.S., *Emerging food packaging technologies: Principles and practice*, Elsevier Inc., Cambridge, UK, 2012.
21. Momin, J.K., Jayakumar, C., Prajapati, J.B., Potential of nanotechnology in functional foods. *Emir. J. Food Agric.*, 25, 1, 10–19, 2013.
22. Primožič, M., Knez, Ž., Leitgeb, M., [Bio] Nanotechnology in food science—Food packaging. *Nanomaterials*, 11, 2, 292, 2021.
23. Sharma, C., Dhiman, R., Rokana, N., Panwar, H., Nanotechnology: An untapped resource for food packaging. *Front. Microbiol.*, 8, 1735, 2017.
24. Siegrist, M., Stampfli, N., Kastenholz, H., Keller, C., Perceived risks and perceived benefits of different nanotechnology foods and nanotechnology food packaging. *Appetite*, 51, 2, 283–90, 2008.
25. Sumit, G., Nanotechnology in food packaging a critical review. *Russ. J. Agric. Soc.*, 10, 10, 14–24, 2012.
26. Cerqueira, M.A., Vicente, A.A., Pastrana, L.M., Nanotechnology in food packaging: Opportunities and challenges, in: *Nanomaterials for food packaging*, pp. 1–11, 2018.
27. Chellaram, C., Murugaboopathi, G., John, A., Sivakumar, R., Ganesan, S., Krithika, S. *et al.*, Significance of nanotechnology in food industry. *APCBEE Proc.*, 8, 109–13, 2014.
28. Enescu, D., Cerqueira, M.A., Fucinos, P., Pastrana, L.M., Recent advances and challenges on applications of nanotechnology in food packaging. A literature review. *Food Chem. Toxicol.*, 134, 110814, 2019.
29. Jaiswal, L., Shankar, S., Rhim, J.-W., Applications of nanotechnology in food microbiology. *Methods Microbiol.*, 46, 43–60, 2019.
30. Kuswandi, B., Nanotechnology in food packaging, in: *Nanoscience in Food and Agriculture 1*, pp. 151–83, Springer, 2016.
31. Jiang, J., Ruan, M., Jiang, Y., A novel nanoscale catalyst system composed of nanosized Pd catalysts immobilized on Fe_3O_4@SiO_2–PAMAM. *Nanotechnology*, 19, 7, 19–30, 2008.
32. Glezer, A.M., Structural classification of nanomaterials. *Russ. Metall.*, 2011, 263, 2011.
33. Kodama, R.H., Magnetic Nanoparticles. *J. Magn. Magn. Mater.*, 200, 1–3, 359–372, 1999.
34. Chavali, M.S. and Nikolova, M.P., Metal oxide nanoparticles and their applications in nanotechnology. *SN Appl. Sci.*, 1, 607, 2019.

8

Nanoencapsulation of Probiotics in Food Packaging

Gurleen Kaur[1], Rajinder Kaur[1], Nitu Rani[2] and Sukhminderjit Kaur[1]*

[1]*Department of Biotechnology, Chandigarh University, Mohali, Punjab, India*
[2]*Department of Agricultural Sciences, Chandigarh University, Mohali, Punjab, India*

Abstract
Probiotics have wide applications in the functional foods and nutraceuticals. The health benefits of the probiotic to the human body are conferred by its viability in the gastrointestinal tract. Poor survival of probiotics in the gastrointestinal tract is a major concern due to the adverse conditions prevailing in the intestine. Overcoming these physical barriers by encapsulating the probiotic in a protective matrix is considered as a promising option to enhance its survival. However, as to assure the viability of probiotics during the encapsulation process and to ensure its controlled release in the desired gut area, several aspects are needed to be taken care. Nanotechnology has been extensively involved in nanoencapsulation of food components like food additives and flavor enhancers and enhances the bioavailability of nutritional components. In addition to enhancing the survival rate of probiotic in gastrointestinal tract, encapsulation technique also enhances the organoleptic and physico-chemical properties of the probiotic-based food material. Advantages of nanoencapsulation in food technology extend from prolongation of shelf life, storage and preservation, food processing, pathogen detection, and intelligent packaging. Spray drying, electrospinning, emulsification, freeze, and vacuum-drying techniques are some of the few technologies used for the nanoencapsulation of probiotics in the food industry. This chapter describes application of nanoencapsulation of probiotics for use in functional foods and nutraceuticals.

Keywords: Probiotic, nanotechnology, nanoencapsulation, storage, intelligent packaging

**Corresponding author*: sukhminderjit.uibt@cumail.in

Annu, Tanima Bhattacharya and Shakeel Ahmed (eds.) Nanotechnology in Intelligent Food Packaging, (175–202) © 2022 Scrivener Publishing LLC

8.1 Introduction

The living microbial consortium that is incorporated into the food in the adequate amount for possible improved health benefits is termed as probiotics. Different bacterial species belonging to genera *Lactobacillus*, *Bifidobacterium*, *Leuconostoc*, *Enterococcus*, *Streptococcus*, *Pediococcus*, and *Propionibacterium* are commercially being used as probiotics in different foods [1]. The consumption of probiotic supplemented food is associated with curation of diarrhea, reduction of cancer cells, controlling bad cholesterol, and intestinal infections. Beyond nutritional benefits, probiotics are associated with bacteriocin's formation and usually prevent the adherence of bacteria that are pathogenic in nature to the intestinal wall by competing them out. For the intended benefits of probiotics, the recommended level of probiotic count in the food product or as dietary supplements should be 10^{-6} to 10^{-7} cfu/ml during the consumption duration [2]. The probiotic strains are commercially produced as lyophilized cultures that can be incorporated into different diary and non-dairy food products. The loss of functionality and viability during the commercial food processing and long storage period is a serious matter of consideration for the food or edible industry. Various factors determine the viability of probiotics during food processing and storage, which are regarded as the critical factors for determining probiotic viability in the foods. Fermentation conditions including temperature and oxygen concentration, freezing and thawing operations, composition of growth media, water activity, and storage temperature are deciding factors for probiotic survival. It is usually seen that most of probiotic foods contain less probiotic count compared to the recommended unit [3]. Different technologies are being tested for enhancing the survival rate, functionality, and viability of the probiotics in the food systems. Microencapsulation technique and nanoencapsulation have been used for protecting the microbial cells. During the devouring of probiotic food, different determinants such as moisture, consistency, aroma, and temperature of food affect the probiotics which further disrupt their activity and functioning in the gastrointestinal system. Accordingly, significant encapsulation techniques must be employed to support and assure probiotics using different constituents like proteins, polysaccharides, fibers, and gums [1]. Encapsulation techniques possess profuse benefits such as distribution at target site, controlled dispensation, and prevention of functionality of probiotics [3]. Nanoencapsulation of probiotics help to enhance the viability and stability within the human gastrointestinal tract (Figure 8.1). This chapter provides detailed description of different aspects related

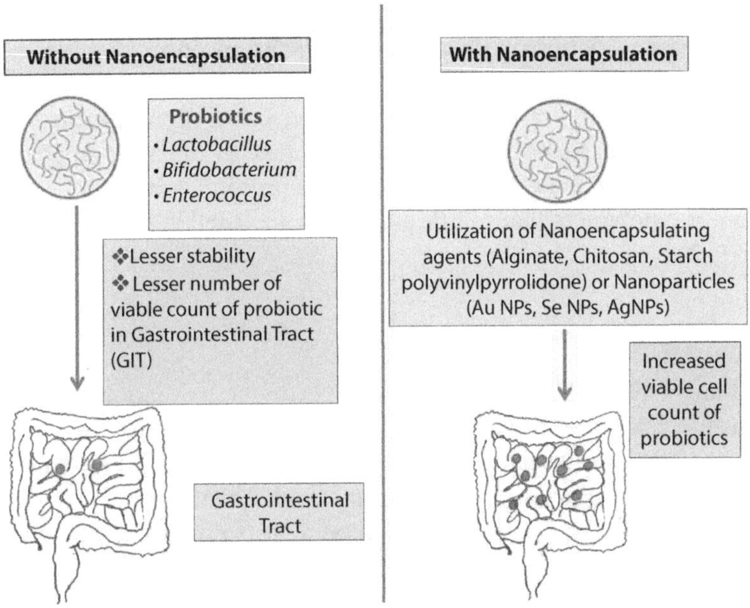

Figure 8.1 Enhancement of viability of probiotic microorganism by nanoencapsulation.

to nanoencapsulation of probiotics, techniques applied, and nanomaterials that can be used for the nanoencapsulation of probiotics. The recent applications of the nanoencapsulation of probiotics have also been explained. The advantages and disadvantages of the probiotic nanoencapsulation have been discussed briefly.

8.2 Nanomaterials for Encapsulation of Probiotics

Natural and synthetic polymers are widely studied for the nanoencapsulation of probiotics (Figure 8.2). As the probiotic nanoencapsulated formulations are intended for human use, the nanomaterials should be proved GRAS (generally recognized as safe), biocompatible, biotolerant, and biodegradable. Lipid, protein, and polysaccharide-based nanomaterials are used for efficient drug delivery and for the oral bioavailability enhancement of considerable biomolecules and various drugs. Among polysaccharides, chitosan, alginate, and gum arabic are being used for the delivery of probiotics to human intestine [4]. Probiotic bacteria must be protected within the gastrointestinal system because its endurance and pioneering is determined by distinct aspects such as bile acids, pH of gastric system, and digestive enzymes. Currently, researchers are working continuously on the advancement of nanomaterials required for the encapsulation of probiotics. Biodegradable nature and small size are the absolute benefits of adopting probiotic nanoencapsulation. Within the body, probiotic bacteria are efficiently approachable and dynamically uptake by different range of cells [5]. Therefore, for the nanoencapsulation of probiotics, different synthetic and natural polymers that can be applicable are listed in the following and given in Table 8.1.

8.2.1 Chitosan

Chitosan, a natural polysaccharide, is pertinent for nanoencapsulation because of active immune stimulation, long shelf life, biocompatiblility, robust coagulation strength, and non-toxicity [6]. Using chitosan, nanoencapsulation was conducted by Ebrahimnezhad *et al.* [5] on *L. acidophilus* as probiotic bacteria to examine its viability in gastrointestinal environment. It was inspected that, after nanoencapsulation of *L. acidophilus*, its viability and survival were enhanced in the presence of gastrointestinal juices.

8.2.2 Gum Arabic

Human cardiovascular system remains healthy and active by the consumption of cluster of polysaturated fatty acids referred to as omega-3 fatty acid present in fish oil. During its storage and processing, omega-3 fatty acids are prone to oxidation and results in substantial degradation. One of the carriers or wall materials used for the nanoencapsulation of fish oil is gum arabic carrier, and the emulsifier used is Tween 80. Such emulsions can be integrated in the fermented milk that results in increased viability of the probiotic bacteria *Lactobacillus plantarum* [7]. Gum arabic is a negatively charged highly branched arabinogalactan polysaccharide. It is biocompatible and provides suitable number of negatively charged sites for reaction with other biomolecules and, therefore, is widely used for ionic gelation reactions.

8.2.3 Whey Proteins

Coating of probiotic cells with whey proteins have been well known to enhance their survival. *Lactobacillus rhamnosus* encapsulated in pectin-whey protein aids to improve the resistance mechanisms of bacteria to bile salts and gastrointestinal conditions. Whey proteins offer a low cost nanoencapsulating agent for delivery of probiotics through oral route [8].

8.2.4 Polyvinylpyrrolidone

Polyvinylpyrrolidone is a linear nonionic polymer, amphiphilic in nature possessing good adhesive and cohesive characteristics, physiologically inert, biocompatible, provide curative dermal benefits, and used as a nanoencapsulating agent. In electrospray atomization method, polyvinylpyrrolidone can be utilized as a nanoencapsulating matrix to demonstrate best survival rate of probiotic freeze-dried powder of *Bifidobacterium.* Polyvinylpyrrolidone coating can be used to demonstrate protective aspects by acting as a potent barrier against oxygen and humidity [9]. Polyvinylpyrrolidone via electrospinning was used by Khan *et al.* [10] to assist the encapsulation of *Enterococcus mundtii* QAUEM2808. In this particular study, *E. mundtii* functionalized nanocomposite scaffold of polyvinylpyrrolidone was done by electrospinning. The probiotic was utilized for the treatment of pathobiomic disease. An increase in probiotic survival rate was indicated by scaffolds in the shelf-life evaluation of 28 days. This particular research was conducted to control infection and dermal burn healing in mice. The antagonistic interpretation indicated that bioscaffold disintegrated in accelerated wound fluid that was crucial for *E. mundtii* activation to offend pathogens causing infection. As a result, encapsulated probiotic strain supported for rapid healing of burned wound and control of caused infection [10].

8.2.5 Cellulose Acetate Phthalate

Cellulose acetate phthalate is soluble at basic pH but insoluble at acidic pH which distinguishes it as significant for the nanoencapsulation of probiotics so that dissolution of cellulose acetate phthalate occurs within the gut and not in stomach [11]. By using spray drying technique, *B. lactis* and *L. acidophilus* can be nanoencapsulated by using cellulose acetate phthalate as a wall material [12].

8.2.6 Alginate

Naturally occurring marine polysaccharide extracted from seaweed commonly known as alginate is inevitably investigated and used for the process of nanoencapsulation of probiotics. The nanoencapsulation of probiotic bacteria within alginate is achievable because of its expeditious, adaptable, and harmless characteristics. The benefit of utilizing alginate is its clear formation of gel matrices surrounding the probiotic microorganisms [13]. Mandal *et al.* [14] encapsulated *L. acidophilus* using alginate as an encapsulating agent that increased its viability and survival in stimulated gastrointestinal system.

8.2.7 Skim Milk

For the nanoencapsulation of probiotic microbes, dairy matrices can act as a beneficial nanoencapsulating material. When the probiotics passes through the gastrointestinal tract, the digestibility of casein and whey proteins is distinct; skim milk can aid to protect the viability of probiotics. Certainly, skim milk increases the viability of probiotic microorganisms. By applying spray drying procedure, skim milk can be used as nanomaterials for the nanoencapsulation of *L. acidophilus*. Skim milk favored an increment in the viability of *L. acidophilus* [15]. Under stimulated gastrointestinal conditions, skim milk can be employed for nanoencapsulation of *L. gastricus* BTM7 to improve viability and stability [16].

8.2.8 Starch

Among carbohydrates, starch is dense, insoluble, and naturally occurring as distinct particles as granules. Starch is among the most potent encapsulating agent in supporting growth of probiotics in acidic medium.

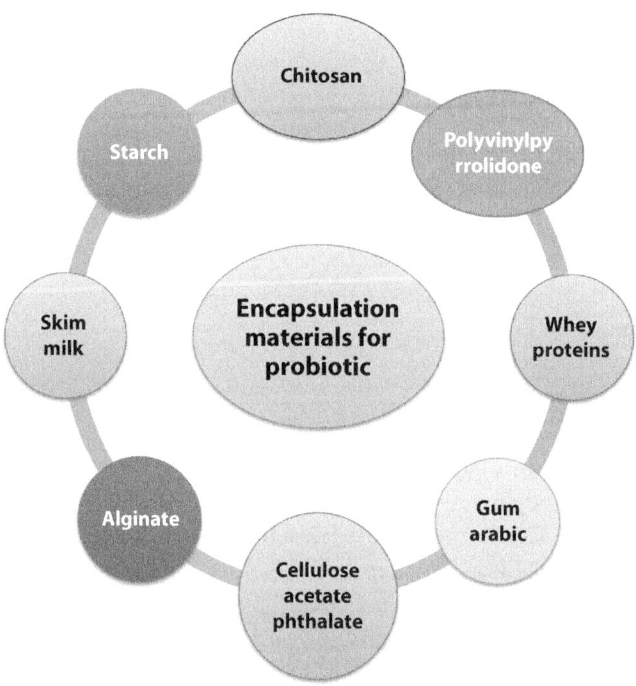

Figure 8.2 Nanomaterials for encapsulation of probiotics.

Starch can be utilized as encapsulating matrix for lipophilic flavors. Certain studies explain that resistant starch is a potent encapsulating agent for probiotics to reside in large intestine. The nanoencapsulation of *Bifidobacterium* and *L. rhamnosus* using starch could increase the survivability within the gastrointestinal tract. Even the oxygen toxicity can be blocked with nanoencapsulation using starch [17]. By applying spray drying method, gelatinized starch can be used for the nanoencapsulation of probiotic *Bifidobacterium* to prolong viability during storage conditions [18].

8.3 Packaging Material for Nanoencapsulated Probiotics

Nowadays, the principal focus of different studies is in renewable materials because of their substantial impact on agronomy, ecosystem, and economy. Plastic is synthetic in nature and generally exhibit resistance to decomposition. Because of non-decaying aspect, synthetic packaging has generated serious environmental issues and so the attention has shifted to utilization of biopolymers and protein-based components. The integration of antimicrobial

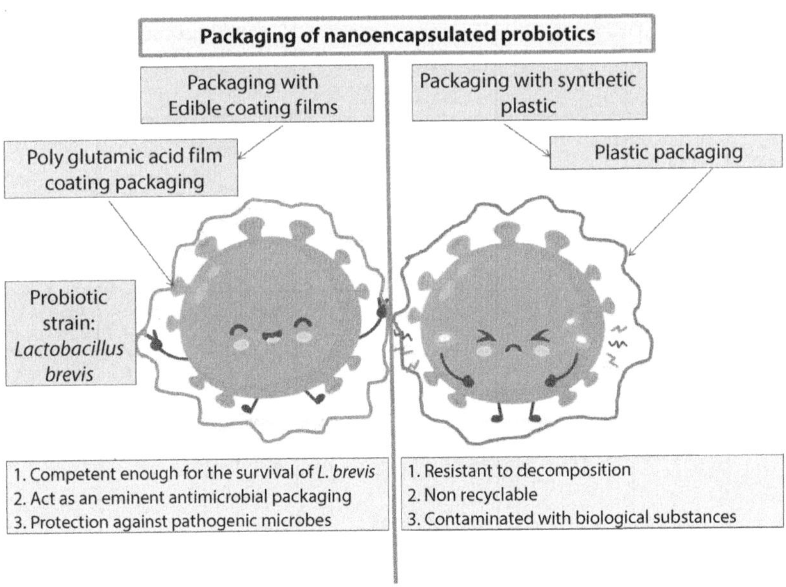

Figure 8.3 Packaging material for nanoencapsulated probiotics.

Table 8.1 Nanomaterials that can be utilized for the nanoencapsulation of various probiotics.

Sr. no.	Nanomaterial	Probiotic strain	Application	Reference
1.	Cellulose acetate phthalate, alginate, skim milk, and starch	*L. acidophilus* *L. bulgaricus* *L. paracasei* *Bifidobacterium PL1*	Increased survivability	[20]
2.	Seasame oil emulsion	*L. delbrueckii* sp. *Bulgaricus*	Increased viability	[21]
3.	Whey proteins and calcium alginate	*L. plantarum* CIP A159 *L. plantarum* 299v *L. plantarum* 800	Survival in stimulated gastric and intestinal fluid	[22]
4.	Alginate	*L. acidophilus* *L. paracasei* *B. lactis Bi-07* *B. longum* *L. plantarum* *L. rhamnosus*	Enhanced survival and stability in acidic and bile conditions	[23]
5.	Alginate, Lecithin and starch	*L. casei* *L. lactis* *B. adolescentis* NCIMB2204	Increased survival rate and shelf life	[24]
6.	Whey cheese	*L. acidophilus* LAFTIL10 *B. animalis* Bo *L. casei* LAFTIL26	Increased viability under intestinal stimulated conditions	[25]
7.	Vegetable fat	*L. acidophilus* *B. lactis*	Survival in stimulated gastric and intestinal fluid	[26]
8.	Whey protein	*L. paracasei* subsp. *Paracasei* DC412	Enhanced viability	[27]

(Continued)

Table 8.1 Nanomaterials that can be utilized for the nanoencapsulation of various probiotics. (*Continued*)

Sr. no.	Nanomaterial	Probiotic strain	Application	Reference
9.	Egg yolk	*L. paracasei* subsp. *Paracasei* DC412	Increased survival rate	[27]
10.	Gum arabic or xanthan	*L. paracasei* subsp. *Paracasei* DC412	Enhanced viability	[27]
11.	Skim milk	*L. paracasei* subsp. *tolerance* *L. delbrueckii* subsp. *bulgaricus*	Good viability of cells	[28]
12.	Whey	*Bifidobacterium Bb-12*	Increased viability during storage and tolerance to NaCl	[29]
13.	Alginate	*E. fecium* (FIdM3) *Staphylococcus succinus* (MAbB4)	Improved survival in acidic and bile conditions	[30]
14.	Polyvinylpyrrolidone	Probiotic Yeast	Enhanced storage	[31]
15.	Inulin, Maltodextrin, Oligofructose,	*B. animalis* spp. Lactis	Same viable cell count	[32]
16.	Hi-maize, dextrin and maltodextrin	*L. plantarum*	Improved viability of probiotic cell	[33]
17.	Hi-maize starch and alginate	*L. casei* 01	Greater survival rates in stimulated gastric and bile fluids	[34]

(*Continued*)

Table 8.1 Nanomaterials that can be utilized for the nanoencapsulation of various probiotics. (*Continued*)

Sr. no.	Nanomaterial	Probiotic strain	Application	Reference
18.	Alginate and inulin	*L. reuteri* *L. salivarius* *Pediocucus acidilactici*	Increased viability against bile salts	[35]
19.	Chitosan	*L. acidophilus*	Increased viability	[5]
20.	Alginate	*L. bulgaricus*	Protection against cell mortality in both bile tolerance and gastric juice	[36]
21.	Chitosan	*L. gasseri* *B. bifidum*	Improved survivability in stimulated gastrointestinal conditions	[37]
22	Gum arabic	*L. plantarum*	Increased viability	[7]
23.	Carrageenan and sodium alginate	*L. acidophilus*	Improved probiotic survival in gastrointestinal tract and prepared ice cream	[38]
24.	Gum arabic and Maltodextrin	*L. acidophilus* NCDC 016	Inlet air temperature tolerance	[39]
25.	Sodium alginate and polyvinyl alcohol	*L. paracasei* KS-199	Improvement and upsurge of probiotics viability	[40]

material within such components improves the shelf life of edible material. In addition, even antibacterial coating and films possess the efficiency for active packaging that further improves the viability of probiotics. In a recent research, conducted by Karimi *et al.* [19], the main focus was on the production of altered PG (poly glutamic acid) film by utilizing polylysine that is a food safety preservative having heat stability, good antimicrobial aspect, and less toxicity for the protection of probiotic bacterium and to prevent the growth of pathogenic microbe [19]. Probiotic must be prevented against stressful conditions and their shelf life should be maintained and improved. Gamma amino butyric acid is a biologically active non-proteinogenic amino acid that is produced by *L. brevis*. Gamma amino butyric acid aids in enhancing immunity and regulation of blood pressure and acts as antidepressant agent. The researchers developed the edible composite film of poly glutamic acid, utilizing various concentrations of polylysine. Literature reports that the coating with poly glutamic acid films proved to be competent in survival of *L. brevis* and can also act as an eminent antimicrobial packaging material against pathogenic microbes. Such edible films can be employed for the nanoencapsulation of probiotics (Figure 8.3) [19].

8.4 Techniques Employed for Nanoencapsulation of Probiotics

8.4.1 Electrospinning

The technique of electrospinning is a fabrication approach, administering high voltage conditions to polymer solution generating voluminous nanofibers for the encapsulation of probiotics and therefore, exhibiting better immobilization ability, decreased sensorial impact, and vigorous stability. Recently, Ragavan and Das conducted a study on probiotic yeast in which they employed electrospinning technique (Table 8.2) for encapsulating *Saccharomycopsis fibuligera* VIT-MN04 with exopolysaccharide, wheat bran fiber, and polyvinylpyrrolidone [3]. Exopolysaccharide act as a promising probiotic carrier and the use of exopolysaccharide demonstrates beneficent corporeal effects like decrease in cholesterol level, antitumor, antiulcer, and ant-inflammatory effects [41]. Gut microbiota is substantially influenced by dietary fiber. Among dietary fibers, wheat bran fiber is considered as the dominant source of fiber constituting sterols, carotenoids and arabinoxylans flavonoids as bioactive substances. Decrease in the plasma cholesterol level and controlled blood glucose level is supported by wheat bran fiber as the eminent health benefit [31]. The encapsulating agent utilized here for the probiotic yeast is polyvinylpyrrolidone. Such electrospinning technique contributes toward the decisive production of symbiotic supplements required in food industry.

In another recent research, an alternate method was employed to encapsulate probiotics using electrospun alginate nanofibers. Yilmaz *et al.* [40]

encapsulated *L. paracasei* KS-199 by fabricating sodium alginate and polyvinyl alcohol-based nanofibers mats. To improve the stability of probiotic cells, seaweed polysaccharide particularly alginate has been consumed for the purpose of nanoencapsulation of probiotics [42]. Alginate has proved to be beneficial in terms of providing confirmed endurance, whereas polyvinyl alcohol and sodium alginate has offered with thermal and chemical stabilization, reliability, and enhanced affinity. The nanofibers mats could be employed in the heat processing industries because of their higher melting temperature. In order to acquire superior bioactive properties, improvement and upsurge of probiotics viability forms the fundamental aspect for the production of functional food.

8.4.2 Spray Drying

The utmost applicable method of nanoencapsulation employed in food industry is spray drying. In this technique, an emulsion is produced wherein the active constituent is integrated in the nanoencapsulating component. Hydrocolloid solvent frequently used is dextrin, gelatin, and modified starch. Dried form solution is acquired and impediment to oxygen and combative agents is provided. Spinning wheel or fluid nozzle is used for the atomization of liquid admixture within a vessel. Further, the evaporation process is conducted by interacting with air possessing high temperature. Principally, concurrent type spray dryer are employed in food industries where the produce invade within the dryer flowing in similar route as the drying air [13]. In order to carry out nanoencapsulation of probiotic strains, various materials can be utilized such as cellulose acetate phthalate, alginate, skim milk, and starch for *L. acidophilus*, *L. bulgaricus*, *L. paracasei*, and *Bifidobacterium PL1*, respectively. Utilizing spray dried technique, Desmond *et al.* conducted nanoencapsulation studies on *L. paracasei* NFBC 338 using gum acacia and skim milk mixture (Table 8.2). *L. paracasei* treated with gum acacia illustrated 10 times increased survivability [20]. Lactic acid bacteria have gained much significance as probiotic bacteria however bile acid and stomach solution influences the activity of probiotic bacteria. *L. acidophilus* JCM 1132 was integrated in the intrinsic phase of water in oil in water emulsion to provide protection against cytotoxic bile acid and to enhance its survivability [43].

8.4.3 Nanoemulsification

The process of emulsion implicates two different phase ascribed as dispersed phase and continuous phase. This method permits the mix up of two liquids that have immiscible characteristic in the presence of surfactant/emulsifier wherein droplets are developed in dispersed phase sustaining in continuous phase when lipid is integrated in the aqueous media. The kind of emulsion determines the size of droplet. Nanoemulsion method includes the droplet size ranging from 20 to 200 nm also referred to as submicron emulsion and demonstrates narrow size distribution. Being nonequilibrium system, nanoemulsions are not produce immediately. Machine-driven systems such as ultrasonic generators, homogenizers with higher pressure, and high shear stirrer are utilized for the formation of nanoemulsions as they are able to generate powerful and extreme disruptive forces [44]. Nanoemulsion techniques can be employed for the nanoencapsulation of probiotics. Even under the strident intestinal and stomach conditions, the viability of microbial probiotic cells can be encouraged by adopting the nanoencapsulation of microbial cells in emulsion droplets using emulsification process. Seasame oil emulsion can also be practiced for the nanoencapsulation of probiotics. *L. delbrueckii* ssp. *Bulgaricus* was encapsulated by Hou *et al.* [12] within seasame oil emulsion. The results illustrated 10^4 times rise in survivability in the gastrointestinal tract and an increment in the viability of lactic acid lactic was observed from 0.023% to 5.45% during storage conditions [21]. Mantzouridou *et al.* [27] conducted the research on encapsulation of *Lactobacillus paracasei* subsp. *Paracasei* DC412 utilizing three different emulsifiers, namely, gum arabic or xanthan, egg yolk, and whey protein isolates to provide complete protection in the human gastrointestinal system and storage conditions by entrapping the probiotic microbes. Aqueous phase egg yolk emulsion resulted in survival of more number of probiotic cells. The integration of inulin improved the viability of probiotic *L. paracasei* subsp. *Paracasei* DC412 in against enzymatic destruction [27]. Abovementioned examples of encapsulating macromolecules can be employed for the nanoencapsulation of probiotic microorganisms.

8.5 Recent Advances in Nanoencapsulation of Probiotics

8.5.1 Augmentation of Nutritional Qualities of Bitter Gourd Juice Powder With Probiotics

Bitter melon or bitter gourd also known as *Momordica charantia* Linn belonging to the family of Cucurbitaceae is an equatorial and subtropical vine indigenous of Asia and Africa. Bitter gourd is a groundbreaking plant for its adaptability as comestible, beneficial, and for remedial applications. Kalal *et al.* [45] used spray drying method for the nanoencapsulation of probiotic, *L. casei* in bitter gourd juice powder using three different nano-encapsulating elements such as gum arabic, maltodextrin, and a mixture of both. Laboratory spray dryer was used for the process of nanoencapsulation. At room temperature, the feed solution encompassing *L. casei* was magnetically agitated. Peristaltic pump assisted in delivering the solution in the main chamber. The sample of nanoencapsulated *M. charantia* powder was obtained and loaded in polypropylene packets. Bitter melon comprises of bountiful mix of steroidal saponins referred to as charantins which confers to the hypoglycemic and antihyperglycemic properties of the plant (Figure 8.4). Bitter gourd is abundantly nutritious because of higher amount of iron, protein, phosphorous, calcium, and ascorbic acid. Antitumor and antimutagenic properties are among the therapeutic application of seed and extract of pulp of *M. charantia*. The results indicated that the functional properties of probiotic bitter gourd were not affected by encapsulating agents. During storage conditions, the stability of bitter gourd powder was retained. As compared to mixture of both maltodextrin and gum arabic encapsulating agents, maltodextrin nanoencapsulation illustrated low bulk density. In conclusion, the viability and survivability of *L. casei* with maltodextrin nanoencapsulation was reliable in terms of providing complete protection by acting as a thermoprotector. The viability of probiotic cells was much enhanced during storage condition as compared to the original count of cells, therefore demonstrating positive results [45].

8.5.2 Nanoencapsulation of *L. rhamnosus GG* for Antitumor Activity

Currently, nanoparticles have gained much attention and are regarded as one of the most periodically used materials. As compared to miscellaneous conventional materials available, nanomaterials are considered as the material of 21st century because of its demanding peculiarity, attributes, and qualities. Nanomaterials are extensively utilized in pharmaceutical and human health sector. Silver (Ag) nanoparticles are well chosen as a fundamental and pivotal material due to its major necessity in development of products. Silver has anti-inflammatory, antimicrobial, and fungicidal characteristics and possesses a wide range of applications such as antibacterial activity, antimicrobial activity, medical devices, food sector, pharmaceutical industry, and in medical management sectors. In current years, major cause of death in most of the western countries is cancer. Different therapeutic techniques have been employed and available for the treatment of cancer comprising chemotherapy and radiotherapy followed by lethal consequences, therefore affecting patient's life quality. Lactic acid bacteria which are more often used as probiotic bacteria play a very imperative role as a complete therapy practice for intestinal symbiosis and enhance the immunological environment of gastrointestinal system. The silver nanoparticles have been reported with cytotoxic attributes for the treatment of cancer. Taking into account the characteristics and properties of probiotics and silver nanoparticles, recently, study has been conducted by Mousavi *et al.* [46] investigating the *in vitro* anticancerous activity of silver nanoparticles and probiotic *L. rhamnosus* against HT-29 cell line colorectal cancer. The results illustrated active influence against the survivability of cancer cell line HT-29. It has been reported that, with increase in the concentration of *L. rhamnosus* silver nanoparticles, the viability of cancer cell lines efficaciously declined. Therefore, reactive oxygen species was induced within the cell, and stress was caused leading to damage and resulting in cell death [46].

8.5.3 Antibacterial Properties of *L. casei* Nanoparticles

An alternative approach to chemical and physical industrial science is the construction of nanoparticles by consuming biological entities that is being regarded as an ecological and worthwhile process. For the synthesis of nanoparticles, bacteria are observed like a strong candidate retaining eminent properties of short generation time, tremendous stability, and ease of culturing. In human body, a healthful and non-pathogenic probiotic *L. casei* is reported in diverse parts including mouth, urinary tract, stomach, and abdomen, which possess the power to rapidly reduce metal ions. Copper oxide has achieved more consideration as a nanomaterial due to its better stability, antibacterial property, good conductivity, and anticancerous potential against human cancer cell lines. In a literature study, copper oxide nanoparticles have been synthesized using *Lactobacillus casei* subsp. *Casei* [47]. The fabricated nanoparticles of copper oxide were identified and characterized by using X-ray diffractometer, Fourier transform infrared (FTIR) spectroscopy, transmission electron microscopy, and scanning electron microscopy. Broth microdilution antibacterial assay and agar well diffusion method were carried out by researchers to examine the antibacterial properties of nanoparticles against *P. aeruginosa* and *Staphylococcus aureus* and anticancerous properties against AGS gastric cancer cell lines and HT-29 colorectal cell lines. Positive result was exhibited by nanoparticles by justifying both antibacterial and anticancerous properties. The analysis of metabolic activity of cancer cells were done by employing Methyl Thiazolyl Diphenyl-Tetrazolium Bromide (MTT) assay. The mitochondrial function of metabolically active or viable cancer cell to cleave trizolium salt into formazan (blue in color) was assessed. In a dose- and time-dependent manner increased concentration of *L. casei* copper oxide nanoparticles unquestionably decreased growth of cancer cells. Even at lowest concentration of nanoparticles, the cell proliferation was ceased and cytotoxic effect was observed against the viability of cancer cell lines signifying *L. casei* copper nanoparticles as a propitious cancer therapy [47].

8.5.4 Cytotoxic and Antioxidant Potential of *L. kimchicus* Nanoparticles

In the field of biomedicine and biotechnology, gold nanoparticles have exhibited inevitable accomplishments due to their biocompatibility, scattering competence, absorption of light, and surface plasmon resonance. The effectiveness of nanoparticles is diminished due to the utilization of noxious ligand molecule and costly solvents upon the surface of nanoparticles, therefore negatively influencing biomedical applications. Thus, for the development of nanoparticles in reliable, non-noxious, appropriate, and biologically active manner, green synthesis or green chemistry is imperative approach. Markus *et al.* have adopted green synthesis method for the production of gold nanoparticles by using *L. kimchicus DCY51^T* isolated by emplyoing intracellular membrane bound technique from kimchi, a staple food of Korea [48]. For the synthesis of gold nanoparticles, *L. kimchicus DCY51^T* whole cells acted as biocatalyst and the synthesis was screened by change in color through visual observation. The surface characteristic of gold nanoparticles was examined by FTIR spectroscopy. Researchers evaluated *in vitro* toxicity against colorectal adenocarcinoma HT29 cell lines of human and murine macrophage (RAW264.7). MTT assay was conducted to investigate cytotoxicity. After treatment with different concentration of gold nanoparticles, both the cell lines indicated cell viability even after 3 days. To inspect the antioxidant activity of gold nanoparticles DPPH (Diphenylpicryl Hydrazine), radical scavenging activity was performed and analyzed. From this research, Markus *et al.* concluded that intracellular membrane bound synthesis of gold nanoparticles by *L. kimchicus DCY51^T* indicated long lasting stability. Cell viability was observed in case of adenocarcinoma HT29 and murine macrophage RAW264.7 cell lines. The latter one exhibited no cell death and continued to multiply, therefore demonstrating lesser cytotoxic characteristic of gold nanoparticles. Gold nanoparticles have antioxidant activity and illustrated ability of free radical scavengers. The lower cytotoxicity and good antioxidant activity of gold nanoparticles ascribe to protein capping layer, thus preventing agglomeration. Tyrosine, tryptophan, and cysteine amino acids were characterized by FTIR analyses and reported that such protein and amino acid residue were accountable for offering stability to *Lactobacillus* mediated gold nanoparticles. Green synthesis of gold nanoparticles is economical approach for medical imaging and drug delivery system [48].

8.5.5 *L. casei 393* Nanoparticles as Anti-Inflammatory Agents

For animal and human health, selenium is an indispensible trace nutrient crucial for supporting immune regulation and redox homeostasis and possesses antitumor properties. The different chemical forms of selenium include elemental selenium, selenite, selenocysteine, and selenate.

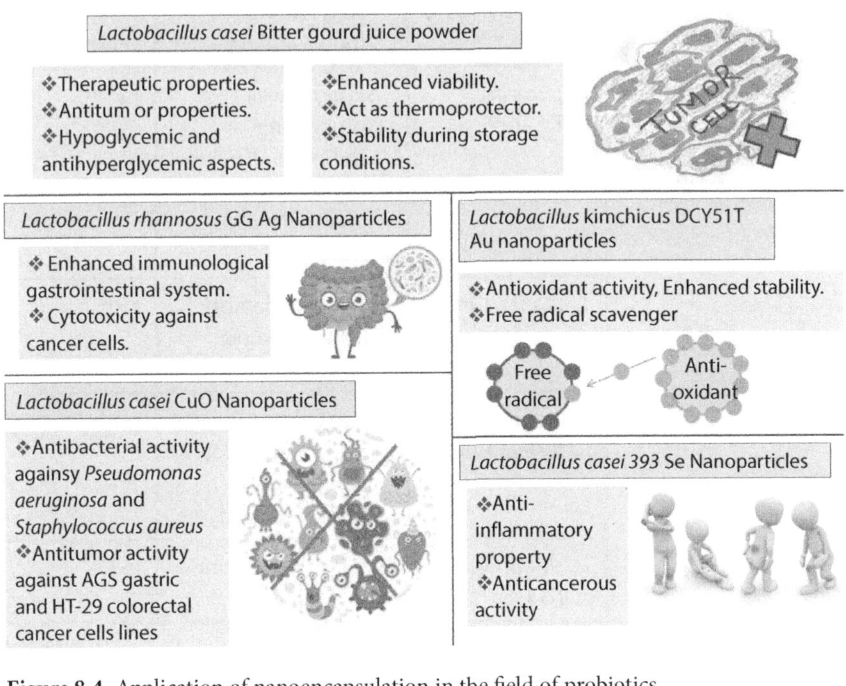

Figure 8.4 Application of nanoencapsulation in the field of probiotics.

Among all the chemical forms, sodium selenite is the most noxious form. However, its transformation can reduce cytotoxicity. Various chemical and physical methods restrict the applications of selenium nanoparticles in the field of biomedicine and food industry. By considering fermentation technology, Xu *et al.* [49] have synthesized novel nanoparticles using *L. casei 393* and analyzed its effects on intestinal barrier dysfunction generated by enterotoxigenic *E. coli* K88 [49]. *L. casei* 393 exhibits anticancerous and anti-inflammatory properties, aided in transformation of sodium selenite to selenium nanoparticles, and indicated intracellular accumulation. Intracellular synthesis of nanoparticles was identified by energy dispersive X-ray, flame emission atomic absorption spectrometer analysis, and transmission electron microscope. For cytotoxic analysis, three different cell lines

Table 8.2 Techniques employed for the nanoencapsulation of probiotics.

Nanoencapsulation technique	Probiotic strain	Importance	Reference
Electrospinning	*Saccharomycopsis fibuligera* VIT-MN04	Production of symbiotic supplements in food industry	[3]
	L. paracasei KS-199	Useful in heat processing industries due to higher chemical and thermal stability	[40]
	L. plantarum	Enhanced viability under adverse conditions	[50]
	Lactobacillus fermentum	Increased viability against stimulated gastric and intestinal fluid	[51]
	L. rhamnosus KLDS	Improved probiotic survivability	[52]

(*Continued*)

Table 8.2 Techniques employed for the nanoencapsulation of probiotics. (*Continued*)

Nanoencapsulation technique	Probiotic strain	Importance	Reference
Spray drying	*L. plantarum*	Effective encapsulation of probiotics and potentially beneficial in food industry	[53]
	L. paracasei NFBC 338	Increased survivability	[20]
	L. acidophilus NCDC 016	Increased encapsulation efficiency	[39]
	L. acidophilus JCM 1132	Resistant to cytotoxic bile acid	[43]
	L. casei	Enhanced storage and thermal protection	[45]
Nanoemulsification	*L. delbrueckii* ssp. *Bulgaricus*	Enhanced viability and storage	[21]
	L.s paracasei subsp. *Paracasei* DC412	Increased survivability	[27]
	L. acidophilus LAFTIL10, *B. animalis* Bo, *L. casei* LAFTIL26	Enhanced viability	[25]
	Bifidobacterium pseudocatenulatum G7	Improved stability and target gut release	[54]
	L. delbrueckii subsp. *bulgaricus*	Enhanced probiotic stability	[55]

were used including IPEC-J2 (porcine jejunal cell line), NCM460 cell line (human normal epithelial cell), and THP-1 (human acute monocyte leukemia cell line). The results observed were with no inhibitory effects on the cell lines, and the proliferation of cell lines was promoted by *Lactobacillus casei 393* selenium nanoparticles. Toxic effects were exerted by *E. coli* k88 on IPEC-J2 (porcine jejunal cell line), resulting in cell deformation and apoptosis. However, when IPEC-J2 cells were treated with *L. casei 393* selenium nanoparticles, the toxic effect was opposed. Thus, it demonstrates the protective influence of *L. casei 393* selenium nanoparticles on enterotoxigenic *E. coli* k88 induced intestinal barrier function damage. The study was also conducted on *in vitro* assessment of antioxidant and antitumor properties of *Lactobacillus casei* nanoparticles on HepG2 cells. In comparison to control group, the growth of HepG2 cells suppressed which resulted in cell apoptosis [49].

8.6 Advantage and Disadvantage of Nanoencapsulation of Probiotics

Probiotics are non-pathogenic living microbes, which, when employed in the diet or food supplements, improve the health of the individual. Probiotics are gaining popularity as an alternative to traditional antibiotics in the fight against bacterial resistance [56]. Nanotechnology has the potential to bring major advances in medicine and the primary goal of nanoencapsulation is to secure the core material while allowing it to be released when needed. Nanomedicine, along with the well-known methods of probiotics and prebiotics, demonstrates a beneficial method in establishing an ideal atmosphere in the gastrointestinal system [57]. Nanoencapsulated probiotics are the colloidal-sized particles with size range from 10 to 1,000 nm and are presented either as nanoparticle, nanolayer, nanoemulsion, or nanobead. Given that the particle size of any bioactive compound has a direct impact on its therapeutic effectiveness, nanoencapsulation of probiotics plays an important role in delivering bioactive compounds to different locations within the body more efficiently than other micro and macro formulations [58].

The best method for encapsulating probiotics was chosen primarily to shield the cells from extreme gastrointestinal conditions, allowing them to be released in as metabolically active and in a viable manner within the intestine [59]. The synthesis of probiotic cells with nanoparticles is propitious method for developing effective and versatile DNA vaccines [60]. Probiotic nanoparticles have been used to prevent and treat neuroinflammation, according to Feher and other researchers [61]. In comparison to microencapsulation, nanoencapsulation has the potential to increase bioavailability, enhances slow controlled release, and allows selective targeting of bioactive compounds. Therapeutic agents are more effective, specific, and targetable when they are nanoencapsulated. In particular, nanoencapsulation technologies have magnificent constant release properties and sub-cellular volume and are biologically compatible with cells and tissue. They also permit for changes in drug biological availability and enhance the pharmacokinetic aspect of a variety of pursuit [4].

Furthermore, nanoencapsulation of microbes reduces their toxicity, overcomes resistance, and lowers the cost of using them because a smaller amount of the active is needed. Both nanoencapsulation systems have limitations for their employment in edible and food industries due to their higher manufacturing charges and absence of permitted materials [62]. The other disadvantage with nanoencapsulation involves the requirement of highly skilled individuals, and also, it is difficult to obtain the continuous and uniform film. There are, however, a number of issues that need to be resolved. The procedure is not appropriate for all medications. In addition, post-preparatory steps, such as purification and preservation, which are especially essential for nanocapsules, and residual solvent analysis need to be thoroughly investigated. Other difficulties remain, including the development of an incomplete or discontinuous film, insufficient stability of certain active components, unreproducible or predictable release characteristics, and other constraints which render the final product economically unviable [63].

Nanoemulsions, which are extremely thermodynamically stable dispersions of water and oil stabilized by an interfacial layer of surfactant and co-surfactant molecules with 100 nm as droplet size, are one of the techniques used in nanoencapsulation [64]. Nanoemulsions can be produced using a variety of methods, including microfluidization, high-pressure homogenization, and low-energy emulsification. Nanoencapsulation formulations possess many advantages for the delivery of medications, biologics, and diagnostic agents. Apart from the fact that a higher percentage of surfactant is used for the formulation of nanoemulsion, still, there is absence of toxicological examination of the prepared nanoemulsion that could be a large research field in the world to come. Similarly, the advantages and disadvantages of various nanoencapsulation formulation methods are listed in the Table 8.3.

Table 8.3 Advantages and disadvantages of nanoencapsulation formulations [65].

Delivery system	Advantage	Disadvantage
Nanoemulsions (liquid)	✓ Transparent systems ✓ Carrier of hydrophobic compounds ✓ Quick absorption excellent shelf stability ✓ Helpful in taste making ✓ Enhances bioavailability ✓ Less energy needed	✓ Quick release ✓ Lesser stability in acidic condition ✓ Need large concentration of surfactant ✓ Use of nontoxic surfactant
Nanoparticles	✓ Boosts compounds aqueous solubility ✓ Large surface to volume ratios ✓ Regulated and control release of insoluble actives ✓ Declined toxic or side effects	✓ Lack of stability of some actives ✓ High production costs ✓ Chances of recrystallization and lesser encapsulation load
Liposomes (liquid)	✓ Possess ability for the water soluble drug's encapsulation or to solubilizes lipophilic drugs in their bilayer ✓ More stable aspect of components in foodstuff with higher water content ✓ Production at large scale	✓ Short life ✓ Accelerated release
Nano fiber (solid)	✓ More porosity and large scale production Competence to bear heat sensitive compounds ✓ Higher gas permeability	✓ Limitation in the utilization in electrospinning due to biopolymer solubility.

8.7 Conclusion

Nowadays, in the context of nutrition and health, probiotics are the major considerations as they offer miscellaneous health benefits including immune stimulation, healthy gut microflora maintenance, pathogen inhibition, antimicrobial properties, and healthy advantages to the gastrointestinal system. However, the stability and viability of probiotics is not maintained when it is being processed through various sectors in the food industry. Nanoencapsulation is expeditiously proliferating technology having dynamic aspects in the field of edible and medical industry. A revolutionary influence has been brought within food industry due to nanotechnology. Different nanoencapsulating agents or matrices like chitosan, gum arabic, whey proteins, polyvinylpyrrolidone, cellulose acetate phthalate, skim milk, starch, and alginate can be employed for the nanoencapsulation of probiotics. Food sector is involved in financing much in nanosystem to establish and carry out nanoencapsulating process to increase the yield in terms of greater viability and stability of probiotics. Researchers are much more interested in generating efficacious nanocarriers to improve the stability and viability of probiotics within gastrointestinal system. Even the studies are being conducted in utilization of nanoparticles such as AgNPs, SeNPs, CuNPs, and AuNPs for the better enhanced probiotic viable count and improved resistance of probiotics against bile acids and salts. Nanoencapsulation has actually emerged as an indispensable approach to mitigate the limitations associated with utilization of probiotics in food and pharmaceutical industries.

References

1. Kaur, S., Kaur, R., Rani, N., Sharma, S., Joshi, M., Sources and selection criteria of probiotics, in: *Advances in Probiotics for Sustainable Food and Medicine*, pp. 27–43, 2021.
2. Mortazavian, A., Razavi, S.H., Ehsani, M.R., Sohrabvandi, S., Principles and methods of microencapsulation of probiotic microorganisms. *Iran. J. Biotechnol.*, 5, 1–18, 2007.
3. Ragavan, M.L. and Das, N., Nanoencapsulation of *Saccharomycopsis fibuligera* VIT-MN04 using electrospinning technique for easy gastrointestinal transit. *IET Nanobiotechnol.*, 14, 766–773, 2020.
4. Durazzo, A., Nazhand, A., Lucarini, M., Atanasov, A.G., Souto, E.B., Novellino, E., Capasso, R., An updated overview on nanonutraceuticals: Focus on nanoprebiotics and nanoprobiotics. *Int. J. Mol. Sci.*, 21, 2285, 2020.
5. Ebrahimnezhad, P., Khavarpour, M., Khalili, S., Survival of Lactobacillus acidophilus as probiotic bacteria using chitosan nanoparticles. *Int. J. Eng. Sci.*, 30, 456–463, 2017.
6. Kim, J., Muhammad, N., Jhun, B.H., Yoo, J.-W., Probiotic delivery systems: A brief overview. *Int. J. Pharm. Investig.*, 46, 377–386, 2016.

7. Moghadam, F.V., Pourahmad, R., Mortazavi, A., Davoodi, D., Azizinezhad, R., Use of fish oil nanoencapsulated with gum arabic carrier in low fat probiotic fermented milk. *Food Sci. Anim. Resour*, 39, 309, 2019.
8. Gerez, C.L., Font de Valdez, G., Gigante, M.L., Grosso, C.R.F., Whey protein coating bead improves the survival of the probiotic Lactobacillus rhamnosus CRL 1505 to low pH. *Lett. Appl. Microbiol.*, 54, 552–556, 2012.
9. Libran, C.M., Castro, S., Lagaron, J.M., Encapsulation by electrospray coating atomization of probiotic strains. *Innov. Food Sci. Emerg. Technol.*, 39, 216–222, 2017.
10. Khan, M.A., Hussain, Z., Ali, S., Qamar, Z., Imran, M., Hafeez, F.Y., Fabrication of electrospun probiotic functionalized nanocomposite scaffolds for infection control and dermal burn healing in a mice model. *ACS Biomater. Sci. Eng.*, 5, 6109–6116, 2019.
11. Gbassi, G.K. and Vandamme, T., Probiotic encapsulation technology: From microencapsulation to release into the gut. *Pharmaceutics, 4,* 149–163, 2012.
12. Favaro-Trindade, C.S. and Grosso, C.R.F., Microencapsulation of L. acidophilus (La-05) and B. lactis (Bb-12) and evaluation of their survival at the pH values of the stomach and in bile. *J. Microencapsul.*, 19, 485–494, 2002.
13. Chavarri, M., Maranon, I., Carmen, M., Encapsulation technology to protect probiotic bacteria. *Probiotics.*, pp. 501–540, IntechOpen, London, 2012.
14. Mandal, S., Puniya, A.K., Singh, K., Effect of alginate concentrations on survival of microencapsulated *Lactobacillus casei* NCDC-298. *Int. Dairy J.*, 16, 1190–1195, 2006.
15. Maciel, G.M., Chaves, K.S., Grosso, C.R.F., Gigante, M.L., Microencapsulation of Lactobacillus acidophilus La-5 by spray-drying using sweet whey and skim milk as encapsulating materials. *Int. J. Dairy Sci.*, 97, 1991–1998, 2014.
16. Singh, M., Sharma, D., Chauhan, R., Goel, G., Skimmed milk-based encapsulation for enhanced stability and viability of Lactobacillus gastricus BTM 7 under simulated gastrointestinal conditions. *Probiotics Antimicrob.*, 11, 850–856, 2019.
17. Iravani, S., Korbekandi, H., Mirmohammadi, S.V., Technology and potential applications of probiotic encapsulation in fermented milk products. *J. Food Sci. Technol.*, 52, 4679–4696, 2015.
18. Oriordan, K., Andrews, D., Buckle, K., Conway, P., Evaluation of microencapsulation of a Bifidobacterium strain with starch as an approach to prolonging viability during storage. *J. Appl. Microbiol.*, 91, 1059–1066, 2001.
19. Karimi, M., Yazdi, F.T., Mortazavi, S.A., Shahabi-Ghahfarrokhi, I., Chamani, J., Development of active antimicrobial poly (l-glutamic) acid-poly (l-lysine) packaging material to protect probiotic bacterium. *Polym. Test.*, 83, 106338, 2020.
20. Desmond, C., Ross, R.P., O'callaghan, E., Fitzgerald, G., Stanton, C., Improved survival of Lactobacillus paracasei NFBC 338 in spray-dried powders containing gum acacia. *J. Appl. Microbiol.*, 93, 1003–11, 2002.

21. Hou, R.C.W., Lin, M.Y., Wang, M.M.C., Tzen, J.T.C., Increase of viability of entrapped cells of Lactobacillus delbrueckii ssp. bulgaricus in artificial sesame oil emulsions. *Int. J. Dairy Sci.*, 86, 424–428, 2003.
22. Gbassi, G.K., Vandamme, T., Ennahar, S., Marchioni, E., Microencapsulation of Lactobacillus plantarum spp in an alginate matrix coated with whey proteins. *Int. J. Food Microbiol.*, 129, 103–105, 2009.
23. Ding, W.K. and Shah, N.P., An improved method of microencapsulation of probiotic bacteria for their stability in acidic and bile conditions during storage. *J. Food Sci.*, 74, M53–M61, 2009.
24. Donthidi, A.R., Tester, R.F., Aidoo, K.E., Effect of lecithin and starch on alginate-encapsulated probiotic bacteria. *J. Microencapsul., 27*, 67–77, 2010.
25. Madureira, A.R., Amorim, M., Gomes, A.M., Pintado, M.E., Malcata, F.X., Protective effect of whey cheese matrix on probiotic strains exposed to simulated gastrointestinal conditions. *Food Res. Int.*, 44, 465–470, 2011.
26. Lara Pedroso, D., Thomazini, M., Heinemann, R.J.B., Favaro-Trindade, C.S., Protection of *Bifidobacterium lactis* and *Lactobacillus acidophilus* by microencapsulation using spray-chilling. *Int. Dairy J.*, 26, 127–132, 2012.
27. Mantzouridou, F., Spanou, A., Kiosseoglou, V., An inulin-based dressing emulsion as a potential probiotic food carrier. *Food Res. Int.*, 46, 260–269, 2012.
28. Jalali, M., Abedi, D., Varshosaz, J., Najjarzadeh, M., Mirlohi, M., Tavakoli, N., Stability evaluation of freeze-dried Lactobacillus paracasei subsp. tolerance and Lactobacillus delbrueckii subsp. bulgaricus in oral capsules. *Res. Pharm. Sci.*, 7, 31, 2012.
29. Castro-Cislaghi, F.P., Silva, C.D.R.E., Fritzen-Freire, C.B., Lorenz, J.G., Sant'Anna, E.S., *Bifidobacterium* Bb-12 microencapsulated by spray drying with whey: Survival under simulated gastrointestinal conditions, tolerance to NaCl, and viability during storage. *J. Food Eng.*, 113, 186–193, 2012.
30. Sathyabama, S. and Vijayabharathi, R., Co-encapsulation of probiotics with prebiotics on alginate matrix and its effect on viability in simulated gastric environment. *LWT*, 57, 419–425, 2014.
31. Patel, S., Cereal bran fortified-functional foods for obesity and diabetes management: Triumphs, hurdles and possibilities. *J. Funct. Foods*, 14, 255–269, 2015.
32. Paim, D.R., Costa, S.D., Walter, E.H., Tonon, R.V., Microencapsulation of probiotic jussara (Euterpe edulis M.) juice by spray drying. *LWT*, 74, 21–25, 2016.
33. Chotiko, A. and Sathivel, S., Three protective agents for pectin-rice bran capsules for encapsulating Lactobacillus plantarum. *Food Biosci.*, 16, 56–65, 2016.
34. Pankasemsuk, T., Apichartsrangkoon, A., Worametrachanon, S., Techarang, J., Encapsulation of Lactobacillus casei 01 by alginate along with hi-maize starch for exposure to a simulated gut model. *Food Biosci.*, 16, 32–36, 2016.

35. Atia, A., Gomaa, A., Fliss, I., Beyssac, E., Garrait, G., Subirade, M., A prebiotic matrix for encapsulation of probiotics: physicochemical and microbiological study. *J. Microencapsul.*, 33, 89–101, 2016.
36. Chen, H.Y., Li, X.Y., Liu, B.J., Meng, X.H., Microencapsulation of Lactobacillus bulgaricus and survival assays under simulated gastrointestinal conditions. *J. Funct. Foods*, 29, 248–255, 2017.
37. Chavarri, M., Maranon, I., Ares, R., Ibáñez, F.C., Marzo, F., del Carmen Villaran, M., Microencapsulation of a probiotic and prebiotic in alginate-chitosan capsules improves survival in simulated gastro-intestinal conditions. *Int. J. Food Microbiol.*, 142, 185–189, 2010.
38. Afzaal, M., Saeed, F., Arshad, M.U., Nadeem, M.T., Saeed, M., Tufail, T., The effect of encapsulation on the stability of probiotic bacteria in ice cream and simulated gastrointestinal conditions. *Probiotics Antimicrob.*, 11, 1348–1354, 2019.
39. Arepally, D. and Goswami, T.K., Effect of inlet air temperature and gum Arabic concentration on encapsulation of probiotics by spray drying. *LWT*, 99, 583–593, 2019.
40. Yilmaz, M.T., Taylan, O., Karakas, C.Y., Dertli, E., An alternative way to encapsulate probiotics within electrospun alginate nanofibers as monitored under simulated gastrointestinal conditions and in kefir. *Carbohydr. Polym.*, 244, 116447, 2020.
41. Fung, W.Y., Yuen, K.H., Liong, M.T., Agrowaste-based nanofibers as a probiotic encapsulant: Fabrication and characterization. *J. Agric. Food Chem.*, 10, 8140–7, 2011.
42. Liu, H., Gong, J., Chabot, D., Miller, S.S., Cui, S.W., Zhong, F., Wang, Q., Improved survival of Lactobacillus zeae LB1 in a spray dried alginate-protein matrix. *Food Hydrocoll.*, 78, 100–108, 2018.
43. Shima, M., Matsuo, T., Yamashita, M., Adachi, S., Protection of Lactobacillus acidophilus from bile salts in a model intestinal juice by incorporation into the inner-water phase of a W/O/W emulsion. *Food Hydrocoll.*, *23*, 281–285, 2009.
44. Paredes, A.J., Asensio, C.M., Llabot, J.M., Allemandi, D.A., Palma, S.D., Nanoencapsulation in the food industry: Manufacture, applications and characterization. *J. Food Bioeng. Nanoprocess.*, 1, 56–79, 2016.
45. Kalal, A.Y., Hiregoudar, D.S., Nidoni, U.D., Nanoencapsulation of probiotic bitter gourd juice powder. *Int. J. Agric. Sci.*, 6, 9–20, 2017.
46. Mousavi, S.M.A.A., Mirhosseini, S.A., Panahi, M.R.S., Hosseini, H.M., Characterization of biosynthesized silver nanoparticles using lactobacillus rhamnosus GG and it's *in vitro* assessment against colorectal cancer cells. *Probiotics Antimicrob.*, 12, 740–746, 2020.
47. Kouhkan, M., Ahangar, P., Babaganjeh, L.A., Allahyari-Devin, M., Biosynthesis of copper oxide nanoparticles using *Lactobacillus casei* subsp. casei and its anticancer and antibacterial activities. *Curr. Nanosci.*, 16, 101–111, 2020.

48. Markus, J., Mathiyalagan, R., Kim, Y.J., Abbai, R., Singh, P., Ahn, S., Yang, D.C., Intracellular synthesis of gold nanoparticles with antioxidant activity by probiotic Lactobacillus kimchicus DCY51 T isolated from Korean kimchi. *Enzyme. Microb. Technol.*, 95, 85–93, 2016.
49. Xu, C., Guo, Y., Qiao, L., Ma, L., Cheng, Y., Roman, A., Biogenic synthesis of novel functionalized selenium nanoparticles by Lactobacillus casei ATCC 393 and its protective effects on Intestinal barrier dysfunction caused by Enterotoxigenic Escherichia coli K88. *Front. Microbiol.*, 9, 1129, 2018.
50. Feng, K., Huang, R.M., Wu, R.Q., Wei, Y.S., Zong, M.H., Linhardt, R.J., Wu, H., A novel route for double-layered encapsulation of probiotics with improved viability under adverse conditions. *Food Chem.*, 310, 125977, 2020.
51. Duman, D. and Karadag, A., Inulin added electrospun composite nanofibres by electrospinning for the encapsulation of probiotics: Characterisation and assessment of viability during storage and simulated gastrointestinal digestion. *Int. J. Food Sci. Technol.*, 56, 927–935, 2021.
52. Ma, J., Xu, C., Yu, H., Feng, Z., Yu, W., Gu, L., Liu, Z., Chen, L., Jiang, Z., Hou, J., Electro-encapsulation of probiotics in gum Arabic-pullulan blend nanofibres using electrospinning technology. *Food Hydrocoll.*, 111, 106381, 2021.
53. Ceja-Medina, L.I., Ortiz-Basurto, R.I., Medina-Torres, L., Calderas, F., Bernad-Bernad, M.J., González-Laredo, R.F., Manero, O., Microencapsulation of Lactobacillus plantarum by spray drying with mixtures of Aloe vera mucilage and agave fructans as wall materials. *J. Food Process Eng.*, 43, 8, 2020.
54. Zhang, Z., Gu, M., You, X., Sela, D.A., Xiao, H., McClements, D.J., Encapsulation of bifidobacterium in alginate microgels improves viability and targeted gut release. *Food Hydrocoll.*, 116, 106634, 2021.
55. Vaishanavi, S. and Preetha, R., Soy protein incorporated nanoemulsion for enhanced stability of probiotic (Lactobacillus delbrueckii subsp. bulgaricus) and its characterization. *Mater. Today: Proc.*, 40, 148–153, 2021.
56. Zheng, M., Zhang, R., Tian, X., Zhou, X., Pan, X., Wong, A., Assessing the risk of probiotic dietary supplements in the context of antibiotic resistance. *Front. Microbiol.*, 8, 908, 2017.
57. Sun-Waterhouse, D. and Waterhouse, G.I., Recent advances in the application of nanomaterials and nanotechnology in food research, in: *Novel approaches of nanotechnology in food*, pp. 21–66, 2016.
58. Ghorani, B. and Tucker, N., Fundamentals of electrospinning as a novel delivery vehicle for bioactive compounds in food nanotechnology. *Food Hydrocoll.*, 51, 227–240, 2015.
59. Anal, A.K. and Singh, H., Recent advances in microencapsulation of probiotics for industrial applications and targeted delivery. *Trends Food Sci. Technol.*, 18, 240–251, 2007.
60. Lin, L.C.W., Chattopadhyay, S., Lin, J.C., Hu, C.M.J., Advances and opportunities in nanoparticle-and nanomaterial-based vaccines against bacterial infections. *Adv. Healthc. Mater.*, 7, 1701395, 2018.

61. Feher, F., Pinter, E., Helyes, Z., Szolcsanyi, J., Nano-size particles of probiotics for preventing and treating neuroinflammation. ARVO Annual Meeting. *Investig. Ophthalmol. Vis. Sci.*, 53, 331, 2012.
62. Krishna, P.G., Kameswaran, S., Ranjani, T.S., Gunavathi, Y., Recent developments in nanoencapsulation of bioactive compounds of microbial sources and their biomedical applications, in: *Rec. Develop. Appl. Microbiol. Biochem.*, pp. 141–152, 2021.
63. Limayem, I., Charcosset, C., Fessi, H., Purification of nanoparticle suspensions by a concentration/diafiltration process. *Sep. Purif. Technol.*, 38, 1–9, 2004.
64. Patel, R.P. and Joshi, J.R., An overview on nanoemulsion: A novel approach. *Int. J. Pharma. Sci. Res.*, 3, 4640, 2012.
65. Blanco-Padilla, A., Soto, K.M., Hernandez Iturriaga, M., Mendoza, S., Food Antimicrobials Nanocarriers. *Sci. World J.*, 1, 1–11, 2014.

9

Incorporation of Nanocarriers as Antimicrobial Agents in Food Packaging

Shamkumar P. Deshmukh[1]*, Krishna K. Pawar[2] and Dattatray K. Dalavi[3]

[1]Department of Chemistry, D.B.F. Dayanand College of Arts and Science, Solapur, M. S., India
[2]School of Nanoscience and Technology, Shivaji University, Kolhapur, M. S., India
[3]Department of Chemistry, Bhogawati Mahavidyalaya, Kurukali, M. S., India

Abstract
Food deterioration has become a critical concern in recent years due to the proliferation of bacteria in food and its by-products. Therefore, the shelf life of natural food may be increasing using bioactive antimicrobial agents. In contrast, even though some antimicrobial bioactive compounds are promising, the instability reduces antimicrobial activity. Hence, the search for antimicrobial agents and their crucial physicochemical properties enhances the food's shelf life. In this context, nanocarriers are the vital ingredient for protecting antimicrobial food agents from severe environmental conditions. The bioactive food constituents are avoided from harsh processing and environmental conditions while also allowing them to safely reach their target organs and cells. Therefore, the incorporation of nanocarriers in food as antimicrobials agents is favorable for enhanced antimicrobial activity. The lipid, biopolymer, carriers inspired by nature, carriers based on specialized equipment, metal, and non-metal-based nanomaterials are effective antimicrobial agents in food packaging.

Keywords: Food packaging, antimicrobial agents, shelf life, nanocarriers, biopolymer, lipid, nanomaterials, bioactive

**Corresponding author*: shamkumar0@gmail.com

Annu, Tanima Bhattacharya and Shakeel Ahmed (eds.) Nanotechnology in Intelligent Food Packaging, (203–234) © 2022 Scrivener Publishing LLC

9.1 Introduction

Widespread foodborne diseases due to various pathogenic agents as viruses, bacteria, and fungus lead to anxiety. According to the World Health Organization (WHO), almost 600 million people were infected by food contamination in 2015, with 420,000 mortality documented globally [1]. With expanding foreign trade, the risk of disease transmission from infected food items and foodborne illnesses rises. As a result, it is critical to improve global steps to prevent the risk of foodborne diseases, and packaged foods are a key focus region for achieving a goal. Packaging is important in the food that provide numerous advantages including as improved life span, simpler storage, and safety from physical damage in transportation and handling [2]. Besides the benefits of standard packaging, it also has several significant benefits, for instance, limiting the development of microbial pathogens in food stuff. Antimicrobial (organic or inorganic) agents may be incorporated into the packaging system by creating active packaging. Making use of organic antimicrobials materials in packaging to improve shelf-life, effectiveness, and safety is a popular strategy for reducing the threat of foodborne microorganisms. In this context, food safety the Codex Alimentarius Commission (2004) reported the concentration of hazards in a portion of food that offers the required intensity of safeguard. It encompasses all food-associated accomplishments for instance handling food, refining, storage, distribution, prevention, identification, and response to foodborne illness. As a result, food manufacturers must maintain a vastly efficient sanitizing method to reduce the amount of pathogenic and spoilage on surfaces and during their formulation, handling, and packaging phases to confirm food quality. Physical techniques such as high-pressure spraying, steam, mechanical cleaning, hot air disinfection, drying, and irradiation are used to reduce the microbial load in the food company.

Instead, a wide range of substances such as halides, solvents, polyphenols, metals, and tetra ammonium salts are used. Furthermore, biological routes, for example, enzymes, are also primarily used [3, 4]. These techniques are not continuously appropriate for foods and certain coats. There is also a consumer desire for foods that are chemical-free or utilize less chemicals in their production. Nanotechnology, a relatively new choice, is investigated as an innovative technology with tremendous potential to provide potential solutions to the food industry's food safety issues. Nanotechnology can prevent the evolution of microorganisms and their wastage. One method makes use of an antimicrobial agent that has been converted to nanoscale material, whereas the other method employs a nanotechnology-based device to produce traditional food antimicrobials. Nanomaterials help in keeping food healthy by prohibiting harmful microorganisms microbes from adhering to food surfaces, modifying molecular transport mechanisms, and killing present microorganisms [5]. Nanocarriers have a maximum area ratio and are smaller than microcarriers. It can act as a carrier to thorough the cell membrane for potential antimicrobial activity. Such small particles, on either side, could pass easily through cell membranes and enter into specific receptors, releasing their cargo [6].

The biocompatible materials with optimized size show a keen interest in improving the quality of foods. Particularly, nanocarriers enhance the carrier's encapsulation efficiency and monitor the dissolution rate under different stimulations for target delivery. The development of appropriate nanocarriers has become increasingly important with non-toxic nanomaterials. Furthermore, stable, biocompatible, and biodegradable carriers are needed. Forming nanoscale emulsions, liposomes, protein particles, and other delivery systems, for example, expose a more accessible active surface area of these delivery systems, which is helpful for absorption via the mucus-adhering process inside the body's digestive system [7]. The ingredients mentioned above are used to formulate these encapsulation processes. The encapsulation technique reduced the frequency of essential oils release and resulted in long-lasting antimicrobial activity. The

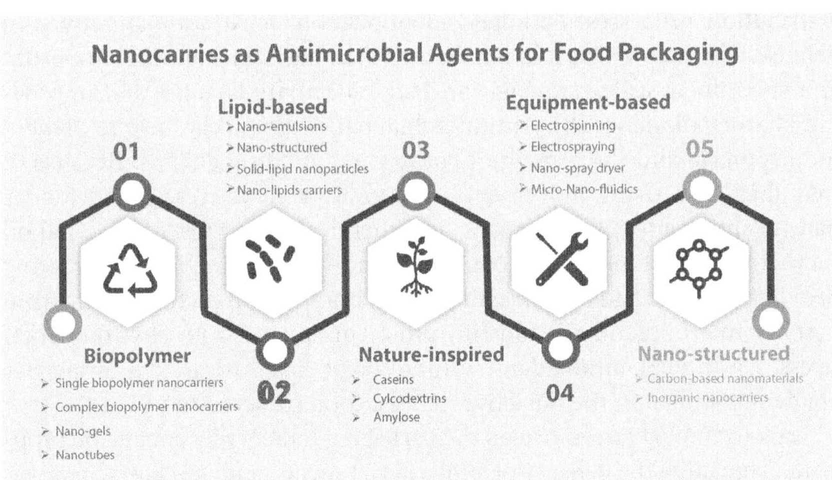

Scheme 9.1 Represents the nanocarriers as antimicrobial agents for food packaging.

regulated disclaimer active packaging loaded with nano encapsulated antimicrobial agents efficiently preserves food quality and extends life span. Consequently, active packaging is a crucial aspect of food packaging. We can divide food nanocarriers into lipid, biopolymer, nature-inspired, and special equipment-based nanocarriers.

9.2 Need of Nanocarriers as Antimicrobial Agents in Food Packaging

Food security is a significant interest for human health because a vast population suffers from digestive or other diseases due to contaminated food worldwide. As a result, food must be preserved by inactivating or inhibiting harmful microbial growth in the food. Chemical preservatives are commonly used to keep food. Food preservation chemicals are well-known for having detrimental effects on human health when consumed more than daily limits. Therefore, there is an alternative method required to overcome it. Nowadays, nanotechnology could pave the ways to develop new practical materials, micro-nanoscale manufacturing, creation, and method and equipment design in the food industry. Nanoencapsulation is an emerging area due to its larger surface area, bioavailability, and targetability. Furthermore, the protection, quality, bioavailability, and nutritious value of foods can be affected by nanoscale. Nanotechnology's crucial links to the agricultural and food industries are increasing food security and distribution, plant ability to soak up nutrients, flavour and nutrition, distribution processes, pathogen identification, food compatibility, and defense. The nanoencapsulation frameworks have outstanding properties at a subcellular scale, allowing for drug bioactivity changes and improving pharmacokinetics [8]. Antimicrobial nanocarriers can help to prevent food spoilage and the growth of pathogenic microorganisms. Because of this, there is a rising movement on natural preservatives, giving food a natural appearance. The aromatic and chemical properties of essential oil serve a variety of vital functions in plants in nature, including attracting beneficial insects and (ii) attracting pollinators, (ii) protecting plants from pests or microorganisms, and (iii) protecting plants from environmental stress. Antifungal, antioxidant, antimicrobial, and insecticidal properties are demonstrated in the bioactive EOs for food preservation [9].

Nanotechnology transformed the packaging industry by enhancing properties, regulating the delivery of antibacterial agents, and developing innovative packaging through nanosensors. Standard methods for producing active packaging include integrating antimicrobial nanocarriers into the wrapping framework, infusing the kit with antibacterial agents using antimicrobial packets/pads, immobilizing antimicrobials to the product and consolidating natural polymers with antimicrobial activity in the packaging framework. Regardless of the manner of AP processing, antimicrobials can be supplied into the environment and later come into touch with meals, or they can be discharged directly into food [10]. If antimicrobials have a high release rate, then most of them would just be released in the first few storage periods, resulting in reduced packaging antimicrobial effects. Conversely, an extended release rate may present a consistent delivery of antimicrobials through space. For example, food stuffs coated with a fine layer of incorporated biopolymers. It is one of the most effective ways to manufacture slow-release active packaging. Therefore, nanocarriers have a role in food packaging for the monitored antimicrobial vehicles to improve the shelf life of food items.

Furthermore, nanoencapsulation improves the solubility of disinfectant agents due to surface area, resulting in sustained delivery [11]. Antimicrobials can be delivered to the site of infection using nanocarriers, which secure them from harsh environments, improve their stability, and guide them to the area of diseases. Each encapsulation device has its own set of benefits and drawbacks. The nanoencapsulation processes have outstanding continuous release. The subcellular scale and tissue biocompatibility allow for changes in drug bioavailability and improve the pharmacokinetic profile of various active ingredients [12]. Therefore, nanocarriers are the new strategy to boost the food products.

9.3 Biopolymers and Their Nanocomposites as Antimicrobial Nanocarriers for Food Packaging

9.3.1 Biopolymers

Food wrapping has long been used to prevent needlessly close human contact with food. In recent years, a variety of packaging technologies have been utilized to interact with food in order to accomplish desired outcomes [13]. Polymer products are the leading materials for food packaging owing to their properties, recyclability, and mixture possibilities. In comparison, the long life of polymer packaging materials is one of the issues of their application. An ever-increasing rate of packaging waste and environmental concerns has necessitated the introduction/incorporation of biodegradable/compostable materials. Bio-based polymers have arisen as a credible option to petroleum-based polymeric products, reducing reliance on the severely depleted crude oil supply in recent days. For instance, the antioxidant efficacy has also been recently explored for chitosan biopolymer and PVA-based material loaded with *Ocimum tenuiflorium* for active food packaging applications [76]. Packaging made from biodegradable agricultural and marine raw materials is known as biopolymer-based packaging [14].

Biopolymers are polymers made up of covalently bound monomeric units. Biopolymers can decompose or degrade by naturally occurring processes and leaves chemical by-products harmless to the environment. The biopolymers are categorized by the process of manufacturing, such as natural and synthetic. Natural biopolymers are proteins and carbohydrates used in food packaging. Carbohydrates include starch, cellulose, and chitosan, whereas proteins include gelatine, gluten, and alginate [15]. In synthetic biopolymers, microbial production or fermentation and chemical synthesis are two significant types widely used in recent days.

Furthermore, biopolymers are synthesized chemically via petroleum and biomass. It includes polyvinyl alcohol (PVA), polylactic acid (PLA), polycaprolactone (PCL), and polyglycolic acid (PGA). The detailed classification of biopolymer is as shown in Figure 9.1.

Biopolymers as food packaging materials have inadequate mechanical strength and barrier characteristics when compared to the conventional non-biodegradable materials, notably those made from petroleum. As a result, there has been a lot of study into improving the film parameters of biopolymer-based packaging films. Even though researchers have made substantial gains in their features, they still have poor physical, thermal, and mechanical properties. It has been a struggle to improve biopolymer's

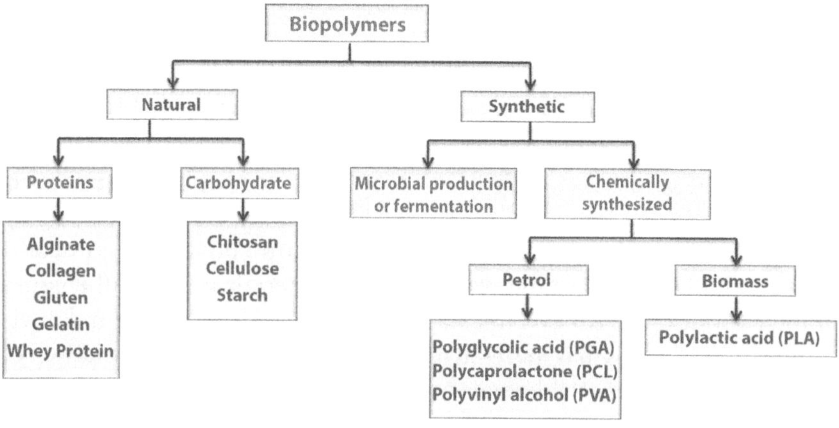

Figure 9.1 Classification of biopolymers based on the natural and synthetic mode.

mechanical and barrier properties for decades without detracting from their inherent properties. In addition, new age requirements like antimicrobial, antioxidant properties, and microencapsulation of aroma materials have increased the problems of researchers and industries. Thus, the strategy to develop the biopolymer-based packaging material with enhanced mechanical and barrier properties with new-age properties like antimicrobial, antioxidant with fragrance-enriched biopolymers, or composites is in high demand.

9.3.2 Nanofillers

A long drive study of biopolymers revealed that the incorporation of nano/micro fillers has a significant impact on the properties of biopolymers. Thus, incorporating of nanocarriers in the biopolymers enhances the physicochemical properties of packaging material. Typically, the size of the nanofiller is lying at 1 to 100 nm, resulting in a drastic increase in the surface area. It leads to quantum trapping, energy quantization, molecular motion, and electromagnetic fields becoming very active at the nanoscale. On the other hand, incorporating of the nanofiller particle has a significant impact on surface interactions with the matrix, which further works on adhesion, antimicrobial, anti-antioxidant, and other final product properties [16].

Nanofillers are classified into organic and inorganic materials, synthesized using various chemical and biological techniques. Among them, metal oxides, carbon-based zeolites, nanocellulose, metal nanoparticles, and their metallic alloys are the few examples widely used to fulfill modern requirements such as high mechanical strength, adhesion, antimicrobial, anti-antioxidant, and other properties [17].

9.3.3 Bio-Nanocomposites

Composites are made up of two materials: a matrix and a reinforcing agent, each having their own set of physical, biological, and chemical characteristics. Nanocomposites have superior mechanical, optical, antibacterial, and barrier characteristics than traditional composites, making them ideal for food packaging. Nanocomposites have been used in environmental applications because of their superb structural strength. Mainly, composites are classified into three types. Among various kinds of bio-nano composites (laminated, fibrous, and particulate composites), the particulate-based composites crystalline and amorphous nanostructures or beeds are used. Again, there are three major types of particulate nanocomposite: polymer, metal, and ceramic, which are used depending on matrix materials [18]. However, one prominent issue still arises while attempting food packaging is food losses and safety due to microbial activity. Antimicrobial bio-nano composites are the modern generation of packaging products that helps to reduce the growth of contaminant microorganisms, prolong the shelf life of food, keep product consistency, and protection during storage. To design and incorporate an antimicrobial agent in packaging material, selecting an adequate antimicrobial agent for the proper packaging is necessary. Further, the agent should maintain a proper balance between the release of the antimicrobial and microbial growth.

Antimicrobial characteristics of nanocomposite materials have long been known and employed in a number of industries, including packaging, as growth inhibitors, antimicrobial agents, and carriers in packaging films. Since bio-nano composite antimicrobial agents and systems have an apparent large surface area, its improved surface reactivity makes it more effective against microbes than any classical composite or material. Also, it is necessary to study microbial interactions with the packaging biopolymer matrix through hydrogen bonds, hydrophobic, or electrostatic interactions, with or without antimicrobial agents. As discussed above, bio-nano composites' antibacterial nanofillers or nanostructures are classified as natural antimicrobial agents, metal oxides, carbon-based, zeolites, enzymes, cellulose, metal nanoparticles, and their metallic alloys [18].

9.3.3.1 Metal and Metal Oxide Bio-Nanocomposites

Metal and metal oxide nanoparticles, as well as other materials, are commonly used in packaging. Nanoparticles in food packaging materials have well-recognized benefits due to their longer product life, including improved barrier, collecting dangerous chemicals, UV irradiation, flexibility, and functionality. Most of these products also need to hold up under normal use, which includes being inert, relatively affordable to manufacture, light, and reusable. They must also withstand the conditions of producing food and filling, while also being sluggish during storage and transportation with a resistance to physical and mechanical powers. Metal nanoparticle-based nanomaterials, such as silver, platinum, gold, stainless steel, and copper, have been widely utilized as composite mixes in a diversity of biopolymers. They have biocidal effects against gram-positive and gram-negative bacteria [19].

In recent years, silver nanoparticles have mostly been thoroughly investigated as antibacterial agents. Silver nanoparticles are antimicrobial surface coatings because of their long-term bactericidal activities and minimal toxicity to eukaryotic organisms. Scientists utilized polyethyleneimines to cultivate *E. coli* cells, agar plates to foster the growth, and silver zeolites to kill the bacteria. It has been applied as an antibacterial agent against *Escherichia coli* and *Staphylococcus aureus* to decelerate polymer dissolution rates [20]. Copper, gold, and other metals with biopolymers against *Escherichia coli, Salmonella, Cronobacter*, and more have also been extensively researched for food packaging. In other words, antimicrobial activity may be highly associated with metal or metal oxides directly interacting with microbial cells, such as interrupting transmembrane electron transport or disrupting/penetrating the cell envelope, or burning cell components. Reactive oxygen species (ROS) and heavy metal ions in solution are examples of indirect products which might be dangerous.

9.3.3.2 Bio-Nanocomposites Made with Polysaccharides

Non-toxic polysaccharides are prevalent macromolecules created by nature. Their activity categorises them as storage, structural, or gel-forming polysaccharides. They are identified as unique materials for developing antimicrobial bio-based packaging materials due to their abundance and good biodegradability. On the other hand, they are selectively permeable to CO_2 and O_2, restricting the availability of ripening fruits and vegetables. Secondly, they have the potential to transmit active components, which is a favorable condition [21].

9.3.3.2.1 Starch

Starch is made up of linear amylose and highly branches amylopectin and is a natural, hydrophilic, and biodegradable polymer. It contains potatoes, wheat, maize (corn), rice, cassava, and other foods. It is also used in a variety of food applications since it is edible, cheep, plentiful, and biodegradable. The microorganism count of coated strawberries was diminished, and the longevity was extended from 14 to 28 days using starch-potassium sorbate formulations. The potassium sorbate starch-based films resist *S. aureus, Penicillium* spp., *Candida* spp., and *Salmonella* spp. at various pH levels and can extend chilled cheese's shelf life [22]. After mixing starch and sodium alginate, antimicrobial polyelectrolyte films were created, with inhibitory effects on *E. coli* and *S. aureus. S. aureus* and *E. coli* are shown to have restricted by modified starch-based materials such as 6-bromomethyltriazole-6-deoxy starch (BMTS), 6-hydroxymethyltriazole-6-deoxy starch (HMTS), 6-carboxyltriazole-6-deoxy starch (CBTS), and 6-chloromethyltriazole-6-deoxy starch (CMTS) [23]. Combining various quantities of clove and cinnamon essential oils with starch films produced in an antibacterial agent in food products that was effective against Penicillium commune, fungi, and Eurotium amstelodami.

9.3.3.2.2 Chitosan

Chitosan, a cationic polysaccharide obtained from different fungi, would be another impressive polysaccharide. It consists of linear 2-amino-deoxy-d-glucose linked by 1,4 glycosidic bonds obtained from chitin. Because of its antibacterial activity, biodegradability, biocompatibility, and nontoxicity, as well as the fact that it can be obtained from renewable resources, chitosan has a wide range of applications in food packaging. B. subtitles were efficiently suppressed by wheat starch-based antimicrobial films combined with chitosan produced for antimicrobial packaging material [24]. Because of its antimicrobial properties, chitosan film infused with rosemary essential oil has been reported for food packaging. Surface spoilage bacteria in processed beef were restricted by antimicrobial films made by combining acetic and propionic acids into a chitosan matrix. The antibacterial activity of the chitosan-based organoclay sheet against food poisoning bacteria has been reported (gram-positive bacteria) [25].

9.3.3.2.3 Cellulose and its Derivate

Cellulose is the abundantly available, organic, and biodegradable polymers found in nature. It is a linear homopolysaccharide of anhydroglucose with highly crystalline and fibrous. Their composite films are gaining much attention due to their recyclability and degradability. Various concentrations of antibacterial PLA film were used to create it. Mentha piperita EO, Bunium percicum EO, and nanocellulose had the best antibacterial efficacy against Staphylococcus aureus, Enterobacteriaceae, and Pseudomonas aeruginosa [26]. The cellulose matrix film based on pediocin developed with high antimicrobial efficiency can prevent the pathogenic bacteria's development during ham slices' storage for up to 15 days. The cellulose-based nisin displays the decrease in growth of *L. monocytogenes* after 14 days, enhancing antibacterial property meat packaging applications [27]. The convenient nisin-grafted carboxylated cellulose nanofibers developed for long-term antibacterial active food packages against various gram-positive bacteria like *B. subtilis*. The immobilized nisin reduced the levels of Listeria innocua and *Staphylococcus aureus* in both cheese and ham and was reported as bioactive food packaging material [28].

9.3.3.3 *Protein-Based Bio-Nanocomposites*

Proteins are macromolecules made up of amino acids joined together by peptide bonds. Plant proteins like soy protein, pea protein, and gluten are classed as plant proteins, whereas animal proteins like collagen, casein, keratin, and whey are classified as animal proteins. Because of various groups attached to the carbon atom, amino acids produce a wide range of proteins with varying functions, sizes, and shapes. Bio-nano composite films were created from fish gelatine, alginate, gluten, and whey protein for food packaging due to their low cost, non-toxic nature, biodegradability, bioactivity, stabilizing, gel forming, and thickening capabilities [29].

9.3.3.3.1 Alginate

It is a linear copolymer consisting of D-mannuronic and L-guluronic acid monomers isolated from brown seaweeds of the Phaeophyceae class (alginic acid salts). The alginate can alter with divalent and trivalent cations for alginate film. Alginate is specifically used for biofilm formation due to its biocompatibility, biodegradability, and nontoxicity. The Citrus paradisi seed extract was used as an antibacterial agent in the alginate matrix for the packaging material of potatoes, which inhibits the *Listeria monocytogenes* and *Escherichia coli* as targeted bacteria [30]. Essential oils comprising oregano, pimento berry, and lemongrass combined with nutmeg, lemongrass, citral, reinforced methylcellulose, and an alginate blend are utilized in the packaging of fresh broccoli.

In contrast, the film acts antibacterial agent versus *L. monocytogenes, Escherichia coli,* and *Salmonella typhimurium.* The alginate-based films containing cinnamon, winter savory, and oregano essential oils showed antibacterial effects *Listeria monocytogenes* and *Salmonella typhimurium* appear in packaging of bologna and ham. In the packaging of smoked Salmon, lysozyme-based alginate films were found effective against *L. monocytogenes* and *Salmonella anatum.* Essential of lemongrass and alginate is proven to be antibacterial composite, used for packaging fruit like pineapple. It is found to lowers the total count, firmness, and sensory scores of yeast and molds [31].

9.3.3.3.2 Collagen

It is an elastic polymer present in the cell membrane of many connective animal tissues, and it is the most used antibacterial major ingredient in bio-based encapsulation. In addition, multiple bacteria discovered that collagen-based nanocomposite coatings provided a good oxygen barrier at 0% humidity levels. Recently, grapefruit seed extract added collagen hydrogel films were prepared and applied for packaging material against *Listeria monocytogenes* and *Escherichia coli* in potato packaging [30]. However, collagen-based films were rarely used due to their average performance toward packaging with antimicrobial properties.

9.3.3.3.3 Gluten

Gluten is formed primarily by two sets of proteins called gliadin and glutenin. The gluten-based packaging materials have a lustrous surface, similar to propylene, and are water resistant under certain conditions. When thyme essential oil was mixed with biodegradable and edible gluten-based films, the *in vitro* antibacterial activities of the most concentrated samples were observed to be enhanced. Gluten inhibits the growth of *L. plantarum* in packing materials [32]. The natamycin-based wheat gluten-based methylcellulose films used to package fresh Kashar cheese had an effect on *A. niger* and *P. roquefortii* [33].

9.3.3.3.4 Gelatin

Biodegradability is a critical factor of gelatin protein, produced through the partial breakdown of collagen and commonly utilized to make biodegradable films lead to the promising material. Gram-negative bacteria are inhibited by the gelatin used in the packing of cod patties. In comparison, enzymatically synthesized antibacterial chitosan-gelatin polymer contributed to prolonging the shelf life of meat products. Citric acid and chitosan are incorporated as antibacterial agents in gelatin/chitosan composite films to inhibit *E. coli* [34].

9.3.3.3.5 Whey Protein

Protein molecules are soluble and remain to be soluble even after caseinate crystallises. The prepared cheese uses the bovine serum albumin, immunoglobulins, lactoglobulin, and lactalbumin to represent a mixture of various proteins. Whey protein isolates (WPI) have a large concentration of hydrophilic amino acids and create a thin, transparent film with low mechanical intensity, and enhanced water vapor permeability, making them ideal for packing. In the WPI matrix, the cellulose nanofibers with WPI found maximal mechanical stability, shorter elongation, and increased elasticity [35]. The combination of oregano essential oils to whey protein–based edible films used as an application in food packing. The whey protein films incorporated with lactoperoxidase extend the shelf life of smoked salmon by effectively reducing the growth of *Listeria monocytogene*. Antimicrobial whey protein–based edible films containing p-aminobenzoic used against *L. monocytogenes* and *E. coli* O157:H7 for packaging material of Bologna slices. The lysozyme, nisin-based whey protein films act as an antimicrobial against *L. monocytogenes* and *Salmonella Typhimurium*. Montmorillonite (MMT) and microbial transglutaminase (1%–3% wt.%) are added to fish myofibrillar protein (FMP) films to enhance the antimicrobial activity in food packaging. Also, MMT with a natural polymer matrix has been widely introduced for the antimicrobial and mechanical strength of biopolymer-based packets [36].

9.4 Lipid-Based Nanocarriers

Nanoencapsulation system is based on lipids as it is a suitable carrier. Additional nanocarriers consist of construction blocks of carbohydrates or proteins that are appropriate candidates for the complete process due to complex chemical and thermal processes. However, lipid-based nanocarriers are in the nutrition trade with higher capture efficiency and lower tolerance. Lipid-based nanoencapsulation structures are so well carriers that they are helpful for various applications. Various lipid-based nanocarriers for food therapeutic agents have been identified and tested in the last decade. Bioactive components include polyphenols, fatty acids, lipophilic vitamins, nutraceuticals, flavors, and preservatives. In this context, lipophilic is prepared oil in water. Nanoliposomes, nanoemulsions, solid lipid nanoparticles (SLNs), and nanostructured lipid carriers (NLCs) are emerging in the food avenue [37].

A liposome is derived from two Greek words: Lipos (fat) and Soma (body or structure), and it refers to a fat envelope containing the aqueous load. Liposomes are lipid vesicles with colloidal interior chambers that are tiny and spherical. Liposomes have one or more concentric bilayers due to their operating conditions and chemical properties. The total number of lamellae diameters and dimensions distinguishes liposomes. The liposome formulation model comprises interactions among the aqueous movement and amphiphilic lipids. To position the acoustic phase of the inner and exterior media, the polar head sequence of phospholipids is oriented, and the bilayers are associated with hydrophobic tails. The existence of lipid and aqueous functional groups allows water-soluble, lipid-solutions, and amphiphile compounds to be trapped, supplied, and released. It has some key features as controlled supply, non-toxic characteristics, developed fat-based amphiphilic bioactive solubility rates, and improved nutrient bioavailability [37]. Nanoliposomes are also known as nanoscale bilayer liposomes; meanwhile, the phrase liposome encompasses a wide range of vesicles with average sizes ranging from nanometers to micrometers. Nanoliposomes are a larger surface area, appropriate stability, and nano-sized scales as 100 nm. The main components of these carriers are lipids and phospholipids. Because of their amphiphilic character, they may entangle and release a wide spectrum of hydrophilic and hydrophobic molecules, providing a consolidated benefit. Furthermore, they are extremely compatible with the skin's surface and can function as active material penetration agents into specific areas [38]. To consolidate foods and produce nutritional supplements, nanoliposomes are incorporated into food products. The phospholipid vesicular in nanocapsules can be single or multi-layered. Hydrophilic nanocarrier components are contained inside the core polar cavity, whereas hydrophobic components in the non-polar layer can be encased at the outside border. Chitosan on the outer surfaces of liposomes was used to investigate the stability of nanoliposomes, resulting in changes in the surface charges of the nanocarrier [39]. Most enzymes encapsulated in liposome/nanoliposome systems were used in the dairy sector. Therefore, future research studies could focus on other areas of food such as beverages and baking products on liposomal or nanoliposomal enzymes. Most nanofibers are for active packaging, but several instances were reported on antimicrobial electrospun fibers. The various physicochemical properties such as good hydrophilicity, biodegradability, and antibacterial properties in cationic cellulose derivatives are useful in the pharmaceutical, food, and textiles industries. Electrofibers obtained the polymer blend of Polyquaternium-4 Cellulose (PQ-4), Dimethyl Ammonium Chloride Copolymer, and PVA. A polymeric cellulose mixture produced electrofiber. These nanofibers demonstrated effective antibacterial activity against *E. coli* and *S. aureus* [40].

9.4.1 Nanoemulsions

Nanoemulsions are sub-micron emulsions as their average radii are minimal, about 50–500 nm. They are described as two immiscible phase combinations as water-in-oil (W/O) or oil-in-water (O/W) with a suitable emulsifier. These nanocarriers have embodied active molecules to improve their solubility and bioavailability in various food stuffs. Nanocarriers are nanostructures in simple and multiple nanoemulsions, Pickering nano-emulsions, nanoemulsion gels, etc. The bioactive component is stuck in simple nanoemulsions and stabilized by surfactants/

biopolymers. The nanoemulsion with the base material was loaded in the interior phase for multiple nanoemulsions. The next step will be distributed during the second stage as an internal, medium, and external phase [41]. Different techniques like microflow, ultrasonics, and homogenization methodologies integrate immiscible liquids fluids into one phase. There are also two groups of nanoemulsions: one or two emulsions (O/W or W/O) and double (W/O/W or O/W/O) emulsions [42]. We may either use high intensity for double emulsions or methods for low energy. Ultrasound and homogenization superior methods that are consistent and smaller in size of droplets. High-energy processes create extreme destructive forces that reduce particle size and shape emulsions, while minimal energy techniques encourage natural emulsification by combining all emulsion materials. The particles are vital to the stability of the emulsion; the tiny particles are somewhat more stable for the emulsion. In addition, nanoencapsulation with nanoemulsions can be observed by both hydrophobic and hydrophilic bioactive components. In general, the performance, physiochemical properties, encapsulation, and release aspects of various emulsions are generated using multiple emulsification processes. O/W nanoemulsions can encompass and transmit badly soluble diet disinfectants, enhancing the active compound's physicochemical characteristics and enhancing its effective transmission in food stuffs. The nanometric droplet dimension has the potential of improving functional compound interactions with microorganism's cell membranes, impacting the lipid membrane's stability and resulting in intracellular component leakage. This generic activity mechanism slows the growth of tolerant bacteria. To maximize antimicrobial activity and utilize concentrations low enough to just slightly alter the consistency of a food product, they must ideally improve mass transfer rating in the area of action to limit bacteria growth while remaining within food standard parameters [43]. The sunflower petroleum-based nanoemulsions included carvacrol, limonene, and cinnamaldehyde, which were homogenized under high pressure and stabilized using specific emulsifiers. After 2 hours of carvacrol emulsion, *E. coli* and *L. delbrueckii* were entirely inactive when 2logs were used to minimize the inhabitants of *S. cerevisiae*. Following 24 hours, *S. cerevisiae* was fully inactivated. Emulsions from limonene and cinnamaldehyde demonstrated less antibacterial activities than carvacrol and showed full inactivation of *E. coli*, *L. delbrueckii*, and *S. cerevisiae* after 24 hours of concentrated oil nanoemulsion. A dramatically improved bactericidal effect over shorter scales was observed due to increased product availability compared to nonencapsulated essential oil. The bacteriostatic action of emulsifiers, which in liquid media, slightly dissolves the essential oil (lecithin and pea proteins) [44]. Lemongrass oil (LO) was therefore embedded by high-pressure homogenization in carnauba-shellac wax (CSW) and ultrasonic nanoemulsions and microfluidization alginate nanoemulsion. The overall residents of *E. coli* O157:H7 and *L. monocytes* after 2 hours of lemongrass oil-loaded CSW-based nanoemulsion decreased by 8.18 L CFU/g. After 5 months, as food stuff, CSW nanoemulsions were unloaded and loaded from storage. A decrease of 0.8 and 1.4 logs CFU was permitted for apples of, respectively, aerobic bacteria. The coatings hampered yeast and mold development. In addition, the layer *Salmonella typhimurium* and *E. coli* inhibited their growth O157:H7, respectively, on apples and prunes. Furthermore, different physicochemical qualities have been preserved in nanoemulsions [45].

9.4.2 Solid Lipid Nanoparticles and Nanostructured Lipid Carriers

SLNs consist of lipid bonds that maintain a crystal structure in the surrounding environments or at room temperature as they collide with delivery cargoes. The spherical carriers with a grain volume distribution of 50 to 1,000 nm protect the active ingredients of hydrophilia/lipophilia. There are a few benefits of using SLN compared to vesicular lipid systems:

- Modulate bioactive release by providing extended dynamic release because of its slow dissolution and solid structure;
- No biological toxicity because they come from food inspired by the natural ingredients;
- Increased lipophilic component entanglement performance;
- Protecting bioactive against extraordinary safety;
- Harmful products such as humidity, deterioration, or oxidation of the chemical/light.

SLNs are cost-efficient, scalable, and reproducible, with no need for organic solvents, sonication, hot or cold heat, pulp homogenization, solvent dissolution, micro-emulsion, or double-emulsion well as high-speed stirring. The high-pressure scalable homogenization was argued to become a more acceptable, reliable, and efficient technique for generating SLNs with superior nanoparticles than high-shear and ultrasonic methods. Apart from the manufacturing process, these structures are type of lipid, trap, and particle size. Three key factors govern the effectiveness of bioactive release. For example, the higher the hot homogenization, the bigger the size of the SLNs and the lipid melting point.

Moreover, a suitable surfactant type/concentration selection has been another significant factor that significantly affects SLN physical and chemical attributes [46]. In contrast to the above therapeutic benefits, the unwanted transition of fat crystalline phase throughout storing is the chief limiting factor of fruitful distribution from SLNs: the particulate assembly/extension, a low incorporation capability to bioactive load distribution caused by the rigid structure of fats, likely gelation-phenomenon, and bioactive expulsion. Nanofabrication techniques and inputs of NLCs are comparable to the SLNs; the NLC formation is the addition to the primary solid-lipid grid of at least one liquid-lipid movement. That leads to crystallinity imperfection and makes superior capacity for bioactive loading [47]. As verbal movers for carotenoids, various lipid formulations have been introduced (e.g., nana-liposomal vectors, nanoparticles, nanosomes, NLCs, niosomes, and cubosomes). The impact of lipid release methods on the bioavailability of bioactive has been examined in various studies, and their negative impacts have been confirmed for the conventional approaches and their consumer satisfaction [48]. Carotenoids can also be produced by adjusting basic substance, design components, method variable quantity, and various lipid nano-vectors for the binding even process engineering. Those carriers require a thorough encapsulation theoretical understanding, the trained delivery mechanism of the carotenoids encapsulated. Physicochemical characteristics of nano feeding systems prepared, new technological, and each method's shortcomings [49].

9.5 Nature-Inspired Nanocarriers

The emerging trend in nanotechnology is to ensure the design, functionalization, and use of nanometric level natural biopolymers for food encapsulation approaches. The exploitation of biopolymer nanocarriers in a variety of food delivery systems maintains better biocompatibility and biodegradability than those used as synthetic materials. The naturally inspired nanocarriers such as casein, cyclodextrin, and amylose are employed to encapsulate some food bioactive ingredients and nutraceuticals. Such nanostructures would be recognized as safe, non-toxic, biocompatible, and biodegradable, with suitable physicochemical properties.

9.5.1 Casein

Casein is found in milk and other milk goods. It is a significant protein found naturally in cow's milk, about 80% of its complete protein, in the form of spherical micelles with a nanosized diameter of about 50–200 nm [50]. The most common caseins are αS1, αS2, β, and k related to phosphoproteins and can bind calcium and calcium phosphate. They have hydrophobic interactions, which has led to caseinates and caseins being utilized to encapsulate various hydrophobic bioactive compounds such as hydrophobic vitamins, fatty acids, and Curcumin. Caseins have been used as nanocarriers in pasta and fried foods to increase their nutritional value due to high lysine-containing casein supplements having low lysine-containing amino acids and high sulfur-containing amino acids of wheat and vegetable proteins present in that product. The nanocarriers of casein have physicochemical properties like copolymers with more stable hydrophobic and hydrophilic parts. These natural nanocarriers offer high thermal stability, a massive potential for bioactive food ingredients. These protein complexes transfer all the critical nutrients from the mother to the kid. Cutting-edge the casein micelle nanocluster model, the colloidal calcium phosphates are strewed as tiny "cherrystones" at an average size of 18.6 nm [51]. Moreover, sodium caseinate has been used as a food additive for stabilizing processed foods. However, the use of calcium caseinate increases calcium content and decreases sodium levels in their products. Casein has been proved to be one of the most dependable proteins, economical, nutritional, and safe nanocarriers [11, 52, 53].

9.5.2 Cyclodextrins

Nature-inspired nanocarriers have received remarkable notice due to their safety and economical delivery of nutrients. Cyclodextrins are cyclic oligomers usually utilized as food flavorings in the food industries to improve shelf-life, sensory abilities, and sequestration of constituents. The cyclodextrin can be divided into α, β, and γ containing 6, 7, and 8 glucopyranose units. Generally, as in more centers, the van der Waals and hydrophobic forces hold guests and hosts together with a partial or complete cavity adjustment [50]. However, cyclodextrin establishes an inclusion complex that can alter guest reactivity and improve bioavailability. In addition, cyclodextrins are hollow molecular nanocarriers of a specific size to incorporate many different components into their cavity with the help of a proper "guest" structure. The inclusion complex is designed so that the CD molecule and the guest molecule, its stoichiometry ratio will be differ compared to 1:1. The natural nanoparticles of cyclodextrins are used in the nano-encapsulation of bioactive food components and nutraceuticals. Various food bioactivities are packed into nature-induced nanocarriers such as quercetin, salmon oil, vitamin D3, α-bisobolol, lycopene, and menthol [11, 54, 55].

9.5.3 Amylose

Amylose is a straightforward starch created of α-D-glucose parts, bonded through with α (1→4) glycosidic bonds. Different bioactive substances as ligands can be incorporated within amylose nanostructures as amylose nanocarriers. When ligands are present, the amylose undergoes a conformational shift, transforming the double helix amylose into a characteristic V-shaped amylose nanocarrier. The hydrophobic core chamber of typical V-shaped amylose nanostructures is capable of chemical bonding with lipophilic binding molecules like fatty acids. Inclusion complexes between amylose content maize and fats in chia seed oil are formed using the alkaline technique. Figure 9.2 reveals the formation of inclusion complex using the polymeric guest and amylose.

It has been demonstrated that these complexes retain a high level of linolenic and linoleic acids. The complexation process enhanced the oxidative stability of important fatty acids in chia oil. Such complexes might be a potential delivery method for chia seed oil-derived omega-3 and omega-6 fatty acids in thermally processed foods [56]. The natural nanocarriers of

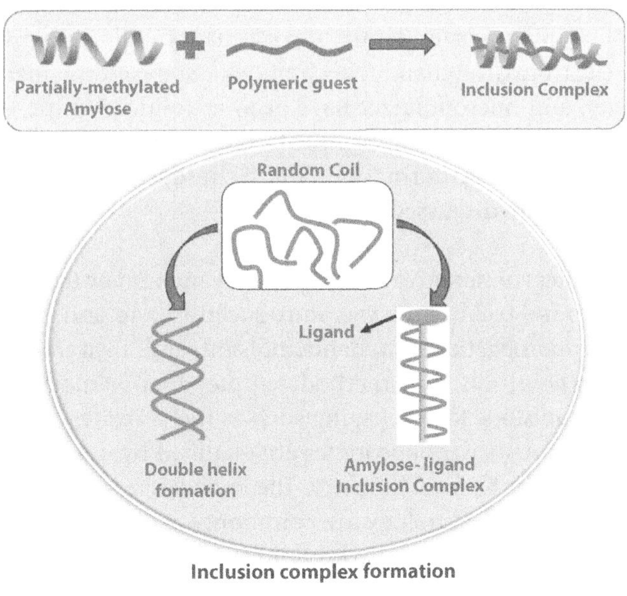

Figure 9.2 The inclusion complex formation of amylose.

amylose are utilized in the nano-encapsulation of bioactive food components. There are some examples of different food bioactive loaded into nature-induced nanocarriers such as genistin, linoleic acid, vitamin D3, linalool, β-Carotene, and menthol [11].

9.6 Equipment-Based Synthesis of Nanocarriers

Equipment-based synthesis is another strategy for the expansion of nanocarriers. A large variety of films is developed using the following polymers such as polyvinylpyrrolidone, polyvinyl alcohol, polylactic acid, and polycaprolactone. However, the use of antimicrobial nanocarriers in various polymers has introduced a new way that assures the safety and lifetime of packaged food. In recent years, the incorporation or coatings or film formation of antimicrobial nanocarriers with polymers have been carried out using electrospinning, dipping, spray dryer, solvent casting, fluidized-bed, and electrospraying methods due to its production scale and ease of process [11, 57]. These methods are classified into two approaches such as top-down and bottom-up. As the name suggests, top-down methods bifurcate the large particles to smaller ones until the nanoscale is achieved. On the other hand, bottom-up grow nanocrystals or nanostructures using chemical reactions for the desired properties.

Furthermore, new practical techniques including membrane technology, ball milling, vortex fluidic device, nanofluidic systems, solution blowing, electrohydrodynamic techniques, sonication, high-pressure homogenizer, and microfluidizer have been introduced in the food packaging. Each of these techniques has its mechanism and preferences of nanofiller/nanoemulsion formation. The techniques and their nano synthesis/formation are discussed below.

1. Nanofillers or nanocarriers prepared by membrane technology
   Membrane technology was introduced to add and mix preferred nano/microparticles in nanoemulsion and then reduced particles in emulsion. This method can prepare organic and inorganic nanoemulsions for packaging such as metal oxides, polymers, and proteins. Also, nanoliposomes are qualified by this method, which is helpful in the food industry. The nanocarriers or nanofillers prepared by this technology are commonly applied in the dairy trade due to nanoemulsion's hydrophilic and hydrophobic properties in packaging [58]. The hydrophobic packaging of polymers and metal oxides-based nanocarriers facilitates safety, antimicrobial property,

and antitoxin property and multiplies the lifetime. In addition, emulsification of oil in water or vice versa is carried out using this hydrophilic and phobic membrane.

2. Nanofillers or nanocarriers prepared by ball milling
   Ball miller is a well-known technique of nano synthesis commonly used for few decades. This technique provides synthesis and mixture of nanocarriers with polymers matrix. The method is one pot, solvent-free, surfactant-free with large-scale production of packaging material. The antimicrobial materials like metal nanoparticles (Ag), metal oxides nanoparticles (ZnO, TiO_2), and organic polymers were predominantly synthesized for packaging layers or films. Also, biopolymers like starch have been widely utilized to get the antimicrobial properties, mixture material for matrix, and the stability and transparency of packaging material [59].
3. Nanofillers or nanocarriers prepared by vortex fluidic device
   The vortex fluidic device technique is a simple method to form a thick and thin coating or film using the dynamic Stewartson/Ekman layer method. This method is helpful to coat or formation of films of emulsions and oil-based matrices with nanocarriers. The film's thickness or coating can be controlled by the number of rotations and angle of deviations of the tube. The advantage of this method is the one-pot surfactant-free simple method. The matrix and nanocarriers such as fish oil, enzymes, proteins, phospholipids, and fullerene are widely used for the coating [60].
4. Nanofillers or nanocarriers prepared by micro/nanofluidic systems
   This method is a new age bottom-up method of nanomaterial synthesis for packaging. This multi capillary system is helpful to form various nanomaterials such as metal, metal oxides, and biopolymers. Especially, biopolymers such as chitosan, polylactic acid, and zein are fabricated using a nanoprecipitation consequence and antisolvent process [60]. The emulsion-based packaging of the matrix with antimicrobial nanocarriers is heavily developed by this method.
5. Nanofillers or nanocarriers prepared by nanospray dryer
   Spray drying is a method for producing both continuous and intense nanomaterials. The feed solution is pumped into the drying chamber and atomized using a nozzle, where the atomized droplets are exposed to inert gas. Material characteristic feed solution parameters including feed concentration, feed solution kinetics, fluidity, active element ratio, carrier agent type, and concentration

are critical elements of this technique. Spray drying has been utilized to nano summarize several bioactive dietary components such as probiotics, essential fatty acids, enzymes, and peptides with a robust antimicrobial agent [60].

6. Nanofillers or nanocarriers prepared by electrospinning technique
   Electrospinning is a new and widely used top-down approach for one dimension nanomaterials in nanofibers with high-power voltage supply, needle, and collector cylinder, as shown in Figure 9.3. This method can synthesize any polymeric mixture for the packaging matrix. However, the synthesis and use of nanomaterials with the polymeric mixture was found useful for uplifting the properties like antimicrobial, antitoxin, and hydrophobic nature of the material. Electrospinning is a nonthermal synthesis process, it is more suitable for the fabrication of biological coatings and films. Due to this, electrospinning gained the advantage and was widely used to prepare proteins and polysaccharides-based biopolymers as a matrix. The modified version of electrospinning combines the spray system to form electrospraying. The use of electrospraying rather than electrospinning increased due to its high performance of polymer formation.
7. Nanofillers or nanocarriers prepared by sonication
   Sonication is one of the oldest and widely used methods of nanomaterial formation. This process relies on ultrasonic waves' effect on the source's solution or material, as shown in Figure 9.4. The phenomenon regarding sonication is called cavitation, which initiates and enhances the chemical reaction at a high vibration frequency. The physical forces are created. The biopolymers have broken when

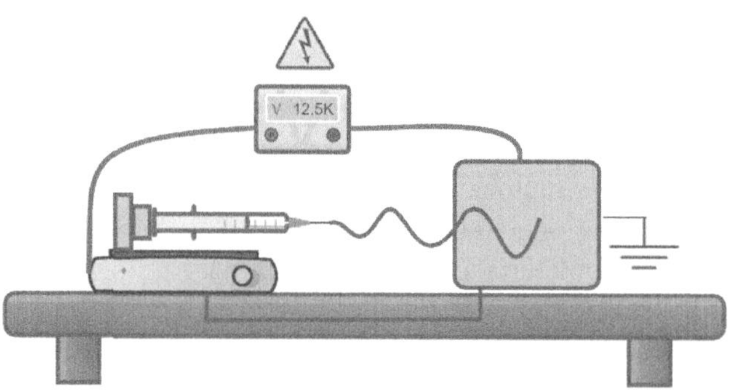

Figure 9.3 The electrospinning technique used for the nanocarriers-matrix synthesis.

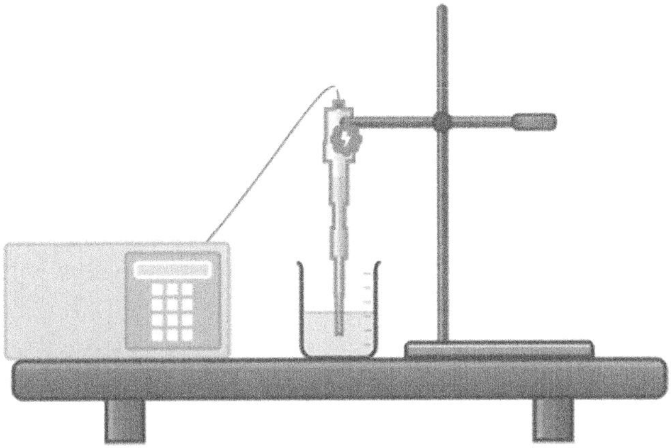

Figure 9.4 The sonication technique of nanocarriers and matrix synthesis.

water jets are used at the solid-liquid interface, and then particles or aggregates form in the nanoscale range. The materials like metal oxide, metal nanoparticles, and polymer-metal mixture are prepared by this method.

9.7 Nanostructured Materials

9.7.1 Carbon-Based Nanomaterials

A new class of carbon-containing antimicrobial agents such as colloidal or nanomaterials of graphene, graphene oxide, reduced GO, and quantum dots of graphene recently used as emerging antimicrobial graphene materials [61]. Carbon nanotubes recently modified with polymers have been developed as disinfectant and intelligent packaging devices. There are carbon nanotubes, i.e., thick nanotubes and circular nanotubes. When added to a polymer matrix, both have extremely high elastic modulus and tensile strength [62]. The CS-CAP-ZnO is a mechanically balanced nanocomposite. Its water vapor and oxygen communication rates were significantly lower than commercial polymeric films. These films are bio-degradable, and they can protect them from UV rays and have excellent antimicrobial activity. As a result, there is only a minor infestation of food-spoiling bacteria and molds on the fruit. Hence, the CS-CAP-ZnO is used for film for green root vegetable and is used mainly to wrap black grapes to protect them from UV radiations and atmospheric agents [63].

The electrocatalytic activity of mesoporous carbon materials has been enhanced by combining them with an electrode. These are utilized to determine hazardous ractopamine in a selective and sensitive electrochemical manner [64]. Prussian blue has also been added to mesoporous carbon for signal amplification. For ultra-sensitive and precise measurement of metolcarb, a 3-D molecularly imprinted electrochemical sensor was created [65]. Due to the synergistic impact of graphene nanosheets and Ag-nanoparticles, the AgNPs-polydopamine-graphene hybrid material displays strong antibacterial activity [66]. Nitrogen-doped graphene-CuNPs have been used to create a non-enzymatic glucose sensor that is selective, is repeatable, and has good recovery in complicated meals [67].

9.7.2 Inorganic Nanomaterials

Inorganic nanoparticles such as metal and semiconductor nanoparticles have fascinated a good contract of attention and are extensively studied from the point of view of scientists and researchers. The precursors used for the preparation of such nanomaterials are usually metal or metal oxides. These nanomaterials contain nanosilver, Cu or CuO nanoparticles, ZnO nanoparticles, TiO_2 nanomaterials, quantum dots, nanogold, etc. A quantum dot is a tightly populated semiconductor crystal with hundreds or thousands of atoms ranging in size from 1 to 10 nm.

9.7.2.1 Zinc Oxide-Based Nanomaterials

The unique physicochemical of ZnO nanoparticles and their various essential characteristics such as concentration, particle size, morphology, crystal structure, and surface activity are performing a substantial role in their antifungal efficacy. SPI/ZnO nanoparticles film created by manufactured ZnO nanoparticles integrated into soybean protein isolate (SPI), increasing the connections between SPI molecules by hydrogen bonding, resulting in possible application of ZnO nanoparticles in food packaging, according to Siying Tang *et al.* [68]. PVC–ZnO-nano-composite films can be used to extend the shelf life of commodities [62]. Hui *et al.* have demonstrated that cerium doped ZnO nanocrystals had strong antifungal action against pathogenic fungus compared to pure ZnO crystals. It revealed the antifungal activity dependent on crystalline morphology and structure of its nanoparticles. The enhancement in the antifungal action of ZnO nanoparticles has been achieved by optimizing parameters such as concentration, structure, and size of its crystals and even preparation technique.

Furthermore, ZnO nanoparticles have been subjected to both light and surface modification in order to increase their antifungal efficacy [69]. Recent studies demonstrated that ZnO nanomaterials functionalized with polysaccharides [70], essential oils [71], and protein-based biopolymers [71] showed significant potential antimicrobial applications against food-borne pathogens. As a result, bioactive compounds can be combined with ZnO nanoparticles to provide a synergistic antifungal effect.

9.7.2.2 *Silver-Based Nanomaterials*

Silver was applied as an disinfectant to protect food, drinks, wine, and water in ancient times. These food materials are protected using a coating of silver. However, Ag nanomaterials have been a good form of active food packaging based on antimicrobial packaging, which performs a vital part in increasing the shelf life of food stuffs and decreasing the threat of pathogens. They have the potential of antimicrobial, antifungal, antiviral, and anti-yeasts activities [72]. Tianxi Yang *et al.* recently reported that Ag nanoparticles incorporated polyethylene for food packaging applications [73]. In addition, Ag nanoparticles synthesized from the Madhuca latifolia oil cake extract showed robust antibacterial activities [74].

9.7.2.3 *Copper-Based Nanomaterials*

Copper-based nanomaterials are divided into two subcategories: CuO nanomaterials and copper-based nanomaterials. CuO is the biggest subsection of copper-based materials for culinary applications among them. Because CuO nanomaterial has antibacterial properties, it is widely used in the industrial food packaging industry. The antibacterial activity of Cu(II) integrated poly(3-hydroxybutyrate-co-3-hydroxy valerate), chitosan, cellulose, and agar against gram-positive and gram-negative bacteria has been investigated and published [75]. CuONPs/AgNPs hybrid nanomaterials enhanced antibacterial and oxygen-water barrier characteristics, as well as biocidal action [77].

9.7.2.4 *Titanium-Based Nanomaterials*

Titanium dioxide is the most common subclass of titanium-based materials in the food packaging industry. Its photocatalytic action boosts the generation of ROS. Food-grade TiO_2 is relatively inexpensive, safe, non-toxic, and photostable, as well as having antibacterial action [78]. Rutile, brookite, and anatase are the three major polymorphs of TiO_2. TiO_2 nanoparticles combined with polymer matrices were used to create nanocomposites that showed mechanical, thermal, and significant bactericidal and photocatalytic activity. Food packaging films are more compatible with TiO_2 nanocomposites. UV-A light irradiation at intensity of 1.3060.15 mW/cm2 for 2 hours resulted in TiO_2 embedded biodegradable polymer films activated by UV-A rays. As a result, bactericidal activity decreased by 1.69 log CFU/ml [79]. Due to the high aspect ratio of nanotubes and the needle effect, which perforates microbial cells, TiO_2 nanotubes are significantly more efficient against gram-positive and gram-negative bacteria than NPs.

9.8 Conclusions

The practical implementation of nanostructured targeted delivery has thoroughly changed the food packaging industry, in which the higher surface area per unit compound mass enhances its excellent nanoscale activity. The nanoencapsulation bodies like nano-emulsion, nanoliposomes, and SLNs are particularly consistent food methods. It has more food stability in the matrix and digestive system. Nano formulations consist of lipid-based bioactive foods that are more soluble and more bio-accessible through studies. However, both food stuffs and the body have yet to investigate how lipids-based nanoencapsulation systems function. Metal and non-metal–based nanomaterials also play a crucial role in food packaging. At present, the likely harmful health impact of nanomaterials remains a mystery and must be investigated beyond. In addition, nano foods are still at the stage of development of their regulatory framework. Furthermore, the interactions of food components with incorporated nanocarriers; and their physicochemical characteristics are the future scopes in food packaging.

References

1. Hoelzer, K., Switt, A.I.M., Wiedmann, M., Boor, K.J., Emerging needs and opportunities in foodborne disease detection and prevention: From tools to people. *Food Microbiol.*, *75*, 65–71, 2018.
2. Pilevar, Z., Bahrami, A., Beikzadeh, S., Hosseini, H., Jafari, S.M., Migration of styrene monomer from polystyrene packaging materials into foods: Characterization and safety evaluation. *Trends Food Sci. Technol.*, *91*, 248–261, 2019.
3. Ayala-Zavala, J., Del-Toro-Sánchez, L., Alvarez-Parrilla, E., González-Aguilar, G., High relative humidity in-package of fresh-cut fruits and

vegetables: Advantage or disadvantage considering microbiological problems and antimicrobial delivering systems? *J. Food Sci.*, *73*, R41–R47, 2008.
4. Chawla, R., Sivakumar, S., Kaur, H., Antimicrobial edible films in food packaging: Current scenario and recent nanotechnological advancements-a review. *Carbohydr. Polym. Technol. Appl.*, *2*, 100024, 2021.
5. Eleftheriadou, M., Pyrgiotakis, G., Demokritou, P., Nanotechnology to the rescue: Using nano-enabled approaches in microbiological food safety and quality. *Curr. Opin. Biotechnol.*, *44*, 87–93, 2017.
6. Jafari, S.M. and McClements, D.J., Nanotechnology approaches for increasing nutrient bioavailability. *Adv. Food Nutr. Res.*, *81*, 1–30, 2017.
7. Assadpour, E. and Mahdi Jafari, S., A systematic review on nanoencapsulation of food bioactive ingredients and nutraceuticals by various nanocarriers. *Crit. Rev. Food Sci. Nutr.*, *59*, 3129–3151, 2019.
8. Reis, C.P., Neufeld, R.J., Ribeiro, A.J., Veiga, F., Nanoencapsulation, I., Methods for preparation of drug-loaded polymeric nanoparticles. *Nanomedicine*, *2*, 8–21, 2006.
9. Dhifi, W., Bellili, S., Jazi, S., Bahloul, N., Mnif, W., Essential oils' chemical characterization and investigation of some biological activities: A critical review. *Medicines*, *3*, 25, 2016.
10. Bahrami, A., Delshadi, R., Assadpour, E., Jafari, S.M., Williams, L., Antimicrobial-loaded nanocarriers for food packaging applications. *Adv. Colloid Interface Sci.*, *278*, 102140, 2020.
11. Assadpour, E. and Jafari, S.M., Nanoencapsulation: Techniques and developments for food applications, in: *Nanomaterials for food applications*, pp. 35–61, Elsevier, London, UK, 2019.
12. Appendini, P. and Hotchkiss, J.H., Review of antimicrobial food packaging. *Innov. Food Sci. Emerg. Technol.*, *3*, 113–126, 2002.
13. Cha, D.S. and Chinnan, M.S., Biopolymer-based antimicrobial packaging: A review. *Crit. Rev. Food Sci. Nutr.*, *44*, 223–237, 2004.
14. Muñoz-Bonilla, A., Echeverria, C., Sonseca, Á., Arrieta, M.P., Fernández-García, M., Bio-based polymers with antimicrobial properties towards sustainable development. *Materials*, *12*, 641, 2019.
15. Othman, S.H., Bio-nanocomposite materials for food packaging applications: types of biopolymer and nanosized filler. *Agric. Agric. Sci. Proc.*, *2*, 296–303, 2014.
16. Akpan, E., Shen, X., Wetzel, B., Friedrich, K., Design and synthesis of polymer nanocomposites, in: *Polymer Composites with Functionalized Nanoparticles*, pp. 47–83, Elsevier, Amsterdam, The Netherlands, 2019.
17. de Oliveira, A.D. and Beatrice, C.A.G., Polymer nanocomposites with different types of nanofiller, in: *Nanocomposites-Recent Evolutions*, pp. 103–104, 2018.
18. Camargo, P.H.C., Satyanarayana, K.G., Wypych, F., Nanocomposites: Synthesis, structure, properties and new application opportunities. *Mater. Res.*, *12*, 1–39, 2009.

19. Klabunde, K.J., Stark, J., Koper, O., Mohs, C., Park, D.G., Decker, S., Jiang, Y., Lagadic, I., Zhang, D., Nanocrystals as stoichiometric reagents with unique surface chemistry. *J. Phys. Chem. A*, *100*, 12142–12153, 1996.
20. Fernandez, A., Picouet, P., Lloret, E., Reduction of the spoilage-related microflora in absorbent pads by silver nanotechnology during modified atmosphere packaging of beef meat. *J. Food Prot.*, *73*, 2263–2269, 2010.
21. Mikkonen, K.S., Rita, H., Helén, H., Talja, R.A., Hyvönen, L., Tenkanen, M., Effect of polysaccharide structure on mechanical and thermal properties of galactomannan-based films. *Biomacromolecules*, *8*, 3198–3205, 2007.
22. Cazón, P., Velazquez, G., Ramírez, J.A., Vázquez, M., Polysaccharide-based films and coatings for food packaging: A review. *Food Hydrocolloids*, *68*, 136–148, 2017.
23. Tan, W., Li, Q., Wang, H., Liu, Y., Zhang, J., Dong, F., Guo, Z., Synthesis, characterization, and antibacterial property of novel starch derivatives with 1, 2, 3-triazole. *Carbohydr. Polym.*, *142*, 1–7, 2016.
24. Salleh, E. and Muhamad, I., In *Starch-based Antimicrobial Films Incorporated with Lauric Acid and Chitosan. AIP Conf. Proc.*, American Institute of Physics, pp. 432–436, 2010.
25. Rhim, J.-W., Hong, S.-I., Park, H.-M., Ng, P.K., Preparation and characterization of chitosan-based nanocomposite films with antimicrobial activity. *J. Agric. Food Chem.*, *54*, 5814–5822, 2006.
26. Talebi, F., Misaghi, A., Khanjari, A., Kamkar, A., Gandomi, H., Rezaeigolestani, M., Incorporation of spice essential oils into poly-lactic acid film matrix with the aim of extending microbiological and sensorial shelf life of ground beef. *LWT*, *96*, 482–490, 2018.
27. Nguyen, V., Gidley, M., Dykes, G., Potential of a nisin-containing bacterial cellulose film to inhibit Listeria monocytogenes on processed meats. *Food Microbiol.*, *25*, 471–478, 2008.
28. Scannell, A.G., Hill, C., Ross, R., Marx, S., Hartmeier, W., Arendt, E.K., Development of bioactive food packaging materials using immobilised bacteriocins Lacticin 3147 and Nisaplin®. *Int. J. Food Microbiol.*, *60*, 241–249, 2000.
29. Nisperos-Carriedo, M.O., Edible coatings and films based on polysaccharides, in: *Edible coatings and films to improve food quality*, vol. 1, pp. 322–323, 1994.
30. Wang, L.-F. and Rhim, J.-W., Preparation and application of agar/alginate/collagen ternary blend functional food packaging films. *Int. J. Biol. Macromol.*, *80*, 460–468, 2015.
31. Azarakhsh, N., Osman, A., Ghazali, H.M., Tan, C.P., Adzahan, N.M., Lemongrass essential oil incorporated into alginate-based edible coating for shelf-life extension and quality retention of fresh-cut pineapple. *Postharvest Biol. Technol.*, *88*, 1–7, 2014.

32. Dawson, P., Hirt, D., Rieck, J., Acton, J., Sotthibandhu, A., Nisin release from films is affected by both protein type and film-forming method. *Food Res. Int.*, *36*, 959–968, 2003.
33. Ture, H., Eroglu, E., Ozen, B., Soyer, F., Effect of biopolymers containing natamycin against Aspergillus niger and Penicillium roquefortii on fresh kashar cheese. *Int. J. Food Sci. Technol.*, *46*, 154–160, 2011.
34. Uranga, J., Puertas, A., Etxabide, A., Dueñas, M., Guerrero, P., De La Caba, K., Citric acid-incorporated fish gelatin/chitosan composite films. *Food Hydrocolloids*, *86*, 95–103, 2019.
35. Carvalho, R.A., Santos, T.A., de Azevedo, V.M., Felix, P.H.C., Dias, M.V., Borges, S.V., Bio-nanocomposites for food packaging applications: Effect of cellulose nanofibers on morphological, mechanical, optical and barrier properties. *Polym. Int.*, *67*, 386–392, 2018.
36. Uyama, H., Kuwabara, M., Tsujimoto, T., Nakano, M., Usuki, A., Kobayashi, S., Organic-inorganic hybrids from renewable plant oils and clay. *Macromol. Biosci.*, *4*, 354–360, 2004.
37. Akhavan, S., Assadpour, E., Katouzian, I., Jafari, S.M., Lipid nano scale cargos for the protection and delivery of food bioactive ingredients and nutraceuticals. *Trends Food Sci. Technol.*, *74*, 132–146, 2018.
38. Farghaly, D.A., Aboelwafa, A.A., Hamza, M.Y., Mohamed, M.I., Topical delivery of Fenoprofen Calcium via elastic nano-vesicular spanlastics: Optimization using experimental design and *in vivo* evaluation. *AAPS PharmSciTech*, *18*, 2898–2909, 2017.
39. Demirci, M., Caglar, M.Y., Cakir, B., Gülseren, İ., Encapsulation by nanoliposomes, in: *Nanoencapsulation technologies for the food and nutraceutical industries*, pp. 74–113, 2017.
40. Jia, B., Zhou, J., Zhang, L., Electrospun nano-fiber mats containing cationic cellulose derivatives and poly (vinyl alcohol) with antibacterial activity. *Carbohydr. Res.*, *346*, 1337–1341, 2011.
41. Jafari, S.M., *Nanoencapsulation technologies for the food and nutraceutical industries*, Academic Press, London, UK, 2017.
42. Rehman, A., Tong, Q., Jafari, S.M., Assadpour, E., Shehzad, Q., Aadil, R.M., Iqbal, M.W., Rashed, M.M., Mushtaq, B.S., Ashraf, W., Carotenoid-loaded nanocarriers: A comprehensive review. *Adv. Colloid Interface Sci.*, *275*, 102048, 2020.
43. Maswal, M. and Dar, A.A., Formulation challenges in encapsulation and delivery of citral for improved food quality. *Food Hydrocolloids*, *37*, 182–195, 2014.
44. Donsì, F., Annunziata, M., Vincensi, M., Ferrari, G., Design of nanoemulsion-based delivery systems of natural antimicrobials: effect of the emulsifier. *J. Biotechnol.*, *159*, 342–350, 2012.
45. Kim, I.H., Lee, H., Kim, J.E., Song, K.B., Lee, Y.S., Chung, D.S., Min, S.C., Plum coatings of lemongrass oil-incorporating carnauba wax-based nanoemulsion. *J. Food Sci.*, *78*, E1551–E1559, 2013.

46. Mukherjee, S., Ray, S., Thakur, R., Solid lipid nanoparticles: A modern formulation approach in drug delivery system. *Indian J. Pharm. Sci.*, *71*, 349, 2009.
47. Montenegro, L., Lai, F., Offerta, A., Sarpietro, M.G., Micicche, L., Maccioni, A.M., Valenti, D., Fadda, A.M., From nanoemulsions to nanostructured lipid carriers: A relevant development in dermal delivery of drugs and cosmetics. *J. Drug Deliv. Sci. Technol.*, *32*, 100–112, 2016.
48. Shin, G.H., Kim, J.T., Park, H.J., Recent developments in nanoformulations of lipophilic functional foods. *Trends Food Sci. Technol.*, *46*, 144–157, 2015.
49. Dos Santos, P.P., de Aguiar Andrade, L., Flôres, S.H., de Oliveira Rios, A., Nanoencapsulation of carotenoids: a focus on different delivery systems and evaluation parameters. *J. Food Sci. Technol.*, *55*, 3851–3860, 2018.
50. Haratifar, S. and Guri, A., Nanocapsule formation by caseins, in: *Nanoencapsulation technologies for the food and nutraceutical industries*, pp. 140–164, Elsevier, London, UK, 2017.
51. De Kruif, C.G., Huppertz, T., Urban, V.S., Petukhov, A.V., Casein micelles and their internal structure. *Adv. Colloid Interface Sci.*, *171*, 36–52, 2012.
52. Esfanjani, A.F. and Jafari, S.M., Biopolymer nanoparticles and natural nanocarriers for nano-encapsulation of phenolic compounds. *Colloids Surf., B*, *146*, 532–543, 2016.
53. Esfanjani, A. and Jafari, S., Nanoencapsulation of phenolic compounds and antioxidants, in: *Nanoencapsulation of food bioactive ingredients*, vol. 1, pp. 63–101, 2017.
54. Hadaruga, D.I., Ünlüsayin, M., Gruia, A.T., Birau, C., Rusu, G., Hadaruga, N.G., Thermal and oxidative stability of Atlantic salmon oil (Salmo salar L.) and complexation with ß-cyclodextrin. *Beilstein J. Org. Chem.*, 12, 179–191, 2016. https://doi.org/10.3762/bjoc.12.20
55. de Sousa Oliveira, F., de Freitas, T.S., da Cruz, R.P., do Socorro Costa, M., Pereira, R.L.S., Quintans-Júnior, L.J., de Araújo Andrade, T., dos Passos Menezes, P., de Sousa, B.M.H., Nunes, P.S., Evaluation of the antibacterial and modulatory potential of α-bisabolol, β-cyclodextrin and α-bisabolol/β-cyclodextrin complex. *Biomed. Pharmacother.*, *92*, 1111–1118, 2017.
56. Di Marco, A.E., Ixtaina, V.Y., Tomás, M.C., Inclusion complexes of high amylose corn starch with essential fatty acids from chia seed oil as potential delivery systems in food. *Food Hydrocolloids*, *108*, 106030, 2020.
57. Suhag, R., Kumar, N., Petkoska, A.T., Upadhyay, A., Film formation and deposition methods of edible coating on food products: A review. *Int. Food Res. J.*, *136*, 109582, 2020.
58. Grumezescu, A.M. and Holban, A.M., *Biopolymers for food design*, vol. 20, Academic Press, London, UK, 2018.
59. Huang, C.C., Chen, Y.F., Wang, C.C., Effects of micronization on the physico-chemical properties of peels of three root and tuber crops. *J. Sci. Food Agric.*, *90*, 759–763, 2010.

60. Sedaghat Doost, A., Nikbakht Nasrabadi, M., Sadžak, A., Van der Meeren, P., Production of food nanomaterials by specialized equipment, in: *Handbook of Food Nanotechnology: Applications and Approaches*, 2020.
61. Karahan, H.E., Wiraja, C., Xu, C., Wei, J., Wang, Y., Wang, L., Liu, F., Chen, Y., Graphene materials in antimicrobial nanomedicine: Current status and future perspectives. *Adv. Healthc. Mater.*, *7*, 1701406, 2018.
62. Chaudhary, P., Fatima, F., Kumar, A., Relevance of nanomaterials in food packaging and its advanced future prospects. *J. Inorg. Organomet. Polym. Mater.*, *30*, 5180–5192, 2020.
63. Indumathi, M., Sarojini, K.S., Rajarajeswari, G., Antimicrobial and biodegradable chitosan/cellulose acetate phthalate/ZnO nano composite films with optimal oxygen permeability and hydrophobicity for extending the shelf life of black grape fruits. *Int. J. Biol. Macromol.*, *132*, 1112–1120, 2019.
64. Yang, X., Feng, B., Yang, P., Ding, Y., Chen, Y., Fei, J., Electrochemical determination of toxic ractopamine at an ordered mesoporous carbon modified electrode. *Food Chem.*, *145*, 619–624, 2014.
65. Yang, Y., Cao, Y., Wang, X., Fang, G., Wang, S., Prussian blue mediated amplification combined with signal enhancement of ordered mesoporous carbon for ultrasensitive and specific quantification of metolcarb by a three-dimensional molecularly imprinted electrochemical sensor. *Biosens. Bioelectron.*, *64*, 247–254, 2015.
66. Zhang, Z., Zhang, J., Zhang, B., Tang, J., Mussel-inspired functionalization of graphene for synthesizing Ag-polydopamine-graphene nanosheets as antibacterial materials. *Nanoscale*, *5*, 118–123, 2013.
67. Shabnam, L., Faisal, S.N., Roy, A.K., Haque, E., Minett, A.I., Gomes, V.G., Doped graphene/Cu nanocomposite: a high sensitivity non-enzymatic glucose sensor for food. *Food Chem.*, *221*, 751–759, 2017.
68. Tang, S., Wang, Z., Li, W., Li, M., Deng, Q., Wang, Y., Li, C., Chu, P.K., Ecofriendly and biodegradable soybean protein isolate films incorporated with ZnO nanoparticles for food packaging. *ACS Appl. Bio Mater.*, *2*, 2202–2207, 2019.
69. Hui, A., Liu, J., Ma, J., Synthesis and morphology-dependent antimicrobial activity of cerium doped flower-shaped ZnO crystallites under visible light irradiation. *Colloids Surf. A Physicochem. Eng. Asp.*, *506*, 519–525, 2016.
70. Rahman, P.M., Mujeeb, V.A., Muraleedharan, K., Thomas, S.K., Chitosan/nano ZnO composite films: Enhanced mechanical, antimicrobial and dielectric properties. *Arab. J.Chem.*, *11*, 120–127, 2018.
71. Ejaz, M., Arfat, Y.A., Mulla, M., Ahmed, J., Zinc oxide nanorods/clove essential oil incorporated Type B gelatin composite films and its applicability for shrimp packaging. *Food Packag. Shelf Life*, *15*, 113–121, 2018.
72. Deshmukh, S., Patil, S., Mullani, S., Delekar, S., Silver nanoparticles as an effective disinfectant: A review. *Mater. Sci. Eng. C*, *97*, 954–965, 2019.
73. Yang, T., Paulose, T., Redan, B.W., Mabon, J.C., Duncan, T.V., Food and beverage ingredients induce the formation of silver nanoparticles in

products stored within nanotechnology-enabled packaging. *ACS Appl. Mater. Interfaces*, *13*, 1398–1412, 2021.
74. Biswal, A.K. and Misra, P.K., Biosynthesis and characterization of silver nanoparticles for prospective application in food packaging and biomedical fields. *Mater. Chem. Phys.*, *250*, 123014, 2020.
75. Castro Mayorga, J.L., Fabra Rovira, M.J., Cabedo Mas, L., Sánchez Moragas, G., Lagaron Cabello, J.M., Antimicrobial nanocomposites and electrospun coatings based on poly (3-hydroxybutyrate-co-3-hydroxyvalerate) and copper oxide nanoparticles for active packaging and coating applications. *J. Appl. Polym. Sci.*, *135*, 45673, 2018.
76. Annu, Ali, A., Ahmed, S., Eco-friendly natural extract loaded antioxidative chitosan/polyvinyl alcohol based active films for food packaging. *Heliyon*, 7, 3, e06550, 2021.
77. Muthulakshmi, L., Rajalu, A.V., Kaliaraj, G.S., Siengchin, S., Parameswaranpillai, J., Saraswathi, R., Preparation of cellulose/copper nanoparticles bionanocomposite films using a bioflocculant polymer as reducing agent for antibacterial and anticorrosion applications. *Compos. B. Eng.*, *175*, 107177, 2019.
78. Deshmukh, S.P., Mullani, S.B., Koli, V.B., Patil, S.M., Kasabe, P.J., Dandge, P.B., Pawar, S.A., Delekar, S.D., Ag nanoparticles connected to the surface of TiO2 electrostatically for antibacterial photoinactivation studies. *Photochem. Photobiol.*, *94*, 1249–1262, 2018.
79. Bodaghi, H., Mostofi, Y., Oromiehie, A., Zamani, Z., Ghanbarzadeh, B., Costa, C., Conte, A., Del Nobile, M.A., Evaluation of the photocatalytic antimicrobial effects of a TiO2 nanocomposite food packaging film by *in vitro* and in vivo tests. *LWT*, *50*, 702–706, 2013.

10

Toxicological Effects of Nanomaterials Used in Food Packaging

Rahul Singhal[1], Deepti Rawat[2] and Bhawna Kaushik[2]

[1]*Department of Chemistry, Shivaji College, University of Delhi, New Delhi, India*
[2]*Department of Chemistry, Miranda House, University of Delhi, New Delhi, India*

Abstract
In recent years, nanotechnology has gathered enormous interest and paved the way to pioneering food packaging materials in order to prolong the life span and safety of foods and beverages. An extensive range of nanomaterials has been employed possessing antimicrobial activity and oxygen scavenging activity in form of film coatings to provide barrier against moisture, oxygen, and carbon dioxide. Currently, the risk posed by the NPs utilized in food packaging is mainly linked with migration. The most significant routes for exposure are inhalation, cutaneous, and ingestion (oral), inhalation/injection being majorly involved and most studied. This chapter summarizes different nanomaterials used in food packaging, their application benefits, and current scenario of commercialized products available in market. It also sheds light on migration of nanomaterials and study of toxicological impact on humans and environment.

***Keywords*:** Food packaging, toxicology, nanotechnology, NPs, antimicrobial

10.1 Introduction

Packaging of food materials has always been a significant part of the food processing industry as it offers protection to the goods we buy and without it the handling of the material will become inefficient and costly. With advent of nanoscience, food packaging industry has evolved to meet the changing demands of consumers along with ensuring food security and

**Corresponding author*: rahulsinghal@shivaji.du.ac.in

Annu, Tanima Bhattacharya and Shakeel Ahmed (eds.) Nanotechnology in Intelligent Food Packaging, (235–270) © 2022 Scrivener Publishing LLC

sustainability [1]. Owing to their nano dimension, these nano packing materials have extraordinary surface area which, when embedded into the polymer support, can provide the desired properties to the resultant nanocomposites [2]. Such packaging not deals with containing and transporting the food to the consumers but also involves protecting it under unfavorable environment such as humidity, microorganisms, gases, dust, and bad odors. Besides this, it also provides mechanical strength during transport and important product information to the consumers of modern society. Employment of nanomaterials in the field of packaging opens up new avenues to reduce the microbial contamination in food and enhancing their shelf life [3]. Currently, this industry is dominated by materials which can act as a barrier or possess oxygen scavenging and moisture absorbing properties. Nanotechnology-assisted packaging can be generally categorized in two kinds, namely, improved packaging and active packaging, wherein the first one involves the embedding of nanomaterials into the polymer matrix to augment the gas barrier and second one includes direct interaction of nanomaterials with food enabling its protection. Recently, edible coatings are also done on the food surface, which prevents the contact of oxygen, moisture, and other gases. These are beneficial as they can be consumed and thus averting the problem of pollution [4]. Recently, smart packaging has been on the forefront of intelligent packing systems. The concept of smart packing involves the use of sensor devices that not only perceive the physical quantity of the material and change the information into readable signs but are also employed for monitoring of inside environment of food which is regularly checked for oxygen content, temperature, possible pathogenic attack, and shown by the alarm indicators. Nano barcodes are also an integral part of smart packaging as they are used for tagging and security [5, 6]. For many decades, plastics have been intensively employed with polyethylene being the first one to be used widely for food packaging. However, with ever-increasing use of non-biodegradable plastics, many critical human and marine health as well as environmental issues have surfaced due to their inappropriate disposal and waste management [7]. Currently, wide range of nano metals and metal oxides are being employed in the current food packaging market owing to their small size high surface area, high thermal, and mechanical strength and strong bactericidal property. However, some of the nanomaterials have been found to be cytotoxic and even carcinogenic in nature. Their extremely small size makes them easily penetrable through the cell barrier, thus generating free radicals in the tissues eventually causing cellular damage [8]. Keeping in mind the grave toxicity issues, researchers have been searching for more green and sustainable alternatives such as bio-nanocomposite packaging materials.

These materials are environmentally benign with high performance, biodegradability, innocuous, and lightweight nature [9].

10.2 Nanomaterials Employed in Food Packaging

Nanomaterials have been employed in the food packaging industry based on their intricate set of factors such as size, shape, surface are, charge and admirable surface activity. Currently, both organic and inorganic nanostructures have been in use in the food processing market. A few of them have been described here in brief [10].

Nano-clay such as montmorillonite (MMT) is extensively utilized in forming polymers as nanocomposites in food packaging industry, owing to its high stability, economic nature, and easy availability. It is made of two silicon oxide tetrahedral layers consisting of magnesium or aluminium hydroxide on the shared edge forming an octahedral between them. When MMT is intercalated with polymers, infiltration of gases and other substances is restricted depending on its degree of orientation and dispersion [11]. For instance, Silvestre and group established that the shelf life of beer was increased from 77 to 210 days upon assimilation of nanoclay in the bottles [12].

Biodegradable polymers are another suitable candidate as far as nano-packaging is concerned. They are superior to others, owing to their eco-friendly nature as they can be decomposed easily by microorganisms. Bio-nanocomposite films are biodegradable films consisting of a polymer matrix created from proteins and lipids and a filling of organic or inorganic material with a dimension below 100 nm. In the quest to look for a more sustainable and green option, cellulose nanocomposites have gained attention as they are a renewable source and preserve food safety and quality. They usually work in the form of matrices and films or as a reinforcement phase in form of cellulose nanofibrils or nanocrystals. Bacterial cellulose has emerged as an ideal candidate because it can be fabricated in the form of natural nanostructured membranes [13].

Chitosan belongs to a family of linear polysaccharides, which retains antibacterial properties and is used in combination of other antimicrobial agents such as biopolymers and metal oxides such as ZnO [14]. Wang *et al.* proved that the use of chitosan-ZnO nanocomposite on polyvinyl acetate support significantly abridged the activity of *E. coli* and *Candida albicans* [15].

Nano metals such as copper, zinc, silver and gold have also emerged as suitable alternatives for food packaging due to their biocidal quality.

Silver NPs in particular have exhibited tremendous bactericidal properties against various bacteria, fungi, yeast, etc., due to their high surface area and exorbitant thermal stability. These can be embedded in various polymer matrices to get desired effects. However, there have been some raising concerns associated with the toxicity of these NPs if ingested [16, 17]. Nano metal oxides such as TiO_2, ZnO, SiO_2, and Al_2O_3 have also contributed immensely toward packaging food materials. These metal oxides have photocatalytic as well as antimicrobial properties with a tendency to capture ethylene useful in packing fresh food products. For instance, TiO_2 which is commonly employed as a pigment in various food materials has antibacterial properties and generates reactive oxygen species (ROS) after coming in contact with UV light, leading to termination of bacteria *via* oxidation of proteins. ZnO, on the other hand, has been found to be superior to silver NPs due to its lower cost and better UV barrier properties [18]. Carbon nanotubes (CNTs) are also being exhaustively used in food sector because of their unique electronic properties, boost mechanical strength of the polymer matrices employed, and possess antimicrobial properties as their long and thin structures puncture the microbial cells [19].

10.3 Food Packaging Functionality

Food packaging has various associated key benefits from food processing, transport, and the market. It not only protects the food material from damage and bacterial attack but also provides important nutritional information about it to the end users [20].

The food packaging aims to have four desired functions, which have been described below (shown in Figure 10.1):

a) **Preservation:** The primary function of food packaging is to guard it from various environmental aspects that can damage or deteriorate the quality of food. It also helps in circumventing the spills and leaks, thus reducing the wastage of food materials. It also safeguards the delicate food such as eggs or fruits from any physical damage caused by shock during the transport by working as a cushion.
b) **Containment:** With population spurt and growing food scarcity, it is indispensable to reduce the wastage of food. Nano-packaging can diminish the wastage along the way as it prolongs the shelf life of food, in turn, extending its usability.

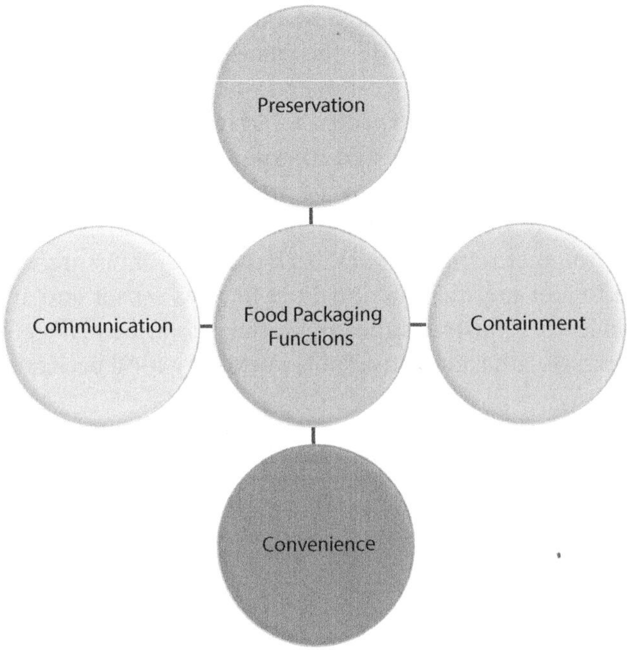

Figure 10.1 Key purpose of food packaging materials.

c) **Convenience:** In the modern era, where people have busy lifestyle, convenience in food handling is a prominent aspect. A good food packaging should have the attributes of easy handling such as facile opening and resealing the food packets as well as disposal.

d) **Communication:** It is essential to offer important nutritional information about the food material to the consumers. An ideal packaging should provide correct information about the ingredients, weight, and manufacturer to the customers.

10.4 Current Market Scenario of Nanomaterials in Packaging Industry

During past few years, nanotechnology has revolutionized the food processing industries. The nanomaterials in packaging industry are in constant process of technological developments with enhanced packaging to make food consumption experience better. According to the market report accomplished by Polaris market Research in the year 2008, the

Nano-Enabled Food and Beverage Market was estimated to be 4.13 billion US dollar, [21], and in 2017, the value expanded to 30.60 billion US dollar. In the next decade, the estimated value will increase at compound rate (CAGR) of 12.7% and expected to arrive at 89 billion by 2026 [22]. The present scenario has witnessed an exponential rise in use of nanotechnology for packaging on global scale and created a scuttle among companies to invest in this sector. Many nanoscale materials have been used to improve polymer reinforcement, barrier against moisture, and processability and to add specific functionalities like gas sensor and antibacterial activity. Table 10.1 illustrates the application of various NPs employed in the commercialized products available in existing food packaging market.

Table 10.1 Nano-packaging material available in current market.

Product	Name of the company	NPs used in packaging	Function/ application	Reference
FresherLonger storage bag	Sharper Image/US	Ag	Antimicrobial activity	[22]
baby milk bottle coated with silver nanoparticles	Baby Dream Co. Ltd., Korea	Ag	Antimicrobial	[23]
Ag NPs coated	A-Do Global Co. Ltd., Korea	Ag	Antimicrobial	[24]
Fresh box food container	BlueMoon Goods, USA Kinetic Go Green, USA Oso Fresh Storage, USA	Ag	Reduce development of bacteria, keeps food fresh, safe and delicious for long	[25]
Baoxianhe nano silver storage box	Quan Zhou Hu Zheng Nano Technology, Co. Ltd, China	Ag	Antimicrobial	[26]
Anson Nano Anti-microbial storage	Anson NanoBiotechnology (Zhuhai) Co. Ltd.	Ag	Antimicrobial	[12]

(Continued)

Table 10.1 Nano-packaging material available in current market. (*Continued*)

Product	Name of the company	NPs used in packaging	Function/ application	Reference
Plastic bag Durethan KU2-2601	Lanxess Deutschland GmbH	Commercial clay nanocomposite Durethan KU2-261 Trade name of nylon6-nanoclay	Prevents penetration of O_2, CO_2 from outside and retains moisture of food on inside; Enhances barrier properties, strength, abrasion, as well as chemical resistance	[27, 28]
Aegis TM OXCE	Honeywell International	Nylon6- nanoclay composite	O_2 scavenging property, used particularly for flavored nano clay alcoholic beverage	
Imperm	Miller Brewing Co. US, Nylon 6 polymer from Mitsubishi Gas Chem Company USA	Nylon 6-nanoclay composite	Multilayer PET bottles for liquor and beer and soft drinks	[29]
Food cling wrap	SongSing Nanotechnology, Taiwan	Nanocomposites containing ZnO	Antimicrobial activity	[30, 31]
McDonald's hamburger clamshells	EcoSynthetix	EcoSphere Biolatex Nanostarch	Biodegradable	[32]

(*Continued*)

Table 10.1 Nano-packaging material available in current market. (*Continued*)

Product	Name of the company	NPs used in packaging	Function/ application	Reference
Water soluble laundry bags	Novamont SA, Italy	Starch/PVOH	Excellent barrier to starch O_2 and CO_2. Keeps flavor and aroma of food inside package and prevent penetration of odors from outside the package	[33]
Joule TM RHB	ColorMatrix	TiN	Reheat additive for PET, superior aesthetic, and reheat performance	[34]

10.5 Nanoparticle Migration in Food

The most important research in packaging is to investigate the migration of nanoparticles from containers into the stuff and evaluate its potential toxicity to understand dynamics inside the human system and ensure safety of nanocomposite packaging. Such a migration can occur through the route of diffusion, dissolution or abrasion of the packing material [35, 36]. Generally, nanomaterials used for food packaging can be categorized as free if they are directly introduced as additives and embedded if they are integrated into food contact substances [37, 38]. This it is important as toxicological effects for both are entirely different with the embedded ones proving to be more dangerous as they migrate from packaged surface and directly interact with the food items [39]. Their toxicological behavior depends significantly on the morphology, size, composition, and charge of the nanostructures [4]. Consequently, it is indispensable to employ various characterization techniques to gain insight into the various important parameters mentioned above as well as the migration rate of the nanoparticles into the foodstuff. Presently, various physicochemical techniques such as inductively coupled plasma mass spectrometry (ICP-MS), size exclusion chromatography, microscopic techniques (such as scanning electron and atomic force microscopy), and filtration have been employed to estimate the NPs present in the samples [40–43].

Numerous studies avail variety of stimulants to study the migration of NPs into the food substances [44]. In fact, European Commission (EC) has established strict threshold levels mentioning the conditions employed for carrying out the migration studies. Food stimulants such water, acetic acid (3% v/v), vegetable oil, and ethanol (10% to 50% v/v) can be utilized relying on the type of packaging material and the foodstuff [5, 45, 46]. Table 10.2 displays the migration of NPs from polymer/plastic packaging to food stuff /food stimulants.

Doimeadios *et al.* studied the migration of silver NPs by employing water and acetic acid (3% v/v) as stimulant. Migration was detected *via* ICP-MS analysis, and it was concluded that no release of AgNPs occurred [47]. Similarly, Gomez and the group evaluated total migration content of silver from food containers under aqueous and acidic conditions *via* ICP-MS analysis. The authors concluded that on elevating the temperature and increasing the time led to enhanced silver migration of 17 ng/g [48]. On similar grounds, Cummins and co-workers examined the migration of nano-sized silver and copper polyethylene nanocomposites to the food and reported that both migrated below the total legislative migration limits of 0.003 to 0.005 mg/dm^2 for Ag and those of copper between 0.024 and 0.049 mg/dm^2 [49]. Motlagh *et al.* studied the outcome of packing Olivier salad in silver NPs embedded polyethylene and how it affected the durability as well as microbial properties of the packaged food material. It was observed that micro-organism count was directly proportional to the storage time and held an inverse relationship with the concentration of silver NPs concentration in the packing material [50]. The influence of size dependent Lin *et al.* investigated the effect of particle size on migration pattern of TiO_2 embedded LDPE packaging films. It was observed that 100 nm TiO_2-LDPE exhibited higher migration rate due to its poor compatibility with the film and food stimulants used [51]. Migration of matrix embedded CNTs (LDPE-incorporated CNTs and polystyrene-incorporated CNTs) was explored by Bott *et al.* Authors utilized various stimulants such as acetic acid (3%) and ethanol (95%) under various temperature and time settings and experimental results affirmed that no amount of carbon black NPs was migrated from the matrices to the food [52]. Schmidt and co-workers studied the migration of organo-modified clay incorporated into polylactic

Table 10.2 Migration studies of NPs from food contact material in different conditions in food.

S. no.	Nanoparticle used	Food contact material	Migration assay conditions	Outcome	Reference
1.	Silver	Polypropylene food containers	Water, acetic acid (3%) 20–70°C	No migration detected	[47]
2.	Silver	Polypropylene and high-density polyethylene (HDPE) containers	Water, acetic acid (3% w/v), 10% (v/v) ethanol, and 95% (v/v) ethanol solutions	17 ng/g	[48]
3.	a) Silver b) Copper	Polyethylene nanocomposites	Temperature of 8.13 and 21.8°C with migration test held for 1.1 or 3.1 days	0.003–0.005 mg/dm^2 0.024–0.049 mg/dm^2	[49]
4.	Clay and silica	Low-density polyethylene	Acetic acid (3%)	No significant migration	[50]
5.	TiO_2	Polyethylene film	3% (w/v) acetic acid, 50% ethanol, 1–8 hours at 25°C, 70°C, and 100°C	1.4 ± 0.02, 6.3 ± 0.5 and 12.1 ± 0.2 $\mu g\ kg^{-1}$ at 25°C, 70°C, and 100°C for acetic acid and 0.5 ± 0.1, 0.6 ± 0.03, and 2.1 ± 0.1 $\mu g\ kg^{-1}$ at 25°C, 70°C, and 100°C for ethanol	[51]

(Continued)

Table 10.2 Migration studies of NPs from food contact material in different conditions in food. (*Continued*)

S. no.	Nanoparticle used	Food contact material	Migration assay conditions	Outcome	Reference
6.	Carbon Black	Low-density polyethylene, polystyrene	Acetic acid (3%), 95% (v/v) ethanol solutions	No migration detected	[52]
7.	Silver	Commercial containers	ethanol (EtOH) 50% v/v and acetic acid (HAc) 3% v/v, 40 and 60°C	1.66–31.46 ng/cm^2 (within permissible limits)	[36]
8.	Silver	a) nano-silver impregnated polymer containers b) coated nano-silver films	Water, 3% (w/v) acetic acid, 40°C	5.66 ± 0.02 mg L^{-1} 28.92 ± 0.01 mg L^{-1}	[55]
9.	Titanium nitride	Low-density polyethylene	Acetic acid (3%), 95% (v/v) ethanol solutions	No migration detected	[56]
10.	Synthetic amorphous silica, silanated	Low-density polyethylene	Acetic acid (3%), 95% (v/v) ethanol, iso octane solutions	No migration detected	[57]

acid (PLA) films at 40°C for 10 days using ethanol as stimulant. ICP-MS analysis was carried out, and it was concluded that clay did not migrate at all. However, a slight amount of nanoclay particles in the range of 50–800 nm was detected via AF^4-MALS analysis plausibly due to a small amount of fragmentation of PLA matrix due to harsh test conditions [53]. Farhoodi and group examined the effect of clay and silica NPs on low-density polyethylene composites. Migration studies were carried out within 10 day of food storage at 40°C. The authors concluded that raising the concentration of NPs directly led to a surge in the swelling index of the composite with acetic acid (3% v/v) as the stimulant [54]. Extensive research should be carried out in this field to clarify whether the NPs migrates from foodstuff packaged in nanocomposite or not, understand the mechanism involved, how diffusion process alters the structure of material in terms of size, shape, and morphology and assess the toxicity to human health.

10.6 Potential Routes for Exposure of NPs

The mode through which nanoparticles can invade into the human system are inhalation, ingestion, and through skin [58] (shown in Figure 10.2). When the NPs enter into the biological system, they interact with various biomolecules which on further dissolution in body fluids accumulate in various parts of body such as kidney, liver, brain, and spleen [59]. In dermal/cutaneous exposure, the NPs penetrate through epidermis of skin. This occurs when NP are used in cosmetics/ointments. Oberdoster *et al.* [60] reported that all skin and epidermis does not stop ultrafine particles from going into the body as they are governed by several factors. They infiltrate and translocate into the lymph nodes [61]. In few cases, these NP may interact with immune system causing oxidative damage via formation of free radicals by the inorganic NP [62]. Inhalation is the most potential route through which NPs enter the system. Volatile NPs present in certain essential oils with diameter <10 nm through the process of diffusion through nasal cavity and get deposited in the respiratory tract. Through the process of diffusion, TiO_2 NPs are deposited into the lungs causing chronic pneumonia and pulmonary infections [63]. These NPs then cross the cell barrier of the blood vessels and move out of the respiratory system employing various mechanisms.

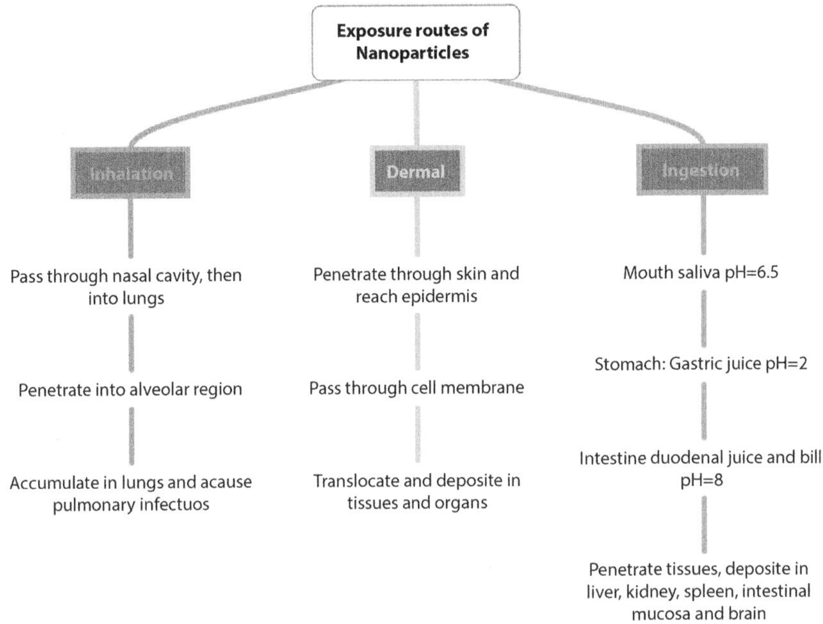

Figure 10.2 Exposure routes of NPs.

Ingestion of NPs present in food penetrates the tissues that are absorbed at the target sites in the body. Inside the stomach, digestion of food takes place in presence of acids and body compounds. Ingested food then penetrates the duodenum, wherein it mixes with bile salts which are discharged by the gall bladder. The ingested NPs damage intestinal epithelial lining and then translocate to blood circulation reaching different organs of the human system like liver, kidney, and brain, thus causing damaging effects. The residence time of NP inside the intestine is dependent upon intricate properties of toxicants such as particle size, shape, concentration or dosage, surface charge, solubility, as well as chemical constituents of nanoparticles.

The toxicity mechanism of nanomaterials is not yet fully understood. Studies have revealed that metal ions may be accountable for deadliness of their corresponding metallic salts. Oxidative stress and inflammation play a significant part in contributing toxicity of nanomaterials, but exact mechanism is yet to be fully understood.

10.7 Toxicological Studies of NPs Used in Packaging

In vitro and *in vivo* studies can be carried out to study the toxic effects of NPs migrated from food packaging material. Initially, *in vitro* toxicity studies are done to assess the potential toxic effects of NPs and helps to calculate the limit of dosage exposure followed by *in vivo* studies to corroborate the behavior in realistic environment inside the body of model animals [64]. In the *in vitro* studies, different cell lines (human and rodent) are commonly employed as experimental models particularly from intestine, liver, lung, or skin cells and exposed to toxic NPs which are left to incubate for definite time [58]. The cytotoxicity and cell viability can be measured by different assays like lactate dehydrogenase assay (LDH), cell-counting, eosin, congo red, and erythrosine blue and trypan blue dye assay [65, 66]. The toxic effects exhibited can be observed through changes in basal biomarkers such as ROS, glutathione content (GSH), inflammation response, DNA damage, cell death, and necrosis. The genotoxicity of NPs is done to identify the rupture of DNA and can be assessed by Ames test, micronucleus assay, and comet. In the *in vivo* toxicity assessment, a small dose of NPs is fed inside the body of mice or rats to examine the histopathological effects and quantify the uptake, distribution and excretion of NPs outside the body of rodents [66]. It is the reliable, acceptable, and prime method to evaluate the toxicity associated with NPs till date.

10.8 Toxicological Effects of NPs

10.8.1 Humans and Animals

10.8.1.1 Clays

Clays have been used promising reinforced material in food packaging to ameliorate the barrier properties, chemical and thermal stability, impedance of flammability, and facile recycling. Integration of clay NPs within the polymer matrix aids numerous benefits such as easy availability, low cost, huge cation exchange capacity, large surface area, more aspect ratio, and improved thermal, mechanical, and optical properties [67, 68]. Clays NPs exhibit toxic effects mainly through inhalation. Since they have been applied in food contact materials, oral route of exposure likely may occur [69, 70]. Potential negative impacts of nano clays related to human well-being are not very well documented in the literature (Table 10.3). *In vitro* analysis has been performed to evaluate their potential toxicological influence.

Toxicities changes with factors like hydrophobicity, concentration, modification. Unmodified clays (bentonite) have been approved by GRAS as it exhibits no toxicity. Jos and co-workers performed cytotoxic studies for unmodified (Cloisite@ Na+) and organically modified of (cloisite 30B) in human hepatocellular cell line HepG2 [71]. It was observed that Cloisite Na^+ did not exhibit any induced cytotoxic or mutagenic effect in concentrations (0–125 μ/ml), whereas Cloisite 30B in concentration (0–25μ/ml) exhibited toxic effect. Toxic mechanism such as oxidative stress, inflammation response, apoptosis, and genotoxicity biomarkers were also assessed. Results depicted depletion in GSH, which is related to ROS scavenging. Alterations in HepG2 cells were observed which resulted in mitochondrial degeneration, dilated endomembrane systems, and heterophagosomes formation.

10.8.1.2 Silver

Silver NPs are widely employed as broad spectrum antimicrobial agents in food packaging by releasing Ag^+ induce the formation release ROS and kills the micro-organisms by oxidative stress. There is possibility of migration of NPs from the containers unintentionally into the food ending up in gastrointestinal tract [36, 72, 73]. The soluble Ag NPs instantly interact with biological medium and react with other components present [74]. Animal studies reveal that high doses of Ag NPs have shown to exhibit toxic effects when ingested (Table 10.3). GIT is quite susceptible to silver induced toxicity; however, adverse effects still remain unresolved. *In vivo* model tests have shown that intake of silver NPs administered through various routes can cross blood brain barrier, inducing neurotoxicity causing inflammation, cell dysfunction, and animal death [75–78]. It has also been observed that silver NPs tend to accumulate in different parts of organs [79–81]. Choi *et al.* treated mice with different particle size of AgNPs (22, 42, 71, 323 nm) at a dose of 1 mg/kg for a period of 14 days by oral administration. No toxicity was ascribed for large sized AgNPs (323nm). Repeated dose toxicity for small size NPs caused organ failure and inflammatory response in mice [79].

Zhande and co-workers [80] reported toxicity studies of <20-nm non-coated Ag NPS and <15-nm AgNP-coated polyvinypyrollidone after 28 days of oral administration in rats at a dose of 90 mg/kg BW of silver and 9 mg/kg BW for $AgNO_3$. High concentrations of silver were observed in liver and spleen for all silver treatment. Biochemical markers and antibody studies no hepatoxicity and immunotoxicity for silver exposure. It was observed that after 8 weeks of post dosing, silver was removed from all organs except brain and testis. Long-term retention of silver in brain and testis can be considered as risk for health of organisms. In a study reported by Kim *et al.*, 8-week-old rats (both male and female weighing 192 and 283 g) were exposed to AgNps at a dose of 30, 300, and 1,000 mg/kg for 28 days. No significant reduction in body weight was observed, but slight liver damage was observed over 300 mg of AgNPs. It can be concluded that silver NPs did not exhibit genetic toxicity but can accumulate and exert toxic impacts when consumed in larger quantities. However, it is unclear whether the levels are in proximity to those actually attainable through ingestion. Ag NPs triggered high toxicity levels due to enhanced levels of ROS, thus generating oxidative stress and causing harm to cell membranes, organelles, and nuclei [83]. Besides toxic behavior toward mammalian cells, Ag NPs also illustrates bactericidal activity and may change the nature of gut bacteria if it reaches the colon. In another report, authors investigated dose-dependent toxic effects of AgNPs and Ag^+ in Sprague–Dawley rats for 13 weeks. Smaller particle size decreased the population growth of Firmicuta phyla and decreased the expression of immunodulatory gene [84]. *In vitro* studies have also disclosed that the impact of silver NPs varies with their size where smaller ones exert high toxicity compared to larger size of NPs [86]. Kim and co-workers [82] studied the size-dependent behavior of three different Ag NPs on cell lines MC3T3-E1 and PC-1. Small-sized Ag NP (10 nm) induced cellular apoptosis than large sized NPs (50 and 100 nm). It was revealed that high dose of NPs was used in cell culture studies.

10.8.1.3 SiO_2

Nanocomposites of SiO_2 possess high thermal, mechanical strength, low cost, and antimicrobial properties to be used in active packaging. It can exert toxic effects through generation of oxidative stress, causing impairment of DNA and apoptosis (Table 10.3). According to the study conducted by Westerhoff on different food grades of SiO_2 was examined in caco-2 cell to study its impact in gastrointestinal tract and environmental fate in waste water treatment plants. Different food grades exhibited particle size in range of 9–26 nm. Disruption of microvilli were observed in GI tract lead to synthesis of ROS [85]. Yang *et al.* suggested the safe application SiO_2 NPs in food additives when used in low concentrations [86, 103]. The chemical composition of nanoparticles plays an important role in determining the toxicity. Presence of metal ions (cobalt, manganese, and titanium) in silica NPs resulted in intracellular oxidation leading to Trojan horse–type mechanism [87]. On contrary, Schubbe *et al.* reported no toxicity associated with

SiO_2 in intestinal epithelial cells [88]. Similarly, in another study, Tarantini and co-workers studied the toxic influence of 15-nm SiO_2 NPs on exposed caco-2 cells. The authors detected apoptosis with increased ROS and chromosomal damage [89]. According to Wang and co-workers, dose-dependent drop in cell viability with enhanced intracellular ROS and decline in GSH was observed in human embryonic kidney (HEK 293) cells [102]. *In vivo* studies SiO_2 NPs increased alanine aminotransferase which caused adverse effect on liver. According to Lee and Van Kesteren *et al.*, oral exposure to rats caused elevated levels of SiO_2 in organs such as liver, lung, and spleen. Small-sized NPs excreted faster (20 nm) because of faster decomposition rates [90, 91]. No toxicity was observed when fed to mice for 90 days and no adverse levels effect observed upto 2,000 mg/kg [92]. Soo and group studied the effect of nano and micron sized silica on mice and concluded that the nano sized silica particle had a toxic effect on the liver of mice on feeding 140 g silica/kg mouse [93]. Toxicity related to SiO_2 NP is not conclusive based on evidences.

10.8.1.4 ZnO

ZnO nanoparticles play a key role as active component in packaging by demonstrating antimicrobial activity. It has received considerable attention in packaging industry, owing to its excellent thermal, mechanical, and barrier properties. They are also employed as potential industrial coatings to protect food from UV light exposure [104, 105]. It is generally recognized as safe for human health as it protects the characteristics of food packaging. After oral ingestion, ZnO NPs enters the circulatory system and gets accumulated in liver and other organelles causing damage through oxidative-mechanism (Table 10.3). Chen and co-workers investigated the cytotoxicity and mechanism of ZnO on human hepatocyte HepG2 cell model using three size of NPs (20, 90, and 200 nm) and compared it with Zn^{2+}. The cytotoxic mechanism, uptake, and elimination behavior of these NPs were investigated through electrochemical vaporization ICP-MS, MTT assay, GSH assay, ROS, and gene expression. It was found out that ZnO NPs exhibit size and dose-dependent cytotoxicity attributed due to released Zn^{2+} which induced oxidative stress and cell death. It was observed that elimination rate of bigger size of Zn NPs was difficult though it possessed low cytotoxicity [106]. The toxicity of nanomaterials is influenced by size, chemical nature, and surface properties. There can be a drastic change in toxicity when nanomaterials in food additives are present in complex environment. Wang and co-workers, for the first time, studied the influence of vitamin C (ascorbic acid) on ZnO NPs. Both *in vitro* and *in vivo* results indicated that when cells were subjected to ZnO NPs at concentration less than 15 mg L^{-1} and ascorbic acid (V_c) at concentration less than 300 mg L^{-1} both in gastric epithelial cell line (GES-1) and neural stem cells (NSCs), no cytotoxicity was observed. On contrary, increasing the concentration of ZnO NPs (~15 mg L^{-1}) and ascorbic acid (~300 mg L^{-1}), the cell viability decreased sharply, indicating significant cytotoxicity due to increased dissolution rate of Zn^{2+} facilitated by ascorbic acid. Further, the authors investigated combined toxicity of ZnO and V_c in mice through oral injection for 30 and 90 days in young and adult mice. High concentration of zinc in liver, kidneys, and modification of related protein after recurrent exposure on mice for 90 days was seen [94]. The authors suggested that the effect of specific food components can affect the toxicity of ZnO. NPs may be spherical or non-spherical solid particles which are largely accumulated when dispersed in aqueous solution [107]. Feeding studies on frogs have revealed that ZnO NPs showed high toxicity than dissolved zinc solution accredited to their greater ability to stimulate oxidative impairment in cells. This study focused on the prominence of establishing the physical form of zinc when ZnO NPs are ingested [95].

10.8.1.5 TiO_2

According to Shi and co-workers, cytotoxicity of TiO_2 is facilitated by generation of ROS after UV exposure, creating oxidative stress and inflammation via DNA damage, peroxidation of lipids and membranes, cell cycle arrest, and mitochondrial dysfunction in pulmonary and inhalation models [108]. Chen and co-workers [96] investigated the influence of both dose and size-dependent toxicity of TiO_2 (25 and 80 nm at 5 g/kg body weight) on adult mice. It was observed that the TiO_2 particles were localized in liver, spleen, kidney, and lung tissues of mice causing hepatic injury and renal lesions on histopathological examination. To study the liver injury, immune response and physiological effects in mice, Hong *et al.* carried out intragastric administration of anatase form of TiO_2 (5 nm) on mice at 62.5, 125, and 250 mg/kg body weight for 30 days [97]. The TiO_2 showed increase in toxicity by dose-dependent behavior [109]. In contrast Chi and co-workers [98] did not report significant toxicity or accumulation of TiO_2 NPs in tissues but high concentrations were detected in faeces which indicates higher elimination rate. From the above research studies, contradictions in animal studies on accumulation and toxicity can be accounted for various reasons like different dosage, particle size, and surface aggregates of NPs. Another reason can be that there may be variation in analytical techniques and animal model

Table 10.3 Research highlights of toxicity studes of various NPs used in food packaging industry.

Reference	NPs	Model	Particle size	Dose	Duration	Observation
[82]	Ag *in vitro*	Cell lines MC3T3-E1 and PC 1.	10, 50, and 100 nm	-	-	• ROS generation • Apoptosis • Necrosis
[84]	Ag *in vitro*	Human GI digestion model	60 nm	10 μg/ml	-	• Accumulation of NPs *via* gastric passage and breaking into original size on reaching intestines
[81]	Ag *in vivo*	Oral exposure of rats	14 nm	9 mg/kg		• Accumulation in liver, kidney, brain, and lungs
[80]	Ag *in vivo*	Rats	<20-nm noncoated and <15-nm Ag-coated with PVP		28 days	• No acute hepatotoxicity nor nonspecific immune responses observed

(*Continued*)

Table 10.3 Research highlights of toxicity studes of various NPs used in food packaging industry. (*Continued*)

Reference	NPs	Model	Particle size	Dose	Duration	Observation
[79]	Ag *in vivo*	Rats	22, 42, 71, and 324 nm	1 mg/kg body weight	14 days	• Organ toxicity (brain, liver, kidney) and inflammatory response • No toxicity observed for particle size 323 nm
[85]	SiO_2 *in vitro*	Human cell line, Caco-2	9–26 nm	2 mg/L	-	• Production of ROS • Rupture of cell membranes
[86]	SiO_2 *in vitro*	Human gastric epithelial cells GE-1 and colorectal adenocarcinoma cells Caco-2	10–50 nm	100–200 μg/ml	48–72 hours	• No induced apoptosis or necrosis observed • Cell cycle arrest
[87]	SiO_2 *in vitro*	Human lung epithelial cells (A549)	20–75 nm Doping with 0.5–1.6 wt %	30 ppm	-	• Oxidative stress due to release of ROS observed in presence of Co and Mn

(*Continued*)

Table 10.3 Research highlights of toxicity studes of various NPs used in food packaging industry. (*Continued*)

Reference	NPs	Model	Particle size	Dose	Duration	Observation
[88]	SiO_2 *in vitro*	Cac-2 cells	32 and 83 nm	100 μg/ml	5, 24, 48, and 72 hours	• No toxicity
[89]	SiO_2 *in vitro*	Cac-2 cells	15 and 55 nm	32 μg/ml	4 days	• Decrease in cell viability with increased reactive oxygen species
[90]	SiO_2 *in vivo*	Rats	20 and 100 nm	100, 500, 1,000, and 2,500 mg/kg (bw/day)	28 days	• High concentration found in liver and spleen
[91]	SiO_2 *in vivo*	Rats	20 and 100 nm	500–100 mg/kg	3 days	• Accumulation in liver, kidney, and lung and spleen • Smaller NPs excreted faster due to decomposition
[92]	SiO_2 *in vivo*	Sprague–Dawley rats	20 and 100 nm	500, 1,000, and 2,000 mg/kg	90 days	• No organ affected

(*Continued*)

Table 10.3 Research highlights of toxicity studes of various NPs used in food packaging industry. (*Continued*)

Reference	NPs	Model	Particle size	Dose	Duration	Observation
[93]	SiO_2 *in vivo*	mice	30 nm and 30 micrometer	140 g/kg	10 weeks	• 30-nm silica exhibited fatty liver • Augmented alanine aminotransferase
[94]	ZnO *in vivo*	Mice	15 nm	15 mg L^{-1} ZnO NPs and 300 mg L^{-1} of ascorbic	-	• Changes in liver and kidneys
[94]	ZnO *in vitro*	Gastric epithelial cell line (GES-1) and neural stem cells (NSCs)	15 nm	15 mg L^{-1} ZnO NPs and 300 mg L^{-1} of ascorbic acid	90 days	• Surge in OS
[95]	ZnO *in vivo*	Xenopus laevis	<50 nm	10–100 mg L^{-1}	96 hours	• Malformations and histological cuts to the gut • Oxidative damage
[96]	TiO_2 *in vivo*	Adult mice	25, 80, and 155 nm	5 g/kg body weight	14 days	• No acute toxicity • In female mice, lesions in liver and kidney were observed

(*Continued*)

Table 10.3 Research highlights of toxicity studes of various NPs used in food packaging industry. (*Continued*)

Reference	NPs	Model	Particle size	Dose	Duration	Observation
[97]	TiO_2 *in vivo*	Mice	Anatase 5 nm	125 and 250 mg/kg	30 days	• Decreased weight • Liver damage • Damage of immune response and haemostasis
[98]	TiO_2 *in vivo*	Rats	21 nm	1,041.4 mg/kg body weight	13 weeks	• No toxicity to exposed cell line
[99]	TiO_2 *in vivo*	Sprague Dawley rats (3 and 8 weeks)	60–200 nm	0, 10, 50, and 200 mg/kg	30 days	• No change in body weight • n adult rats, slight injuries observed in liver and kidney • Heart injuries and liver odema were observed in young rats
[100]	TiO_2 *in vitro*	Human gastric epithelial cells	21 nm (80/20 anatase/ rutile)	150 µg/ml	24 hours	• DNA damage and oxidative stress
[101]	TiO_2 *in vitro*	Caco-2/HT29-MTX cell culture model	30 nm	--	5 days	• ROS generation • Intestinal epithelial cells affected

to determine accumulation and toxicity (Table 10.3). Jia and co-workers reported comparative toxicity studies of TiO_2 NPs in Sprague–Dawley rats 3 weeks (young) and 8 weeks (adult rats) at oral exposure of 0.10, 50, and 200 mg/kg body weight for a span of 30 days. *In vivo* results indicate different adverse effects both on young and adult mice. Young mice were susceptible to liver enema, and heart injuries and non-allergic mast cell activation were found in young rats. Slight injuries in liver and kidney were observed in adult rats [99]. The above observations indicated that age differences affect the dietary intake of of TiO_2 NPs in food. *In vitro* studies reported that TiO_2 NPs exhibit cytotoxicity depending on various factors such as dose, particle size, crystal form, and surface coating. Toxic effects are accredited to generation of ROS which has the potential to interfere and damage the cellular metabolism, efflux pumps, nutrient transport, and alteration of gut microbia due to its antimicrobial activity [100, 101]. Out of three forms of TiO_2, anatase is reported to be the most toxic form. Cell culture study exhibited that mixture of rutile and anatase forms of TiO_2 triggered acute cytotoxic and genotoxic harm than pure forms of TiO_2 NPs [110]. These results contradict the studies that are carried out in animals. The plausible reason is that cells were subjected to pristine TiO_2 in serum-free media used in cell culture but exposed to coated TiO_2 when carried out in animal study [98]. This inconsistency may consequently be due to alterations in the biological fate and toxicity of TiO_2 NPs with different interfacial properties. Literature reveals that presence and absence of serum in cell culture media affects the absorption and toxicity of NPs [111, 112].

10.8.2 Environment

There are basically three routes through which nanomaterials can be exposed to human and environment (Figure 10.3):

a. Leaching of NPs from packaging material into the food;
b. Consumption of nano-food;
c. Disposal of nano food packaging into landfills from which it is released into air, water, and soil.

The increased utilization of NPs will inevitably result in risk to the environment (air, water, and soil). Nanomaterial-based packaging materials are disposed into the environment in form of landfills or residual left from incineration. As a result of physiological and biological factors, the packaging materials ultimately gets decomposed into the environment with release

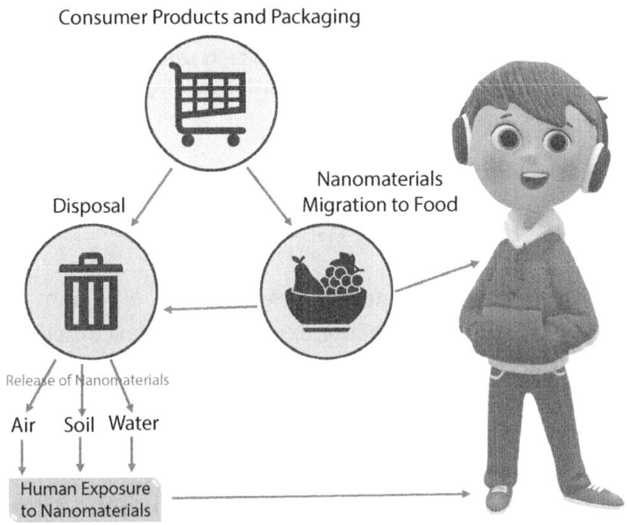

Figure 10.3 Potential routes of exposure of nano-enabled food packaging material to humans and environment.

of nanomaterials, thereby affecting the terrestrial and aquatic system [113]. NPs undergo various transformation during their life cycle and the transformed material differs from the initial one in terms of hazard, transport, and fate in environment [114]. Example: nano graphene oxide decreases soil enzyme activity and was found to be ecotoxic to phytoplankton. Once the packaging material is disposed, the polymeric chain starts degrading in presence extracellular enzymes. The enzymatic activity release oligomers which are assimilated by micro-organisms and metabolized. It has been reported that, when silver NPs are released into the environment, they undergo various transformations such as formation of silver chloride and silver disulfide through the process of aggregation and agglomeration. Due to release of silver ions in aqueous medium, it interacts with microbial target sites causing structural and metabolic disruption [115]. Dwivedi and co-workers reported that small size of Ag NPs increases the process of dissolution, which helps it to persist in the environment for longer period of time [116]. NPs of TiO_2 produces enhanced ecotoxicity due to formation of ROS under UV radiation [117]. CNTs are extremely toxic to aquatic animals, causing respiratory damage. NPs can be removed from water columns and settle in sea sediments bed and eventually may get accumulated in aquatic organisms. The fate and distribution of nanomaterials in the environment is not fully explored and has raised concerns whether they bioaccumulate in the trophic levels of food chain. Literature survey has revealed that majority of work has been carried out on toxicological effects of nanomaterials in human health but scarce survey has been reported for effects on environment.

10.9 Challenge and Future Prospect

The use of engineered nanomaterials has increased exponentially in food sector. New nanomaterials are being explored and developed for usage in packaging material but utmost importance needs to be given to study toxicological impact on human well-being and environment before its usage in consumer market. With potential benefits of nanotechnology in food sector, emphasis on safety aspects of nanomaterials should be laid as it depends on physiochemical factors such as size, chemical composition, surface area, solubility instability in different environments, duration of exposure, concentration, agglomeration, accumulation, metabolism, biotransformation, and media in which nanomaterials are exposed. Therefore, it is of utmost importance to recognize the potential toxicity associated with NPs and efforts should be directed to evaluate toxicity all nanomaterials under a common domain. These efforts have given rise to new branch called nanotoxicology. Many organic polymers are being exploited for packaging but permeable nature of food packaging is the main drawback associated with it. Presently, there has not been a packaging material available that can offer complete resistance to atmospheric gases, and water vapor. The desirable properties required for effective packaging can be attained by blending polymers with NPs or complex multilayer film is generally preferred and can solve problems associated with it.

Last few years have witnessed considerable research in the area of toxic effects of NPs employed in packaging applications. Studies have led into the insights of adverse effects of NPs as well as contradictory results have also been established. The reasons can be attributed to the following: (a) different physiochemical properties of NPs have different impact on biological system and (b) different test methods were established to measure toxicity such as cell culture, microbial, animal, and human feeding studies. (c) Moreover, methods do vary from one laboratory to another, thus making it difficult to compare the obtained results (d) fate and through the environment. Therefore, it is important to standardize the methods to obtain reproducible and transport results under realistic conditions. Studies need to be carried out to mitigate risk of nanotechnology in complex environments and migration of NM needs to be carried out in realistic environment. The NM when discharged into the environment undergoes various transformations which can affect the toxicological profile of NM. Hence, the inadequate knowledge on dynamics of the environment over entire food supply chain can produce uncertainty toward the life cycle assessment (LCA) of NM. All these above factors will help scientist to mitigate sustainable and environmental development of nanomaterials for food industry.

10.10 Conclusion

The significant demand of consumer for longer shelf life of products with safer packaging with improved food safety and quality needs to be satisfied in emerging technologies of nanomaterials in food packaging. The nano-enabled packaging protects the taste, color, and flavor texture a by improving the mechanical barrier and antimicrobial activity of packaging material. It also aids in extending the shelf life; improves food color, taste, flavor, and freshness; and reduces packaging wastes. A plethora of nanomaterials like clay, silver, TiO_2, CNTs, starch, silver, and ZnO have been used as functional additives in packaging. In the current scenario, it is imperative to assess the credible risk associated with nanomaterials on environment and human health. With potential rise of nanomaterials being exploited in food sector, the strict regulations need to be carried out for probable migration of NPs from food packaging along with food stimulants at various experimental conditions. Complete life cycle of nanomaterials containing packaging need to be clarified. There is lack of correlation between the toxicity studies carried out *in vitro* and *in vivo* for different NMs. Thus, it is essential to create the toxicity hub of libraries in order to predict its impact of toxicity of engineered nanomaterials.

References

1. Duncan, T.V., Applications of nanotechnology in food packaging and food safety: Barrier materials, antimicrobials and sensors. *J. Colloid Interface Sci.*, 363, 1, 1–24, 2011.
2. Jildeh, N.B. and Matouq, M., Nanotechnology in packing materials for food and drug stuff opportunities. *J. Environ. Chem. Eng.*, 8, 104338, 2020.
3. Mihindukulasuriya, S. and Lim, L.-T., Nanotechnology development in food packaging: A review. *Trends Food Sci. Technol.*, 40, 2, 149–167, 2014.

4. Emamhadi, M.A., Sarafraz, M., Akbari, M., Fakhri, Y., Linh, N.T.T., Khaneghah, A.M., Nanomaterials for food packaging applications: A systematic review. *Food Chem. Toxicol.*, 146, 111825, 2020.
5. Enescu, D., Cerqueira, M.A., Fucinos, P., Pastrana, L.M., Recent advances and challenges on applications of nanotechnology in food packaging. A literature review. *Food Chem. Toxicol.*, 134, 110814, 2019.
6. Vanderroost, M., Ragaert, P., Devlieghere, F., De Meulenaer, B., Intelligent food packaging: The next generation. *Trends Food Sci. Technol.*, 39, 1, 47–62, 2014.
7. Din, M.I., Ghaffar, T., Najeeb, J., Hussain, Z., Khalid, R., Zahid, H., Potential perspectives of biodegradable plastics for food packaging application-review of properties and recent developments. *Food Additives & Contaminants: Part A*, 37, 4, 665–680, 2020.
8. Yu, Z., Wang, W., Kong, F., Lin, M., Mustapha, A., Cellulose nanofibril/silver nanoparticle composite as an active food packaging system and its toxicity to human colon cells. *Int. J. Biol. Macromol.*, 129, 887–894, 2019.
9. Sharma, R., Jafari, S.M., Sharma, S., Antimicrobial bio-nanocomposites and their potential applications in food packaging. *Food Control*, 112, 107086, 2020.
10. Chaudhary, P., Fatima, F., Kumar, A., Relevance of nanomaterials in food packaging and its advanced future prospects. *J. Inorg. Organomet. Polym. Mater.*, 30, 12, 5180–5192, 2020.
11. Arora, A. and Padua, G., Nanocomposites in food packaging. *J. Food Sci.*, 75, 1, R43–R49, 2010.
12. Silvestre, C., Duraccio, D., Cimmino, S., Food packaging based on polymer nanomaterials. *Prog. Polym. Sci.*, 36, 12, 1766–1782, 2011.
13. Youssef, A.M. and El-Sayed, S.M., Bionanocomposites materials for food packaging applications: Concepts and future outlook. *Carbohydr. Polym.*, 193, 19–27, 2018.
14. Wang, H., Qian, J., Ding, F., Emerging chitosan-based films for food packaging applications. *J. Agric. Food Chem.*, 66, 2, 395–413, 2018.
15. Wang, Y., Zhang, Q., Zhang, C.-l., Li, P., Characterisation and cooperative antimicrobial properties of chitosan/nano-ZnO composite nanofibrous membranes. *Food Chem.*, 132, 1, 419–427, 2012.
16. Rezić, I., Haramina, T., Rezić, T., Metal NPs and carbon nanotubes—Perfect antimicrobial nano-fillers in polymer-based food packaging materials, in: *Food packaging*, pp. 497–532, Elsevier, Academic press, USA, 2017.
17. Garcia, C.V., Shin, G.H., Kim, J.T., Metal oxide-based nanocomposites in food packaging: Applications, migration, and regulations. *Trends Food Sci. Technol.*, 82, 21–31, 2018.
18. Hoffmann, T., Peters, D., Angioletti, B., Bertoli, S., Péres, L., Reiter, M., De Souza, C., Potentials nanocomposites in food packaging. *Chem. Eng. Trans.*, 75, 253–258, 2019.

19. Adeyeye, S.A.O., Food packaging and nanotechnology: Safeguarding consumer health and safety. *Nutr. Food Sci.*, 49, 1164–1179, 2019.
20. Eleftheriadou, M., Pyrgiotakis, G., Demokritou, P., Nanotechnology to the rescue: using nano-enabled approaches in microbiological food safety and quality. *Curr. Opin. Biotechnol.*, 44, 87–93, 2017.
21. Bumbudsanpharoke, N. and Ko, S., Nano-Food Packaging: An overview of market, migration research, and safety regulations. *J. Food Sci.*, 80, 5, R910–R923, 2015.
22. Paul, S.K., Sarkar, S., Sethi, L.N., Ghosh, S.K., Application of Nanotechnology in Food Processing: A Promising Prospective with Considerable Risk, in: *Conference Proceeding*.
23. Rejeski, D. and Lekas, D., Nanotechnology field observations: Scouting the new industrial west. *J. Cleaner Prod.*, 16, 8–9, 1014–1017, 2008.
24. Lin, Q.-B., Li, B., Song, H., Wu, H.-J., Determination of silver in nano-plastic food packaging by microwave digestion coupled with inductively coupled plasma atomic emission spectrometry or inductively coupled plasma mass spectrometry. *Food Addit. Contam.: Part A*, 28, 8, 1123–1128, 2011.
25. Gontard, N., Peyron, S., Lagaron, J.M., Echegoyen, Y., Guillaume, C., Nanotechnologies for active and intelligent food packaging: Opportunities and risks, in: *Nanotechnology in Agriculture and Food Science*, pp. 177–196, 2017.
26. Nakazato, G., Kobayashi, R.K., Seabra, A.B., Duran, N., Use of NPs as a potential antimicrobial for food packaging, in: *Food preservation*, pp. 413–447, Elsevier, Academic press, USA, 2017.
27. Sekhon, B.S., Food nanotechnology–an overview. *Nanotechnol. Sci. Appl.*, 3, 1, 2010.
28. Duran, N. and Marcato, P.D., Nanobiotechnology perspectives. Role of nanotechnology in the food industry: A review. *Int. J. Food Sci. Technol.*, 48, 6, 1127–1134, 2013.
29. Farmer, N., *Trends in packaging of food, beverages and other fast-moving consumer goods (FMCG): Markets, materials and technologies*, Elsevier, Woodhead publishing, Cambridge, 2013.
30. Tsuzuki, T., *Nanotechnology commercialization*, CRC Press, 2013.
31. Chaudhry, Q. and Castle, L., Food applications of nanotechnologies: An overview of opportunities and challenges for developing countries. *Trends Food Sci. Technol.*, 22, 11, 595–603, CRC press, USA, 2011.
32. Ramani, S.V., *Nanotechnology and Development: What's in it for Emerging Countries?*, Cambridge University Press, University of Cambridge, 2014.
33. Duan, Y., Liu, J., Ma, L., Li, N., Liu, H., Wang, J., Zheng, L., Liu, C., Wang, X., Zhao, X.J.B., Toxicological characteristics of nanoparticulate anatase titanium dioxide in mice. *Biomaterials*, 31, 5, 894–899, 2010.
34. Sherman, L.M., Sub-Micron Additives Make Strides(Just Don't Say' Nano'). *Plast. Technol.*, 58, 7, 26–31, 2012.

35. Kuorwel, K.K., Cran, M.J., Orbell, J.D., Buddhadasa, S., Bigger, S.W., Review of mechanical properties, migration, and potential applications in active food packaging systems containing nanoclays and nanosilver. *Compr. Rev. Food Sci. Food Saf.*, 14, 4, 411–430, 2015.
36. Echegoyen, Y. and Nerín, C., Nanoparticle release from nano-silver antimicrobial food containers. *Food Chem. Toxicol.*, 62, 16–22, 2013.
37. Bouwmeester, H., Brandhoff, P., Marvin, H.J., Weigel, S., Peters, R.J., State of the safety assessment and current use of nanomaterials in food and food production. *Trends Food Sci. Technol.*, 40, 2, 200–210, 2014.
38. Khan, A., Wen, Y., Huq, T., Ni, Y., Cellulosic nanomaterials in food and nutraceutical applications: A review. *J. Agric. Food Chem.*, 66, 1, 8–19, 2018.
39. Han, C., Zhao, A., Varughese, E., Sahle-Demessie, E., Evaluating weathering of food packaging polyethylene-nano-clay composites: Release of NPs and their impacts. *NanoImpact*, 9, 61–71, 2018.
40. Dudkiewicz, A., Tiede, K., Loeschner, K., Jensen, L.H.S., Jensen, E., Wierzbicki, R., Boxall, A.B., Molhave, K., Characterization of nanomaterials in food by electron microscopy. *TrAC Trends Analyt. Chem.*, 30, 1, 28–43, 2011.
41. Wrona, M. and Nerín, C., Analytical approaches for analysis of safety of modern food packaging: A review. *Molecules*, 25, 3, 752, 2020.
42. Szakal, C., Roberts, S.M., Westerhoff, P., Bartholomaeus, A., Buck, N., Illuminato, I., Canady, R., Rogers, M., Measurement of nanomaterials in foods: Integrative consideration of challenges and future prospects. *ACS Nano*, 8, 4, 3128–3135, 2014.
43. Wrona, M., Cran, M.J., Nerín, C., Bigger, S.W., Development and characterisation of HPMC films containing PLA NPs loaded with green tea extract for food packaging applications. *Carbohydr. Polym.*, 156, 108–117, 2017.
44. Honarvar, Z., Hadian, Z., Mashayekh, M., Nanocomposites in food packaging applications and their risk assessment for health. *Electron. Physician*, 8, 6, 2531, 2016.
45. Istiqola, A. and Syafiuddin, A., A review of silver NPs in food packaging technologies: Regulation, methods, properties, migration, and future challenges. *J. Chin. Chem. Soc.*, 67, 11, 1942–1956, 2020.
46. Lazić, V., Vivod, V., Peršin, Z., Stoiljković, M., Ratnayake, I.S., Ahrenkiel, P.S., Nedeljković, J.M., Kokol, V., Dextran-coated silver NPs for improved barrier and controlled antimicrobial properties of nanocellulose films used in food packaging. *Food Packag. Shelf Life*, 26, 100575, 2020.
47. Ricardo, A.I.C., García, S.A., Bernardo, F.J.G., Ríos, Á., Martín-Doimeadios, R.C.R., Rapid assessment of silver nanoparticle migration from food containers into food simulants using a qualitative method. *Food Chem.*, 361, 130091, 2021.
48. Artiaga, G., Ramos, K., Ramos, L., Cámara, C., Gómez-Gómez, M., Migration and characterisation of nanosilver from food containers by AF4-ICP-MS. *Food Chem.*, 166, 76–85, 2015.

49. Cushen, M., Kerry, J., Morris, M., Cruz-Romero, M., Cummins, E., Evaluation and simulation of silver and copper nanoparticle migration from polyethylene nanocomposites to food and an associated exposure assessment. *J. Agric. Food Chem.*, 62, 6, 1403–1411, 2014.
50. Motlagh, N.V., Aghazamani, J., Gholami, R., Investigating the effect of nano-silver contained packaging on the olivier salad shelf-life. *BioNanoScience*, 1–10, 2021.
51. Lin, Q.-B., Li, H., Zhong, H.-N., Zhao, Q., Xiao, D.-H., Wang, Z.-W., Migration of Ti from nano-TiO2-polyethylene composite packaging into food simulants. *Food Addit. Contam.: Part A*, 31, 7, 1284–1290, 2014.
52. Bott, J., Störmer, A., Franz, R., Migration of NPs from plastic packaging materials containing carbon black into foodstuffs. *Food Addit. Contam.: Part A*, 31, 10, 1769–1782, 2014.
53. Tang, X. and Alavi, S., Recent advances in starch, polyvinyl alcohol based polymer blends, nanocomposites and their biodegradability. *Carbohydr. Polym.*, 85, 1, 7–16, 2011.
54. Hafttananian, N., Khajavi, M.Z., Farhoodi, M., Jahanbin, K., Pure, A.E., Migration of nano-clay and nano-silica from low-density polyethylene nanocomposites into different food simulants. *J. Food Meas. Charact.*, 1–8, 2021.
55. Metak, A.M., Nabhani, F., Connolly, S.N., Migration of engineered NPs from packaging into food products. *LWT-Food Sci. Technol.*, 64, 2, 781–787, 2015.
56. Bott, J., Störmer, A., Franz, R., A model study into the migration potential of NPs from plastics nanocomposites for food contact. *Food Packag. Shelf Life*, 2, 2, 73–80, 2014.
57. EFSA Panel on Food Contact Material, E., Flavourings and Aids, P., Statement on the safety assessment of the substance silicon dioxide, silanated, FCM Substance No 87 for use in food contact materials. *EFSA J.*, 12, 6, 3712, 2014.
58. Maisanaba, S., Pichardo, S., Puerto, M., Gutierrez-Praena, D., Camean, A.M., Jos, A., Toxicological evaluation of clay minerals and derived nanocomposites: A review. *Environ. Res.*, 138, 233–254, 2015.
59. Farhoodi, M., Nanocomposite materials for food packaging applications: characterization and safety evaluation. *Food Eng. Rev.*, 8, 1, 35–51, 2016.
60. Oberdörster, G., Maynard, A., Donaldson, K., Castranova, V., Fitzpatrick, J., Ausman, K., Carter, J., Karn, B., Kreyling, W., Lai, D., Principles for characterizing the potential human health effects from exposure to nanomaterials: Elements of a screening strategy. *Part. Fibre Toxicol.*, 2, 1, 1–35, 2005.
61. Tinkle, S.S., Antonini, J.M., Rich, B.A., Roberts, J.R., Salmen, R., DePree, K., Adkins, E.J., Skin as a route of exposure and sensitization in chronic beryllium disease. *Environ. Health Perspect.*, 111, 9, 1202–1208, 2003.
62. Kreilgaard, M., Influence of microemulsions on cutaneous drug delivery. *Adv. Drug Delivery Rev.*, 54, S77–S98, 2002.
63. Nel, A., Xia, T., Mädler, L., Li, N., Toxic potential of materials at the nanolevel. *Science*, 311, 5761, 622–627, 2006.

64. Oecd, *Guidance on sample preparation and dosimetry for the safety testing of manufactured nanomaterials*, OECD Paris, France, 2012.
65. Maisanaba, S., Prieto, A.I., Pichardo, S., Jorda-Beneyto, M., Aucejo, S., Jos, A., Cytotoxicity and mutagenicity assessment of organomodified clays potentially used in food packaging. *Toxicol. Vitro*, 29, 6, 1222–1230, 2015.
66. Souza, V.G.L. and Fernando, A.L., NPs in food packaging: Biodegradability and potential migration to food—A review. *Food Packag. Shelf Life*, 8, 63–70, 2016.
67. de Azeredo, H.M.C., Mattoso, L.H.C., McHugh, T.H., Nanocomposites in food packaging–a review, in: *Advances in diverse industrial applications of nanocomposites*, pp. 57–78, 2011.
68. Meneghetti, P. and Qutubuddin, S., Synthesis, thermal properties and applications of polymer-clay nanocomposites. *Thermochim. Acta*, 442, 1–2, 74–77, 2006.
69. Carretero, M., Gomes, C., Tateo, F., Clays and human health, in: *Developments in clay science*, vol. 1, pp. 717–741, 2006.
70. Sharma, A.K., Schmidt, B., Frandsen, H., Jacobsen, N.R., Larsen, E.H., Binderup, M.-L., Genotoxicity of unmodified and organo-modified montmorillonite. *Mutat. Res./Genet. Toxicol. Environ. Mutagen.*, 700, 1–2, 18–25, 2010.
71. Maisanaba, S., Puerto, M., Pichardo, S., Jordá, M., Moreno, F.J., Aucejo, S., Jos, Á., *In vitro* toxicological assessment of clays for their use in food packaging applications. *Food Chem. Toxicol.*, 57, 266–275, 2013.
72. Pugliara, A., Makasheva, K., Despax, B., Bayle, M., Carles, R., Benzo, P., BenAssayag, G., Pécassou, B., Sancho, M.C., Navarro, E., Assessing bio-available silver released from silver NPs embedded in silica layers using the green algae Chlamydomonas reinhardtii as bio-sensors. *Sci. Total Environ.*, 565, 863–871, 2016.
73. Mackevica, A., Olsson, M.E., Hansen, S.F., Silver nanoparticle release from commercially available plastic food containers into food simulants. *J. Nanopart. Res.*, 18, 1, 5, 2016.
74. Loza, K., Sengstock, C., Chernousova, S., Köller, M., Epple, M., The predominant species of ionic silver in biological media is colloidally dispersed nanoparticulate silver chloride. *RSC Adv.*, 4, 67, 35290–35297, 2014.
75. Cha, K., Hong, H.-W., Choi, Y.-G., Lee, M.J., Park, J.H., Chae, H.-K., Ryu, G., Myung, H., Comparison of acute responses of mice livers to short-term exposure to nano-sized or micro-sized silver particles. *Biotechnol. Lett.*, 30, 11, 1893–1899, 2008.
76. Kim, Y.S., Song, M.Y., Park, J.D., Song, K.S., Ryu, H.R., Chung, Y.H., Chang, H.K., Lee, J.H., Oh, K.H., Kelman, B.J., Subchronic oral toxicity of silver NPs. *Part. Fibre Toxicol.*, 7, 1, 1–11, 2010.
77. Shahare, B., Yashpal, M., Gajendra, Toxic effects of repeated oral exposure of silver NPs on small intestine mucosa of mice. *Toxicol. Mech. Methods*, 23, 3, 161–167, 2013.

78. Jeong, G.N., Jo, U.B., Ryu, H.Y., Kim, Y.S., Song, K.S., Yu, I.J., Histochemical study of intestinal mucins after administration of silver NPs in Sprague–Dawley rats. *Arch. Toxicol.*, 84, 1, 63–69, 2010.
79. Park, E.-J., Bae, E., Yi, J., Kim, Y., Choi, K., Lee, S.H., Yoon, J., Lee, B.C., Park, K., Repeated-dose toxicity and inflammatory responses in mice by oral administration of silver NPs. *Environ. Toxicol. Pharmacol.*, 30, 2, 162–168, 2010.
80. van der Zande, M., Vandebriel, R.J., Van Doren, E., Kramer, E., Herrera Rivera, Z., Serrano-Rojero, C.S., Gremmer, E.R., Mast, J., Peters, R.J., Hollman, P.C., Distribution, elimination, and toxicity of silver NPs and silver ions in rats after 28-day oral exposure. *ACS Nano*, 6, 8, 7427–7442, 2012.
81. Loeschner, K., Hadrup, N., Qvortrup, K., Larsen, A., Gao, X., Vogel, U., Mortensen, A., Lam, H.R., Larsen, E.H., Distribution of silver in rats following 28 days of repeated oral exposure to silver NPs or silver acetate. *Part. Fibre Toxicol.*, 8, 1, 1–14, 2011.
82. Kim, Y.S., Kim, J.S., Cho, H.S., Rha, D.S., Kim, J.M., Park, J.D., Choi, B.S., Lim, R., Chang, H.K., Chung, Y.H., Twenty-eight-day oral toxicity, genotoxicity, and gender-related tissue distribution of silver NPs in Sprague-Dawley rats. *Inhalation Toxicol.*, 20, 6, 575–583, 2008.
83. Gaillet, S. and Rouanet, J.-M., Silver NPs: their potential toxic effects after oral exposure and underlying mechanisms–a review. *Food Chem. Toxicol.*, 77, 58–63, 2015.
84. Williams, K., Milner, J., Boudreau, M.D., Gokulan, K., Cerniglia, C.E., Khare, S., Effects of subchronic exposure of silver NPs on intestinal microbiota and gut-associated immune responses in the ileum of Sprague-Dawley rats. *Nanotoxicology*, 9, 3, 279–289, 2015.
85. Yang, Y., Faust, J.J., Schoepf, J., Hristovski, K., Capco, D.G., Herckes, P., Westerhoff, P., Survey of food-grade silica dioxide nanomaterial occurrence, characterization, human gut impacts and fate across its lifecycle. *Sci. Total Environ.*, 565, 902–912, 2016.
86. Yang, Y.X., Song, Z.M., Cheng, B., Xiang, K., Chen, X.X., Liu, J.H., Cao, A., Wang, Y., Liu, Y., Wang, H., Evaluation of the toxicity of food additive silica NPs on gastrointestinal cells. *J. Appl. Toxicol.*, 34, 4, 424–435, 2014.
87. Limbach, L.K., Wick, P., Manser, P., Grass, R.N., Bruinink, A., Stark, W.J., Exposure of engineered NPs to human lung epithelial cells: Influence of chemical composition and catalytic activity on oxidative stress. *Environ. Sci. Technol.*, 41, 11, 4158–4163, 2007.
88. Schübbe, S., Schumann, C., Cavelius, C., Koch, M., Müller, T., Kraegeloh, A., Size-dependent localization and quantitative evaluation of the intracellular migration of silica NPs in Caco-2 cells. *Chem. Mater.*, 24, 5, 914–923, 2012.
89. Tarantini, A., Lanceleur, R., Mourot, A., Lavault, M.-T., Casterou, G., Jarry, G., Hogeveen, K., Fessard, V., Toxicity, genotoxicity and proinflammatory effects of amorphous nanosilica in the human intestinal Caco-2 cell line. *Toxicol. Vitro*, 29, 2, 398–407, 2015.

90. van Kesteren, P.C., Cubadda, F., Bouwmeester, H., van Eijkeren, J.C., Dekkers, S., de Jong, W.H., Oomen, A.G., Novel insights into the risk assessment of the nanomaterial synthetic amorphous silica, additive E551, in food. *Nanotoxicology*, 9, 4, 442–452, 2015.
91. Lee, J.-A., Kim, M.-K., Paek, H.-J., Kim, Y.-R., Kim, M.-K., Lee, J.-K., Jeong, J., Choi, S.-J., Tissue distribution and excretion kinetics of orally administered silica NPs in rats. *Int. J. Nanomed.*, 9, Suppl 2, 251, 2014.
92. Kim, Y.-R., Lee, S.-Y., Lee, E.J., Park, S.H., Seong, N.-w., Seo, H.-S., Shin, S.-S., Kim, S.-J., Meang, E.-H., Park, M.-K., Toxicity of colloidal silica NPs administered orally for 90 days in rats. *Int. J. Nanomed.*, 9, Suppl 2, 67, 2014.
93. So, S.J., Jang, I.S., Han, C.S., Effect of micro/nano silica particle feeding for mice. *J. Nanosci. Nanotechnol.*, 8, 10, 5367–5371, 2008.
94. Wang, Y., Yuan, L., Yao, C., Ding, L., Li, C., Fang, J., Sui, K., Liu, Y., Wu, M., A combined toxicity study of zinc oxide NPs and vitamin C in food additives. *Nanoscale*, 6, 24, 15333–15342, 2014.
95. Bacchetta, R., Moschini, E., Santo, N., Fascio, U., Del Giacco, L., Freddi, S., Camatini, M., Mantecca, P., Evidence and uptake routes for Zinc oxide NPs through the gastrointestinal barrier in Xenopus laevis. *Nanotoxicology*, 8, 7, 728–744, 2014.
96. Wang, J., Zhou, G., Chen, C., Yu, H., Wang, T., Ma, Y., Jia, G., Gao, Y., Li, B., Sun, J., Acute toxicity and biodistribution of different sized titanium dioxide particles in mice after oral administration. *Toxicol. Lett.*, 168, 2, 176–185, 2007.
97. Duan, Y., Liu, J., Ma, L., Li, N., Liu, H., Wang, J., Zheng, L., Liu, C., Wang, X., Zhao, X., Toxicological characteristics of nanoparticulate anatase titanium dioxide in mice. *Biomaterials*, 31, 5, 894–899, 2010.
98. Cho, W.-S., Kang, B.-C., Lee, J.K., Jeong, J., Che, J.-H., Seok, S.H., Comparative absorption, distribution, and excretion of titanium dioxide and zinc oxide NPs after repeated oral administration. *Part. Fibre Toxicol.*, 10, 1, 1–9, 2013.
99. Wang, Y., Chen, Z., Ba, T., Pu, J., Chen, T., Song, Y., Gu, Y., Qian, Q., Xu, Y., Xiang, K., Susceptibility of young and adult rats to the oral toxicity of titanium dioxide NPs. *Small*, 9, 9-10, 1742–1752, 2013.
100. Botelho, M.C., Costa, C., Silva, S., Costa, S., Dhawan, A., Oliveira, P.A., Teixeira, J.P., Effects of titanium dioxide NPs in human gastric epithelial cells *in vitro*. *Biomed. Pharmacother.*, 68, 1, 59–64, 2014.
101. Guo, Z., Martucci, N.J., Moreno-Olivas, F., Tako, E., Mahler, G.J., Titanium dioxide nanoparticle ingestion alters nutrient absorption in an *in vitro* model of the small intestine. *NanoImpact*, 5, 70–82, 2017.
102. Wang, F., Gao, F., Lan, M., Yuan, H., Huang, Y., Liu, J., Oxidative stress contributes to silica nanoparticle-induced cytotoxicity in human embryonic kidney cells. *Toxicol. Vitro*, 23, 5, 808–815, 2009.
103. Yun, J.W., Kim, S.H., You, J.R., Kim, W.H., Jang, J.J., Min, S.K., Kim, H.C., Chung, D.H., Jeong, J., Kang, B.C., Comparative toxicity of silicon dioxide,

silver and iron oxide NPs after repeated oral administration to rats. *J. Appl. Toxicol.*, 35, 6, 681–693, 2015.
104. Sirelkhatim, A., Mahmud, S., Seeni, A., Kaus, N.H.M., Ann, L.C., Bakhori, S.K.M., Hasan, H., Mohamad, D., Review on zinc oxide NPs: Antibacterial activity and toxicity mechanism. *Nano-Micro Lett.*, 7, 3, 219–242, 2015.
105. EFSA Panel on Food Contact Materials, E., Flavourings and Aids, P., Safety assessment of the substance zinc oxide, NPs, for use in food contact materials. *EFSA J.*, 14, 3, 4408, 2016.
106. Chen, P., Wang, H., He, M., Chen, B., Yang, B., Hu, B., Size-dependent cytotoxicity study of ZnO NPs in HepG2 cells. *Ecotoxicol. Environ. Saf.*, 171, 337–346, 2019.
107. Kang, T., Guan, R., Song, Y., Lyu, F., Ye, X., Jiang, H., Cytotoxicity of zinc oxide NPs and silver NPs in human epithelial colorectal adenocarcinoma cells. *LWT-Food Sci. Technol.*, 60, 2, 1143–1148, 2015.
108. Shi, H., Magaye, R., Castranova, V., Zhao, J., Titanium dioxide NPs: A review of current toxicological data. *Part. Fibre Toxicol.*, 10, 1, 1–33, 2013.
109. Nogueira, C.M., de Azevedo, W.M., Dagli, M.L.Z., Toma, S.H., de Arruda Leite, A.Z., Lordello, M.L., Nishitokukado, I., Ortiz-Agostinho, C.L., Duarte, M.I.S., Ferreira, M.A., Titanium dioxide induced inflammation in the small intestine. *World J. Gastroenterol.: WJG*, 18, 34, 4729, 2012.
110. Gerloff, K., Fenoglio, I., Carella, E., Kolling, J., Albrecht, C., Boots, A.W., Förster, I., Schins, R.P., Distinctive toxicity of TiO2 rutile/anatase mixed phase NPs on Caco-2 cells. *Chem. Res. Toxicol.*, 25, 3, 646–655, 2012.
111. Monopoli, M.P., Walczyk, D., Campbell, A., Elia, G., Lynch, I., Baldelli Bombelli, F., Dawson, K.A., Physical- chemical aspects of protein corona: Relevance to *in vitro* and *in vivo* biological impacts of NPs. *J. Am. Chem. Soc.*, 133, 8, 2525–2534, 2011.
112. Lesniak, A., Salvati, A., Santos-Martinez, M.J., Radomski, M.W., Dawson, K.A., Åberg, C., Nanoparticle adhesion to the cell membrane and its effect on nanoparticle uptake efficiency. *J. Am. Chem. Soc.*, 135, 4, 1438–1444, 2013.
113. Caballero-Guzman, A., Sun, T., Nowack, B., Flows of engineered nanomaterials through the recycling process in Switzerland. *Waste Manage.*, 36, 33–43, 2015.
114. Mitrano, D.M., Motellier, S., Clavaguera, S., Nowack, B., Review of nanomaterial aging and transformations through the life cycle of nano-enhanced products. *Environ. Int.*, 77, 132–147, 2015.
115. Hartemann, P., Hoet, P., Proykova, A., Fernandes, T., Baun, A., De Jong, W., Filser, J., Hensten, A., Kneuer, C., Maillard, J.-Y., Nanosilver: Safety, health and environmental effects and role in antimicrobial resistance. *Mater. Today*, 18, 3, 122–123, 2015.
116. Dwivedi, A.D., Dubey, S.P., Sillanpää, M., Kwon, Y.-N., Lee, C., Varma, R.S., Fate of engineered NPs: Implications in the environment. *Coord. Chem. Rev.*, 287, 64–78, 2015.

117. Schaumann, G.E., Philippe, A., Bundschuh, M., Metreveli, G., Klitzke, S., Rakcheev, D., Grün, A., Kumahor, S.K., Kühn, M., Baumann, T., Understanding the fate and biological effects of Ag-and TiO2-NPs in the environment: The quest for advanced analytics and interdisciplinary concepts. *Sci. Total Environ.*, 535, 3–19, 2015.

11

Recent Advances in Micro- and Nanoencapsulation of Bioactive Compounds and Their Food Applications

Mehdi Taib[1], Fouad Damiri[2*], Yahya Bachra[2], Mohammed Berrada[2] and Lahboub Bouyazza[1]

[1]Laboratory of Applied Chemistry and Environment, Hassan 1st University Faculty of Science and Technology, Settat, Morocco
[2]University Hassan II of Casablanca, Faculty of Sciences Ben M'sick, Department of Chemistry, Laboratory of Biomolecules and Organic Synthesis (BIOSYNTHO), Casablanca, Morocco

Abstract
Bioactive compounds have numerous health benefits, but several drawbacks such as the impact on organoleptic properties, instability, and susceptibility to oxidative degradation limit their use in food. Furthermore, fast release, poor solubility, and bioavailability of bioactive compounds greatly restrict their use in food products. Nanoencapsulation is considered to be a promising method for improving bioactive compound dispersion in food, providing protection against degradation, enhance functionality and nutritional value, improve physicochemical functionalities, and control delivery or release. This chapter presents recent advances in nanoencapsulation of bioactive compounds and their food applications. Furthermore, the principles, advantages, and disadvantages of each technique for encapsulation are discussed.

***Keywords*:** Bioactive compounds, microencapsulation, nanoencapsulation, food application

**Corresponding author*: fouad.damiri@outlook.fr

Annu, Tanima Bhattacharya and Shakeel Ahmed (eds.) Nanotechnology in Intelligent Food Packaging, (271–290) © 2022 Scrivener Publishing LLC

11.1 Introduction

The awareness by consumers of the correlation between food and human health has stimulated the trend toward healthy and function-oriented foods. The preparation of healthful foods requires the use of appropriate bioactive ingredients in the food chain to enhance performance characteristics. Bioactive compounds from natural sources, such as anthocyanins, alkaloids, phenolic acids, peptides, flavonoids, carotenoids, fat-soluble, vitamins, and phytosterols, have received considerable attention in the latest decade, because of their functional properties. Nevertheless, these applications in the industry are affected by problems such as poor solubility, physicochemical instability, low bioavailability, sensitivity to environmental factors, and unfavorable flavor of the final product [1–3]. Additionally, the interaction with other food components, their limited shelf-life, and difficulties in transportation and storage all exacerbate the need to stabilize these compounds [4]. As a result, encapsulation is considered a feasible alternative for preserving the quality of bioactive compounds and also enhancing their application in food manufacturing industries.

Nanoencapsulation is a potential approach for protecting and preserving bioactive compounds versus harsh environmental factors, increasing their stability, bioavailability, and targeted release and hence enhancing their use in food processing industries. Based on its size, particles produced by encapsulation have been classified into (i) nanoencapsulation (<1 μm), microencapsulation (1 μm to 5 mm), and (iii) macroencapsulation (>5 mm) [5, 6]. The particle size factor of the bioactive compound directly impacts the bioactive molecule's distribution within a body [7]. Nanocapsules provide a larger surface area that increases solubility compared to microcapsules [8, 7]. The nanoencapsulation method is chosen according to the physicochemical properties of the core and wall materials, as well as the proposed use of the food components. This chapter, therefore, aims to highlight the encapsulation techniques in nanotechnology, the main carrier agents and bioactive ingredients encapsulated, and their application in the food industry.

11.2 The Importance of Encapsulating Bioactive Compounds in Food Science

Nanoencapsulation is the technique whereby base materials fortified with food ingredients or bioactive compounds are loaded into a polymer matrix or "wall material" to form capsules. In the food industry, encapsulation technology can be used for several reasons. Among the greatest benefits of nanotechnology for food products are as follows [7, 9]: 1) improves solubility, bioavailability, and controlled release; 2) mask undesirable flavors; 3) enhance biological activity (antioxidants, antimicrobials, probiotics, etc.), stability, and shelf life; 4) enhance thermal and mechanical properties; 5) protect from gastric and intestinal conditions; and 6) protect against adverse reactions, such as light, oxygen, pH changes, moisture, heat, lipid oxidation, volatile loss, shear, or other extreme conditions. Therefore, encapsulation is important for both protecting and improving the functionality of bioactive substances (Figure 11.1).

According to their size, the particles produced by encapsulation can be classified as follows: (i) nanoencapsulation (<1 μm), (ii) microencapsulation (1 μm to 5 mm), and (iii) macroencapsulation (>5 mm) [5]. The particle size factor of the bioactive compound directly impacts the bioactive molecule's distribution within a body [7, 10, 11]. Nanocapsules offer a larger surface that increases solubility in comparison to microcapsules [7, 8]. Nanocapsules contribute more to sustained drug release and bioavailability, allowing for more precise targeting of active compounds [7]. The difference between microencapsulation and nanoencapsulation is shown in (Figure 11.2).

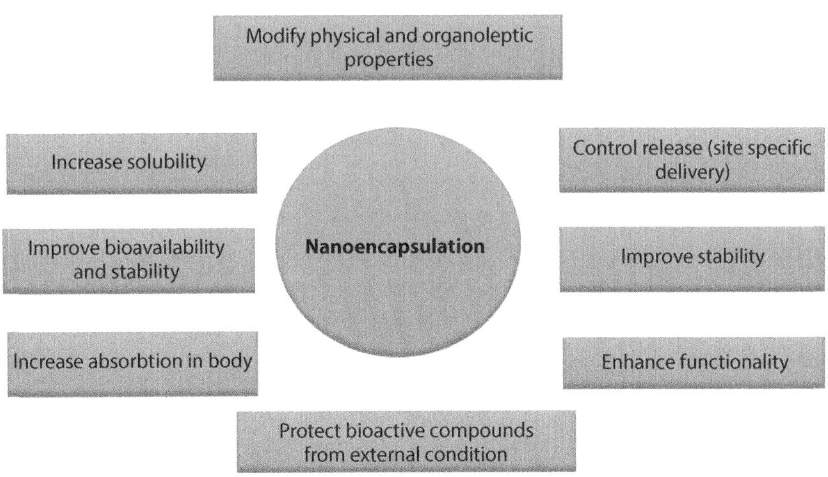

Figure 11.1 The interesting aspect of nanoencapsulation to improve their applicability.

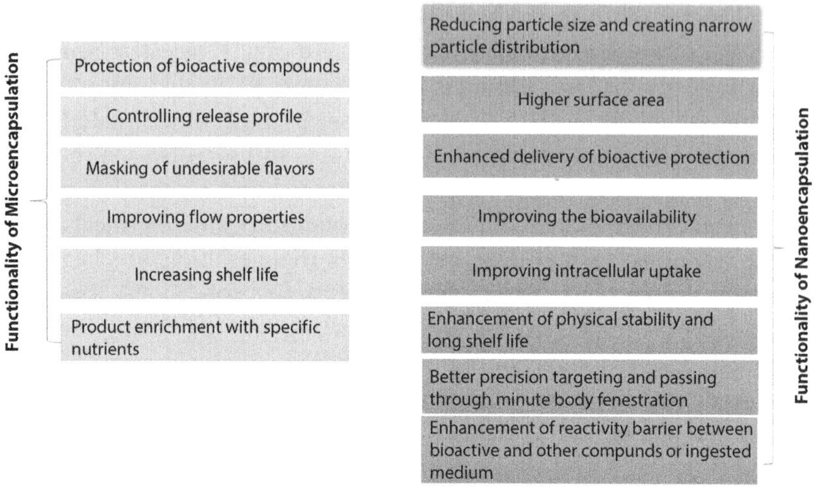

Figure 11.2 The difference between microencapsulation and nanoencapsulation.

11.3 Materials Utilized in Micro and Nanoencapsulation

Among the most important factors of a successful encapsulation process is selecting the proper encapsulating agent. Several parameters should be evaluated before designing an encapsulation system. From a health perspective, encapsulation agents should be regarded as "generally accepted as safe" (GRAS) materials for foods [12]. In an aspect of functionality, encapsulation agents should be inert with respect to the active ingredients, easily soluble, viscous at high concentration, good emulsifier, and compatible with processing conditions and core substances. It should be capable of protecting bioactive compounds against oxygen and humidity. Furthermore, encapsulating agents should be able to resist gastric and intestinal conditions. They should be able to improve the residence time of bioactive compounds within the gastrointestinal tract [13–15]. Furthermore, the availability, ease of production, raw material costs, and regulatory status must be considered for selecting encapsulant materials [7, 12]. However, some biopolymers can be thought of as the principal source of appropriate materials for encapsulating bioactive compounds (Table 11.1).

Table 11.1 Commonly used materials for encapsulation of food applications [7, 9, 16, 17].

Source	Carbohydrate	Protein	Lipid	Polymer
Plant	Pectins Starch Tragacanth gum Cellulose Galactomannans Cyclodextrins Gum Arabic	Gluten (corn) Soy protein, Zein, Gliadin Barley protein Potato protein Amaranth proteins Pulse proteins	Fatty acids Phospholipids Glycerides Waxes	
Marine	Alginate Carrageenan			
Microbial/ animal	Chitosan Gellan Xanthan Dextran	Collagen Gelatin Whey proteins Ovalbumin Caseins		
Synthetic				Polyvinyl alcohol Polycaprolactone Polyglycolides Polyacrylic acid Polyacrylamide

11.4 Nano and Microencapsulation Techniques

The choice of encapsulation technique is essential in the preparation of nano- or microstructured encapsulated systems. From a classical viewpoint, two different types of methods are used to create encapsulate systems "top-down" and "bottom-up" [18]. The method bottom-up (top-down) consists of particle size reduction of large particles into smaller particles with external disturbance forces. The top-down approach encompasses emulsification,

emulsification-solvent evaporation, and extrusion techniques. They can encapsulate both hydrophobic and hydrophilic compounds, but they require sophisticated equipment [18, 19]. These techniques have limited influence on particle size and structure and are only appropriate for specific matrix types [18, 19]. Differently, the bottom-up method involves the development of these large particles by combining small particles through self-organization and self-assembly of the particles. Techniques such as electrospinning, spray drying, supercritical-fluids, nanoprecipitation, and inclusion complexation coacervation are considered bottom-up methods [19]. These technologies use less energy and provide enhanced control of particle size, distribution, and morphology compared to top-down methods [18]. Moreover, in some cases, both approaches can be used to achieve the best results. Table 11.2 outlines the principles, advantages, and disadvantages of a few of the most commonly used technologies for the encapsulation of bioactive food ingredients [7, 12, 20, 21].

Table 11.2 Some methods widely used for encapsulating food bioactive.

Method	Principles	Advantages/limitations
Spray drying	Atomization of bio-actives and biopolymers solutions or suspensions into dry powder using an injector with a hot drying gas stream.	1. Rapid, simple, and relatively inexpensive process. 2. Organic solvents are avoided by using water-based dispersions. 3. Use of relatively high drying temperature. 4. Large particle sizes.
Coacervation	Combination of two phases by electrostatic forces creating a precipitate.	1. The simplest approach to make NPs from biopolymers. 2. The complexity of controlling particle size. 3. The biopolymer used affects the NPs' stability.

(Continued)

Table 11.2 Some methods widely used for encapsulating food bioactive. (*Continued*)

Method	Principles	Advantages/limitations
Extrusion	Feeding a biopolymer solution through a nozzle to form a gelled environment.	1. Encapsulation of a wide range of compounds. 2. Large particle size. 3. It is only suitable for materials with a limited matrix.
Anti-solvent precipitation	The solute precipitation is accomplished via the addition of a nonsolvent.	1. It is a straightforward and low-cost method that does not necessitate the use of any specialist equipment. 2. The usage of organic solvents is required.
Inclusion complexation	A guest molecule (bioactive) is encapsulated by a host (polymer) through physical and chemical forces.	1. Widely used for concealing odors or flavors as well as preserving aromas
Fluid gel formation	Use of shear forces on a solution of biopolymer during the gelation process.	1. The resulting particles are easily distorted and have an uneven shape. 2. It is necessary to carefully control the parameters.

(*Continued*)

Table 11.2 Some methods widely used for encapsulating food bioactive. (*Continued*)

Method	Principles	Advantages/limitations
Electro-spinning	Electro-spinning is a process of preparing nanoparticles with a high voltage potential.	1. The protocol and processing are simple. 2. The particles obtained are small in size. 3. The final product is relatively easy to control in size and shape. 4. Many proteins cannot be used alone in electro-spinning.
Layer-by-layer deposition	Depositing alternating layers of oppositely charged materials around a charged template.	1. It is inexpensive, easily adaptable, and does not require any sophisticated equipment.
Solid dispersion	The active ingredient is dispersed in a hydrophilic inert carrier or matrix.	1. Improve the bioavailability of poorly water-soluble drugs.

11.5 Application to Nanoencapsulation for Producing Bioactive Food Ingredients

Bioactive compounds consist of phenolic compounds, carotenoids, lipids, proteins, vitamins, essential oils, fatty acids, peptides, probiotics, enzymes, and alkaloids. The integration of bioactive substances in food could play a key role in enhancing consumer health and well-being [2]. Nevertheless, their applications in industry are affected by problems such as low solubility, physicochemical instability, loss of biological activity, low bioavailability sensitivity to environmental conditions, and undesirable organoleptic properties of the resulting product [2, 22]. In this sense, nanoencapsulation, by protecting bioactive substances from environmental aggressions, can considerably improve their physicochemical properties as well as promote their functional activity (Table 11.3).

Table 11.3 Nanoencapsulated bioactive ingredients for food products.

BACs	Nanocarriers	Technique	Food product	Dose used	Main effects	Ref.
Ascorbic acid	Gelatin and gum arabic	Complex coacervation/ freeze-drying	Emulsified sausages	30% (w/w)	Improved the thermal stability and provided a gradually release	[8]
Astaxanthin	Polysorbate 20, sodium caseinate, GA: gum Arabic	Nanodispersions	Orange juice, skimmed milk	5, 10, 50 or 100 mg/L	Improved storage stability,	[41]
	oligochitosan/ Zein	Phase separation continued by lyophilization/ magnetic agitation	antioxidant in food systems	1.2 mg/ml	Increased UV-light and storage stability	[42]
Chlorogenic acids	β-cyclodextrin	Nanocomplexation	Bread, cookies, or cheese	237.7 mg/g 237.7 mg/g	Limited formation of preferred aroma volatiles	[28]
Clove EO (CEO)/ Eugenol	Chitosan	Ionic gelation	Pomegranate aril coated with Ch film	-	- Increased shelf life - Fungal decay was delayed	[36]

(*Continued*)

Table 11.3 Nanoencapsulated bioactive ingredients for food products. (*Continued*)

BACs	Nanocarriers	Technique	Food product	Dose used	Main effects	Ref.
Carotenoid extract (high in -carotene	Gelatin	Nanoemulsion	Yogurt	-	Increased dispersibility, high storage stability	[42]
Curcumin	Amaranth protein and pullulan	Electrospinning	*In vitro* digestion process	0.05% - 0.075%	Improved the thermal stability and antioxidant activity	[43]
Lycopene	Tween 20	Emulsification–evaporation	Model beverage	0.015 to 0.085 mg/ml	Improved storage stability	[44]
Fish oils	Lecithin sunflower oil	Nanoliposomes	Bread	5% (w/w)	Improved oxidative stability and sensory properties	[45]
	Sodium caseinate/gum arabic	Layer-by-layer deposition	Fruit juice	40, 50, and 60 mg/L	Acceptable taste scores, High bioaccessibility	[46]

(*Continued*)

Table 11.3 Nanoencapsulated bioactive ingredients for food products. (*Continued*)

BACs	Nanocarriers	Technique	Food product	Dose used	Main effects	Ref.
Olive leaf phenolicsoleuropein	Lecithin cholesterol	Nanoliposomes Ethanol injection	Yogurt	0.7% phenolics/g nanoliposome	Improved antioxidant activity, sensory properties and syneresis	[26]
Quercetin	Capryol PGMC oils or Capmul MCM NF, and Cremophor RH40 as surfactant	Emulsion	Food matrice		Increased antioxidative and antibacterial activities	[47]
Thymus capitatus EO	Aqueous SDS/ soybean oil	Nanoemulsion microfluidizer	Milk	2 mg/ml	Improved physicochemical and microbial (S. *aureus*) quality	[48]
Vitamin D_3	Potato Proteins	Protein nanocomplexes	Beverages	10, 25, 50, 75, and 100 μg/ml	Increased high storage and stability	[37]
Vitamin E	Mustard oil and surfactant Tween 80	Nanoemulsion	Fruit juice	2.0% w/w	Increased shelf life	[39]

11.5.1 Phenolic Compounds

Polyphenols are compounds naturally found in plant foods. These compounds are well-recognized for their antioxidant proprieties and therapeutic effects, including anti-inflammatory effects and the prevention of neurodegenerative diseases, cardiovascular diseases, and type 2 diabetes [23, 24]. The encapsulation of these compounds with nanotechnology has been suggested to enhance the functional properties of foods [9]. Human *et al.* [12] nano encapsulated aspalathin using both synthetic and natural polymers by electrospraying method. The preeminent results were achieved with the use of synthetic polymers and the electrospraying method [advanced encapsulation efficiencies (>50%) and loading capacities (>10%)]. On the contrary, natural polymers showed higher efficiency with the conventional method. In addition, resveratrol was encapsulated in trimethyl chitosan NPs cross-linked with tripolyphosphate and/or alginate to reach improved cellular uptake and controlled release [25]. The developed encapsulation revealed effective controlled release, better distribution, and higher particle size (575 nm) and EE (56.9). Tavakoli *et al.* [26] studied the encapsulation of oleuropein with a considerable EE by the lecithin-cholesterol system (70.7%–88.2%) [26]. The incorporation of oleuropine nanoliposomes in yogurt reduces the syneresis rate, improves the antioxidant activity, and contributes to good sensory properties. Furthermore, Bamaba *et al.* [24] successfully encapsulated total phenolics and blueberry pomace anthocyanins by double emulsion stabilized with whey proteins. Particularly, the impacts of critical physical parameters (time, homogenization pressure, and stirring speed) on the properties of double emulsions were investigated. The results revealed that size dispersity, particle size, ZP, and EE significantly influence the stability and functional performance of nanoemulsions as well as the final products in which they are used. A novel synergistic formulation was formed using low molecular weight chitosan, based on the combination of linalool, thymol, and methyl cinnamates [1]. The nanoencapsulated formulation, with an EE of 35%–40% and a loading efficiency of 6%–7%, demonstrated outstanding antifungal and anti-aflatoxin B1 protection *in vitro* and a food system (Pennisetum glaucum) without compromising sensory properties. Guan *et al.* [27] utilized sucrose fatty acid ester to encapsulate caffeic acid phenethyl ester to improve physiological properties and anti-cancer activities. Additionally, chlorogenic acids were successfully incorporated into β-cyclodextrin to supplement the aroma volatile profile of products such as bread, nutty filling, cookies, and meat stuffing or mushroom [28]. In the same way, Curcumin (Cur) has also been encapsulated in NPs utilizing natural polymers to enhance its solubility and bioavailability [12].

Sassi *et al.* studied the encapsulation of polyphenols obtained from Phoenix dactylifera L pits by gum arabic and egg yolk protein (EYP) to enhance their bioaccessibility and their bioavailability [29]. It was found that encapsulation efficiency was greater when EJP represented a high proportion of the wall material. In addition, the *in vitro* gastrointestinal simulation study revealed that a relatively higher content of EJP increased the resistance of microparticles to intestinal and gastric conditions. In addition, aloe vera skin extract (AVE) was also encapsulated by using polyethylene oxide solutions applying the electrospinning technique [30]. The best results were found with the use of 5 wt% AVE content, showing the possibility of its promising use in active food packaging.

11.5.2 Carotenoids

Carotenoids are important components usually existing in florae, algae, and photosynthetic bacteria. These compounds are interesting functional ingredients in the human diet that also can be applied in processed foods as natural colorants. Apart from their nutritional value, they are also regarded as valuable health-promotional compounds. Nevertheless, their application is still not completely effective due to some weaknesses (susceptibly to environmental and processing stress, low solubility, and bioavailability). In this context, nanoencapsulation has been used to overcome these constraints. Recently, lycopene gained considerable attention due to its potential health benefits in chronic disease prevention. Lycopene NPs were produced and approved as GRAS and are currently utilized as a food additive. They can be combined in baking products and in soft drinks to promote health benefits [31]. Carotenoids obtained from tomato peel were encapsulated in zein NPs to improve their antioxidant capacity and stability using the electrospinning method, displaying the potential of their promising use in food products. In addition, Bezerra *et al.* [32] used pectin as encapsulating agent for the development of nanodispersions of carotenoid extract from Spirulina spp. The nanoencapsulated formulation was characterized for its stability, antioxidant capacity, physicochemical features, and retention of carotenoids when stored at 4°C for 60 days, showing the possibility of their introduction into several food formulations.

11.5.3 Essential Oils

Essential oils are one of the most interesting natural products that have been considered as alternatives to antibiotics [33, 34]. They are extensively used in various countries as medicine, perfumes, cosmetics, and food preservatives. Nevertheless, their use is still ineffective due to various drawbacks (poor solubility, bioavailability, volatility, and stability in the foods). The advancement made in nanotechnology lets one address these obstacles for use of EOs as preservatives. Wu *et al.* [35] studied the encapsulation of carvacrol and thymol Eos using zein NPs. The obtained Eos NPs reduced E. coli counts by 0.8 to 1.8 log CFU/ml. Likewise, depending on the formulation, antioxidant activity was reduced within a range of 24.8% to 66.8%. The encapsulation of clove essential oil (CEO) using chitosan and emulsion/ionic gelation technique was also reported [36]. In comparison to free oil, the nanoencapsulated CEO had a greater effect on *Aspergillus Niger*. Thus, CEO-chitosan NPs could be a promising natural fungicide for agricultural and food applications.

11.5.4 Vitamins

Vitamins can be also protected with nanoencapsulation against harsh environmental conditions that could otherwise affect their integrity during food processing. As an example, vitamin D_3 (VD) nonencapsulated using potato proteins instead of natural material [37]. The formed NPs had better results concerning the samples' shelf life with reduced loss of vitamins. Recently, Santos *et al.* encapsulated vitamin D_3 using carboxymethyl tara gum-lactoferrin complex coacervates as carriers [38]. The results showed that encapsulating vitamin D3 enhanced its stability and supported the controlled release in gastrointestinal simulation *in vitro*. In addition, Dasgupta *et al.* [39] used low energy approach to produced vitamin E nanoemulsions. These vitamin E loaded-nanoemulsions have shown improved antimicrobial and antioxidant activity and might be deployed to improve the storage stability of fruit juice. Furthermore, vitamin C was embedded in casein gel using a spray dryer in order to improve its stability in vitamin gummies [40]. The finding revealed that encapsulating vitamin C enhanced its stability in the gummy under severer environments and accelerated storage conditions.

11.6 Conclusion

Encapsulation represents a great successful approach to enhance the dispersion of bioactive compounds in food. Nanoencapsulation presents various advantages over microencapsulation regarding improved protection, bioavailability, prolonged-release profile, and increased stability of bioactive compounds. The successful application of nanoencapsulation relies on the selection of carriers with the required physicochemical properties to protect the target bioactive compound and proper encapsulation techniques. In the future, the parameters relating to the physicochemical functions of the encapsulated material need to be improved for both encapsulation methods. It should be noted that there is still limited data on the toxicological effects of nanomaterials, and thus, they should be verified before being commercialized. More comprehensive studies on the behavior of the entrapped bio-actives in foods should be performed. In addition, the interaction between the encapsulation material and the ingredient should be also considered.

References

1. Kumar, A., Kujur, A., Singh, P.P., Prakash, B., Nanoencapsulated plant-based bioactive formulation against food-borne molds and aflatoxin B1 contamination: Preparation, characterization and stability evaluation in the food system. *Food Chem.*, 287, 139–150, 2019, https://doi.org/10.1016/j.foodchem.2019.02.045.
2. Akbarbaglu, Z., Peighambardoust, S.H., Sarabandi, K., Jafari, S.M., Spray drying encapsulation of bioactive compounds within protein-based carriers; different options and applications. *Food Chem.*, 359, 129965, https://doi.org/10.1016/j.foodchem.2021.129965, 2021.
3. Damiri, F., Bachra, Y., Bounacir, C., Laaraibi, A., Berrada, M., Synthesis and characterization of Lyophilized chitosan-based hydrogels cross-linked with benzaldehyde for controlled drug release. *J. Chem.*, 2020, Article ID 8747639, 10 pages, 2020, https://doi.org/10.1155/2020/8747639.
4. Arpagaus, C., Collenberg, A., Rütti, D., Assadpour, E., Jafari, S.M., Nano spray drying for encapsulation of pharmaceuticals. *Int. J. Pharm.*, 546, 194–214, https://doi.org/10.1016/j.ijpharm.2018.05.037, 2018.
5. Fang, Z. and Bhandari, B., Encapsulation techniques for food ingredient systems, in: *Food Mater. Sci. Eng.*, pp. 320–348, Wiley-Blackwell, Oxford, UK, 2012, https://doi.org/10.1002/9781118373903.ch12.
6. Damiri Fouad, G.A., Yahya Bachra, N.K., Ouaket, A., Bennamara, A., Berrada, M., A novel drug delivery system based on nanoparticles of magnetite Fe3O4 embedded in an auto cross-linked chitosan, in: *Chitin*, http://dx.doi.org/10.5772/intechopen.94873.
7. Shishir, M.R.I., Xie, L., Sun, C., Zheng, X., Chen, W., Advances in micro and nano-encapsulation of bioactive compounds using biopolymer and lipid-based transporters. *Trends Food Sci. Technol.*, 78, 34–60, https://doi.org/10.1016/j.tifs.2018.05.018, 2018.

8. Rodrigues da Cruz, M.C., Andreotti Dagostin, J.L., Perussello, C.A., Masson, M.L., Assessment of physicochemical characteristics, thermal stability and release profile of ascorbic acid microcapsules obtained by complex coacervation. *Food Hydrocolloids*, 87, 71–82, 2019, https://doi.org/10.1016/j.foodhyd.2018.07.043.
9. Pateiro, M., Gómez, B., Munekata, P.E.S., Barba, F.J., Putnik, P., Kovačević, D.B., Lorenzo, J.M., Nanoencapsulation of promising bioactive compounds to improve their absorption, stability, functionality and the appearance of the Final Food Products. *Molecules*, 26, 1547, 2021, https://doi.org/10.3390/molecules26061547.
10. Yamashita, C., Chung, M.M.S., dos Santos, C., Mayer, C.R.M., Moraes, I.C.F., Branco, I.G., Microencapsulation of an anthocyanin-rich blackberry (Rubus spp.) by-product extract by freeze-drying. *LWT*, 84, 256–262, 2017, https://doi.org/10.1016/j.lwt.2017.05.063.
11. Fioramonti, S.A., Rubiolo, A.C., Santiago, L.G., Characterisation of freeze-dried flaxseed oil microcapsules obtained by multilayer emulsions. *Powder Technol.*, 319, 238–244, 2017, https://doi.org/10.1016/j.powtec.2017.06.052.
12. Abd El-Salam, Mohamed H. et El-Shibiniy, Safinaz, Natural biopolymers as nanocarriers for bioactive ingredients used in food industries, in: *Encapsulations*, pp. 793–829, Elsevier, United States, 2016, https://doi.org/10.1016/B978-0-12-804307-3.00019-3.
13. Jain, A., Thakur, D., Ghoshal, G., Katare, O.P., Shivhare, U.S., Characterization of microcapsulated β-carotene formed by complex coacervation using casein and gum tragacanth. *Int. J. Biol. Macromol.*, 87, 101–113, 2016, https://doi.org/10.1016/j.ijbiomac.2016.01.117.
14. de Souza Simões, L., Madalena, D.A., Pinheiro, A.C., Teixeira, J.A., Vicente, A.A., Ramos, Ó.L., Micro- and nano bio-based delivery systems for food applications: *In vitro* behavior. *Adv. Colloid Interface Sci.*, 243, 23–45, 2017, https://doi.org/10.1016/j.cis.2017.02.010.
15. Penalva, R., Esparza, I., Agüeros, M., Gonzalez-Navarro, C.J., Gonzalez-Ferrero, C., Irache, J.M., Casein nanoparticles as carriers for the oral delivery of folic acid. *Food Hydrocolloids*, 44, 399–406, 2015, https://doi.org/10.1016/j.foodhyd.2014.10.004.
16. Laaraibi, A., Moughaoui, F., Damiri, F., Ouakit, A., Charhouf, I., Hamdouch, S., Jaafari, A., Abourriche, A., Knouzi, N., Bennamara, A., Berrada, M., Chitosan-clay based (CS-NaBNT) biodegradable nanocomposite films for potential utility in food and environment, in: *Chitin-Chitosan - Myriad Funct. Sci. Technol.*, InTech, London, 2018, https://doi.org/10.5772/intechopen.76498.
17. Zuidam, N.J. and Nedovic, V., (Eds.), *Encapsulation Technologies for Active Food Ingredients and Food Processing*, Springer New York, New York, NY, 2010, https://doi.org/10.1007/978-1-4419-1008-0.

18. Joye, I.J. and McClements, D.J., Biopolymer-based nanoparticles and microparticles: Fabrication, characterization, and application. *Curr. Opin. Colloid Interface Sci.*, 19, 417–427, 2014, https://doi.org/10.1016/j.cocis.2014.07.002.
19. Jia, Z., Dumont, M.-J., Orsat, V., Encapsulation of phenolic compounds present in plants using protein matrices. *Food Biosci.*, 15, 87–104, 2016, https://doi.org/10.1016/j.fbio.2016.05.007.
20. Niu, B., Shao, P., Luo, Y., Sun, P., Recent advances of electrosprayed particles as encapsulation systems of bioactives for food application. *Food Hydrocolloids*, 99, 105376, 2020, https://doi.org/10.1016/j.foodhyd.2019.105376.
21. Jones, O.G. and McClements, D.J., Recent progress in biopolymer nanoparticle and microparticle formation by heat-treating electrostatic protein–polysaccharide complexes. *Adv. Colloid Interface Sci.*, 167, 49–62, 2011, https://doi.org/10.1016/j.cis.2010.10.006.
22. Karami, Z., Peighambardoust, S.H., Hesari, J., Akbari-Adergani, B., Cytotoxic effect proteins hydrolysates from wheat germ on human lung carcinoma A549 cell lines in XTT assay. *Food Sci. Technol.*, 16, 43–51, 2020, https://doi.org/10.29252/fsct.16.96.43.
23. Pereira, M.C., Oliveira, D.A., Hill, L.E., Zambiazi, R.C., Borges, C.D., Vizzotto, M., Mertens-Talcott, S., Talcott, S., Gomes, C.L., Effect of nanoencapsulation using PLGA on antioxidant and antimicrobial activities of guabiroba fruit phenolic extract. *Food Chem.*, 240, 396–404, 2018, https://doi.org/10.1016/j.foodchem.2017.07.144.
24. Bamba, B., Shi, J., Tranchant, C., Xue, S., Forney, C., Lim, L.-T., Xu, W., Xu, G., Coencapsulation of Polyphenols and Anthocyanins from Blueberry Pomace by Double Emulsion Stabilized by Whey Proteins: Effect of Homogenization Parameters. *Molecules*, 23, 2525, 2018, https://doi.org/10.3390/molecules23102525.
25. Bin Min, J., Kim, E.S., Lee, J.-S., Lee, H.G., Preparation, characterization, and cellular uptake of resveratrol-loaded trimethyl chitosan nanoparticles. *Food Sci. Biotechnol.*, 2017, https://doi.org/10.1007/s10068-017-0272-2.
26. Tavakoli, H., Hosseini, O., Jafari, S.M., Katouzian, I., Evaluation of Physicochemical and Antioxidant Properties of Yogurt Enriched by Olive Leaf Phenolics within Nanoliposomes. *J. Agric. Food Chem.*, 66, 35, 9231–9240, 2018, https://doi.org/10.1021/acs.jafc.8b02759.
27. Guan, Y., Chen, H., Zhong, Q., Nanoencapsulation of caffeic acid phenethyl ester in sucrose fatty acid esters to improve activities against cancer cells. *J. Food Eng.*, 246, 125–133, 2019, https://doi.org/10.1016/j.jfoodeng.2018.11.008.
28. Budryn, G., Zaczyńska, D., Oracz, J., Effect of addition of green coffee extract and nanoencapsulated chlorogenic acids on aroma of different food products. *LWT*, 73, 197–204, 2016, https://doi.org/10.1016/j.lwt.2016.06.019.
29. Ben Sassi, C., Marcet, I., Rendueles, M., Díaz, M., Fattouch, S., Egg yolk protein as a novel wall material used together with gum Arabic to encapsulate polyphenols extracted from Phoenix dactylifera L pits. *LWT*, 131, 109778, 2020, https://doi.org/10.1016/j.lwt.2020.109778.

30. Solaberrieta, I., Jiménez, A., Cacciotti, I., Garrigós, M.C., Encapsulation of bioactive compounds from Aloe Vera agrowastes in electrospun poly (ethylene oxide) nanofibers. *Polym. (Basel)*, 12, 1323, 2020, https://doi.org/10.3390/polym12061323.
31. Ponce, A.G., Ayala-Zavala, J.F., Marcovich, N.E., Vázquez, F.J., Ansorena, M.R., Nanotechnology trends in the food industry: Recent developments, risks, and regulation, in: *Impact Nanosci. Food Ind.*, pp. 113–141, Elsevier, United States, 2018, https://doi.org/10.1016/B978-0-12-811441-4.00005-4.
32. Bezerra, P.Q.M., de Matos, M.F.R., Ramos, I.G., Magalhães-Guedes, K.T., Druzian, J.I., Costa, J.A.V., Nunes, I.L., Innovative functional nanodispersion: Combination of carotenoid from Spirulina and yellow passion fruit albedo. *Food Chem.*, 285, 397–405, 2019, https://doi.org/10.1016/j.foodchem.2019.01.181.
33. Taib, M., Bouyazza, L., Lyoussi, B., Acorn Oil: Chemistry and Functionality. *J. Food Qual.*, 2020, 1–11, 2020, https://doi.org/10.1155/2020/8898370.
34. Taib, M., Rezzak, Y., Bouyazza, L., Lyoussi, B., Medicinal uses, phytochemistry, and pharmacological activities of Quercus species, evidence-based complement. *Altern. Med.*, 2020, 1–20, 2020, https://doi.org/10.1155/2020/1920683.
35. Wu, Y., Luo, Y., Wang, Q., Antioxidant and antimicrobial properties of essential oils encapsulated in zein nanoparticles prepared by liquid–liquid dispersion method. *LWT - Food Sci. Technol.*, 48, 283–290, 2012, https://doi.org/10.1016/j.lwt.2012.03.027.
36. Hasheminejad, N., Khodaiyan, F., Safari, M., Improving the antifungal activity of clove essential oil encapsulated by chitosan nanoparticles. *Food Chem.*, 275, 113–122, 2019, https://doi.org/10.1016/j.foodchem.2018.09.085.
37. David, S. and Livney, Y.D., Potato protein based nanovehicles for health promoting hydrophobic bioactives in clear beverages. *Food Hydrocolloids*, 57, 229–235, 2016, https://doi.org/10.1016/j.foodhyd.2016.01.027.
38. Santos, M.B., Geraldo de Carvalho, M., Garcia-Rojas, E.E., Carboxymethyl tara gum-lactoferrin complex coacervates as carriers for vitamin D3: Encapsulation and controlled release. *Food Hydrocolloids*, 112, 106347, 2021, https://doi.org/10.1016/j.foodhyd.2020.106347.
39. Dasgupta, N., Ranjan, S., Mundra, S., Ramalingam, C., Kumar, A., Fabrication of food grade Vitamin E nanoemulsion by low energy approach, characterization and its application. *Int. J. Food Prop.*, 19, 700–708, 2016, https://doi.org/10.1080/10942912.2015.1042587.
40. Yan, B., Davachi, S.M., Ravanfar, R., Dadmohammadi, Y., Deisenroth, T.W., Van Pho, T., Odorisio, P.A., Darji, R.H., Abbaspourrad, A., Improvement of vitamin C stability in vitamin gummies by encapsulation in casein gel. *Food Hydrocolloids*, 113, 106414, 2021, https://doi.org/10.1016/j.foodhyd.2020.106414.
41. Anarjan, N. and Tan, C.P., Chemical stability of astaxanthin nanodispersions in orange juice and skimmed milk as model food systems. *Food Chem.*, 139, 527–531, 2013, https://doi.org/10.1016/j.foodchem.2013.01.012.

42. Jiang, G.-L. and Zhu, M.-J., Preparation of astaxanthin-encapsulated complex with zein and oligochitosan and its application in food processing. *LWT*, 106, 179–185, 2019, https://doi.org/10.1016/j.lwt.2019.02.055.
43. Medeiros, I., de Oliveira, G.L.R., de Queiroz, J.L.C., de Carvalho Gomes, C., de Carvalho, F.M.C., de Souza Lima, M.C.J., Serquiz, A.C., de Andrade Santos, P.P., da Silva Camillo, C., Maciel, B.L.L., de Araújo Morais, A.H., Passos, T.S., Safety and bioactive potential of nanoparticles containing Cantaloupe melon (Cucumis melo L.) carotenoids in an experimental model of chronic inflammation. *Biotechnol. Rep.*, 28, e00567, 2020, https://doi.org/10.1016/j.btre.2020.e00567.
44. Akhoond Zardini, A., Mohebbi, M., Farhoosh, R., Bolurian, S., Production and characterization of nanostructured lipid carriers and solid lipid nanoparticles containing lycopene for food fortification. *J. Food Sci. Technol.*, 55, 287–298, 2018, https://doi.org/10.1007/s13197-017-2937-5.
45. Ojagh, S.M. and Hasani, S., Characteristics and oxidative stability of fish oil nano-liposomes and its application in functional bread. *J. Food Meas. Charact.*, 12, 1084–1092, 2018, https://doi.org/10.1007/s11694-018-9724-5.
46. Ghorbanzade, T., Jafari, S.M., Akhavan, S., Hadavi, R., Nano-encapsulation of fish oil in nano-liposomes and its application in fortification of yogurt. *Food Chem.*, 216, 146–152, 2017, https://doi.org/10.1016/j.foodchem.2016.08.022.
47. Das, S.S., Verma, P.R.P., Singh, S.K., Screening and preparation of quercetin doped nanoemulsion: characterizations, antioxidant and anti-bacterial activities. *LWT*, 124, 109141, 2020, https://doi.org/10.1016/j.lwt.2020.109141.
48. Ben Jemaa, M., Falleh, H., Neves, M.A., Isoda, H., Nakajima, M., Ksouri, R., Quality preservation of deliberately contaminated milk using thyme free and nanoemulsified essential oils. *Food Chem.*, 217, 726–734, 2017, https://doi.org/10.1016/j.foodchem.2016.09.030.

12

Applications of Nanosensors as Pathogen Detectors in Packaged Food

Samka Peregrine Maishu* **and Ngwa Celestine Atemenkeh**

Department of Biochemistry, University of Bamenda, Bamenda, Cameroon

Abstract

Foodborne diseases caused by pathogens, for example, *Listeria monocytogenes*, from processed packaged food are on the rise and claiming millions of lives worldwide. Although traditional and conventional detection mechanisms such as microscopy are being set into motion to ensure the safety of such commodities, their effectiveness remain questioned. In respond to the increasing number of deaths from packaged food, there is a surge in the search for biosensors with specificity for fast and easy detection of such pathogens before processing, after processing, and during packaging. Nanomaterials, for example, noble nanoparticles are topping the charts as biosensors in the detection of most prevailing food pathogens due to their specificity. These nanomaterials are easily conjugated with different affinity ligands, antibodies, as well as a multitude of functional groups, thus improving their sensitivity to simultaneous detection of multiple pathogens and their toxins. Therefore, this review aims at bringing out the traditional and conventional methods of detection of a variety of pathogens and their drawbacks and outlining how these drawbacks are being mitigated using biosensors notably nanobiosensor.

Keywords: Pathogen, foodborne disease, conventional and traditional detection system, biosensor, nanobiosensor

Abbreviation

AgNPs	silver nanoparticles
Chi-AgNPs	chitosan/silver nanocomposite

**Corresponding author*: perino7485@gmail.com

Annu, Tanima Bhattacharya and Shakeel Ahmed (eds.) Nanotechnology in Intelligent Food Packaging, (291–312) © 2022 Scrivener Publishing LLC

EIA	enzyme immunoassay
ELISA	enzyme-linked immunosorbent assay
GNPs	gold nanoparticles
IA	immunological assay
ICG	immunochromatography assay
LOD	limits of detection
LPS	lipopolysaccharides
MNP	magnetic nanoparticles
mPCR	multiplex polymerase chain reaction
PCR	polymerase chain reaction
QCM	quartz crystal microbalance
QDS	quantum dots
QPCR	quantitative polymerase chain reaction
RT-PCR	real-time polymerase chain reaction
VBNC	viable but non-culturable bacteria

12.1 Background

Food is any nutritional valued material typically of plant or/and animal origin that contains constituents such as sugars, lipids, growth factors (proteins), vital minerals, and vitamins that, when ingested by an organism, offer life, energy, development and maintenance [1]. Food, despite its many health benefits, can also act as a vehicle for disease transmission and mortality if infected with dangerous microbes and microbial toxins [2]. Each year, rougly two million people, including children, die as a result of food contamination [3]. Food that is properly cared for, processed, and preserved is protected from ailments and damages, thus leading to the provision of safe and high-quality foods in communities [4]. Foodborne infections are a common and evolving public health concern in both industrialized and developing countries [5], making food safety an increasingly important public health issue. Without being examined and traced, foodborne pathogens can spread exponentially, resulting in outbreaks and pandemics [6]. Several methods, such as food hygiene and packaging, have been adopted around the world to slow the transmission and growth of foodborne illnesses and spoilage organisms; yet, new bacteria appear to be able to overcome these barriers [7]. Early intervention via diagnosis using traditional techniques such as culturing and colony counting, etc., is by far the most successful approach for limiting the transmission of foodborne infections, but these methods are difficult to achieve accurate results due to high costs, specialized process conditions, slow data output, labor intensive, time consuming, complexities, and the need for exhaustive techniques to increase specificity [8, 9]. There is a boom in the quest for easy-to-use, economical approaches that are particular, sensitive, stable, and portable to minimize and overcome the disadvantages of traditional methods of food quality management [10]. Nanotechnologically derived materials have lately offered a variety of material alternatives for improving the mobility, discriminating, and responsiveness of sensors and analytical evaluation methods, procedures for quality control in food industry [10]. These nanomaterials may be easily coupled with various affinity ligands, antibodies, and a variety of functional groups, increasing their sensitivity to detect several infections and their poisons simultaneously in food [11]. In the industry, nanomaterials are now being employed as flavor and shelf-life modifiers, and as food pathogen trackers, tracers, sensors, and detectors [12]. To detect infections and pollutants, nanotechnology-based sensors and coating materials are now being included into packaging materials [13]. Nanostructured packaging can be intelligent in the sense that it can respond to external conditions, restore itself, or warn clients about contaminates and disesases [14]. The distribution of high-quality, safe foods in communities is ensured by promising advances of well-developed smart packaging sensitive enough to identify global health organisms like *Salmonella* and *E. coli* [15].

12.2 Package Foods (Ways of Packaging)

The process of transforming raw food ingredients into marketable foods and other forms with a long shelf life is known as food processing. In addition, food processing also includes removing toxins, preventing sickness, preserving food, and improving food homogeneity for better marketing and delivery. Packaging food materials is a nutrition approach that incorporates many strategies such as protection, tamper resistance, and the construction of a unique physical, chemical, and biological environment for food without compromising its nutritional value [12]. To avoid microbial development and spoilage, food preservation via packing relies on maintaining food in a non-reactive and reduced-oxygen environment, which is why the majority of packaging materials are gas-tight [12, 16]. There are a variety of packaging options available, ranging from rigid containers to mineral oil fractionated polymers that have just been available for use in the food business (Table 12.1). Food packaging materials are chosen based on the nature of the food, that is, its content and its ability to survive microbial attacks [17]. Glass, metals, paper, and plastics are

Table 12.1 Packaging materials and their properties.

Packaging materials	Properties	Application/food type and drawbacks	Reference
Glass	• Ordor free and chemically non-reactive with practically all packaged foods. • Impervious to gases and mist, so product integrity is maintained • Rigid and provides good insulation • Non-porous and relatively cheap • Easy to clean and non-toxic • Non-leaching • Reusable and recyclable	Liquids and sauces, for example, alcoholic beverages, milk products, pickles, jams hetchups, and squashes **Drawbacks** • Susceptibility to breakage • Heavy weight	[17, 18].
Metals	• Aluminum containers are non-toxic and non tainting and are impermeable to oxygen, carbon dioxide, moisture, and other scent chemicals. • Most types of corrosion are resistant to aluminium. Air, temperature, moisture, smell, light, microganism, and chemical attacks are all protected by the metal oxide natural layer. • Aluminum foil used in packaging protects foods from temperature changes as it has thermo-insulation features.	The metal aluminum is utilized in packaging liquid soups, herbs, spices, cooking oil, milk products and tomatoes, as well as in lamination. **Drawback** • Because metals are made from non-renewable resources they should be repurposed in order to maintain the environment in its natural state. • Most metal cans are susceptible to rust especially when stored badly	[17, 18, 22].

(*Continued*)

Table 12.1 Packaging materials and their properties. (*Continued*)

Packaging materials	Properties	Application/food type and drawbacks	Reference
Papers	• Biodegradable • Renewable and recyclable • Ultra-light material • Inexpensive for manufacturing	This material is used in packaging vegetables, snack foods, cookies, fruits, flour, and confectionary. Papers are also used as greaseproof for wrapping candy bars and oily foods. **Drawbacks** • Products not preserved for longer time • No strong barrier against water, odor or sun • Simple to tear material • Its production process encourages deforestation	[17, 18, 24].
Plastics	• Strong material • Long lasting material that is highly durable • Light weight material • Air-tight and recyclable • Keep products in their natural states while acting as a shelf life enhancers.	Dairy products, cooking oils, baking flours, **Drawbacks** • Gases, vapors, light, and low-molecular weight molecules can pass through plastic materials.	[17, 18].

among the traditional materials used in food packaging (Figure 12.1) [18]. Many innovative packaging developments are now being seen, including the use of functional additives such as oxygen scavenging agents, carbon dioxide absorbers and emitters, hygroscopic agents that control moisture/water activity to reduce microbial growth, and antimicrobial agents that reduce surface contamination of processed food [19, 20]. Furthermore, packaging materials are increasingly designed to interact with food to

Figure 12.1 Various packaging materials. (a) Glass. (b) Metal. (c) Plastic [18] original/field work pictures.

provide extended shelf life, increased safety, and improved hygiene [21]. Even though all packaging measures have been implemented in industries, designing a packaging system that has no total environmental impact on packaged food products is still difficult [22]. Many microorganisms have been reported to be able to overcome these barriers, particularly facultative anaerobes and anaerobic microbes, as evidenced by the high prevalence of *Pseudomonas* spp. in vacuum packaged food [23].

12.3 Packaged Food Pathogens

Milk is a highly susceptible food to microbial infection. *Listeria monocytogenes, Escherichia coli O157:H7, Staphylococcus aureus*, and *Salmonella* spp. are among the most prevalent infections connected to milk or dairy based products in advanced economies, according to Jakobsen *et al.* [25]. *Listeria monocytogenes*, for example, has been detected in many outbreaks around the world because of compromised milk and dairy products [26]. According to Jackson *et al.* [27], dairy products are the most common source of listeriosis. According to Wuyts and colleagues [28], human salmonellosis had been second quite often identified Zoonotic disease in the Eurozone. *Salmonella* spp. has also been observed to be transferred to people by eating infected dairy products, particularly unpasteurized or under pasteurized milk and cheeses [29].

Meat eating has been linked to a number of epidemics in current history. According to Chaillou *et al.* [30], the most frequent microbes on both raw and packed beef products were *Carnobacterium* spp., *Brochothrix thermosphacta*, *Leuconostoc*, and *Lactobacillus*. Egan *et al.* [31] highlighted that isolates of *Lactobacillus* in Vacuum packed meat were responsible for producing hydrogen sulphide which led to meat spoilage. According to Chai *et al.* [32], *Clostridium perfringens* is responsible for one-third of livestock-associated foodborne illness notably products from chicken recurrences in the United States of America. In Europe, botulism is a sickness caused by exotoxins generated by the bacterium *Clostridium botulinum*, which can be contracted through eating sausages. Armany *et al.* [33] reported *Staphylococcus aureus*, *Listeria monocytogenes*, and *Escherichia coli* as the most prevalent microbes in minced meat, sausage, luncheon, and basterma precisely in Cairo and Giza (Egypt). In addition, *L. monocytogenes* contamination has been recorded in many outbreaks in the United States, particularly in packaged and ready-to-eat beef items. *L. monocytogene* may live and thrive at chilled temperatures, and it can attain high cell counts during retail and storage [34].

Yeasts have been found in alcoholic beverages, bread, cheese, rice, soy sauce, olives, and red-hot chili peppers. This microbe frequently overrun elavated-sugar or increased salt foods, resulting in the deterioration of these products. This microbe also grow on surfaces of animal products notably cheese and meat, and also low pH fruit drinks are also candidates [7]. They grow slowly and release unpleasant smell and carbon dioxide, which can cause food cans and containers to bulge and burst [35]. Different harmful bacteria species were identified in sherbets and juices sold in Dhaka, according to Khan *et al.* [36]. These included *Proteus* sp., *Enterobacter* sp.,

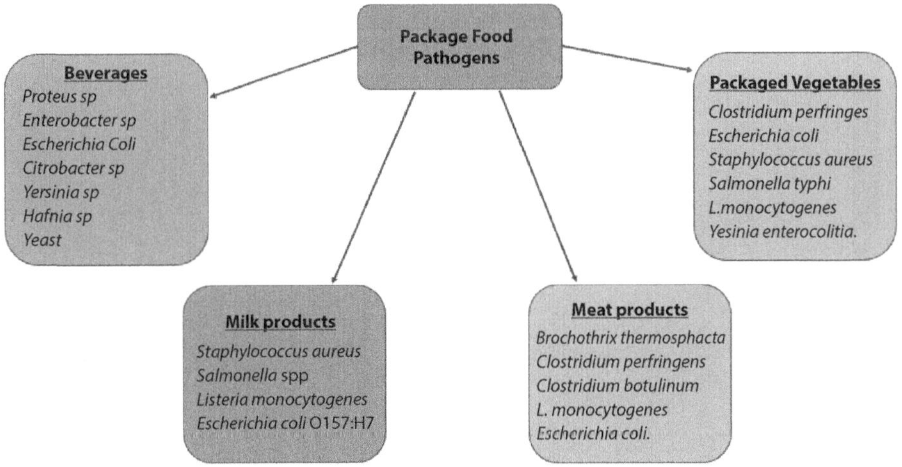

Figure 12.2 Packaged food related pathogens [25, 30, 36, 37].

E. coli, *Shigella* sp., *Citrobacter* sp., *Vibrio* sp., *Yersinia sp.*, and *Hafnia* sp. The beverages have been identified as a potential point of origin of food borne disease transmission.

Vegetables and fruits play a crucial role in digestion especially in humans acting as laxatives and antioxidants. Regardless of the potential health advantages of fresh produce that is vegetables and fruits, they habor microbial pathogens including *Clostridium perfringens*, *E. coli*, *Staphylococcus aureus*, *Salmonella typhi*, *Listeria monocytogenes*, *Yesinia enterocolitia*, *Campylobacter jejuni*, and *Campylobacter coli* at various regions of the globe [37]. When vegetables and fruits are packaged in polythene bags, microbial contamination is said to be decreased. Although, airtight polythene bags have also been reported to hasten deterioration in some situations, especially when a correct concentration of reduced O_2 and elevated CO_2 is not established [38].

12.4 Conventional Detection Techniques for Packaged Food Pathogens

Foodborne illness have now become a major global public health concern. Food safety procedures and detection technology are recommended by public health professionals to avoid the spread of these diseases. Many traditional ways of detecting food-borne infections have been used for decades. Traditional pathogen identification methods include culture and colony

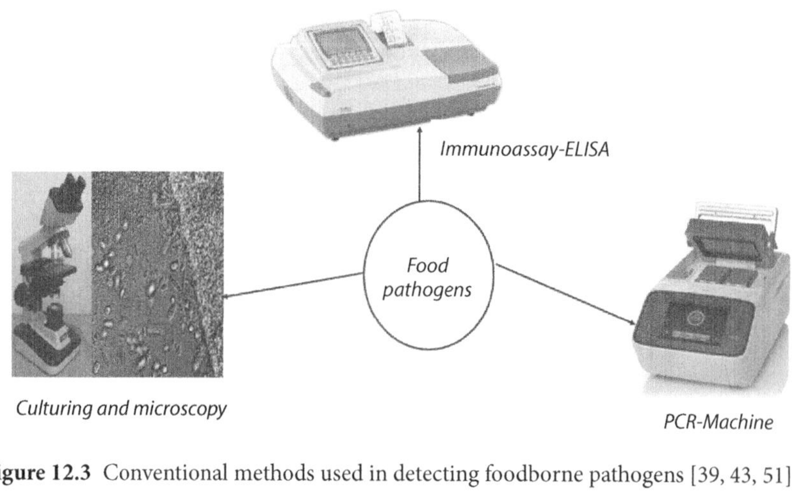

Figure 12.3 Conventional methods used in detecting foodborne pathogens [39, 43, 51].

enumeration, serological/immunological studies, and PCR amplification-based technologies [39]. The aforementioned technologies are based on many detecting characteristics, including sensitivity, speed, selectivity, viable cell discrimination, and *in situ* analytical suitability [40]. These methods provide both quantitative and qualitative data, which is important when dealing with the high prevalence of foodborne diseases.

Collecting and culturing food samples on agar plates, isolating and growing a suspect pathogen, and visual inspection are all attributed to culture and colony counting which is employed to detect bacterial pathogens from foods. This approach is useful for detecting very low quantities of microorganisms; that is single cells because of the unavioidable multiplication that occur during colony expansion [4, 39]. This approach has been used to count enteric bacteria like *E. coli* and *S. typhimurium*, but it has also been demonstrated to count Gram-negative bacteria including *L. monocytogenes*, *B. cereus*, and enterotoxigenic *S aureus.* [40]. Pre-enrichment in buffered peptone water and selective media enrichment, followed by isolation on different media and immunological confirmation respectively, is a basic approach for identifying *Salmonella* in food stuffs according to Ferretti and colleagues [41]. Although this method is regarded as the reference method for pathogen detection, it is indeed labor-intensive, takes a lot of time, and is insensitive, taking up to a week to confirm the specific pathogenic bacteria [4]. Microbial contamination poses a significant risk to this procedure, as it can result in growth inhibition of preferred bacteria including the presence of living (viable) but non-culturable microbes (VBNC) [42, 43].

Immunological assays (IAs) are well known for their ability to detect different pathogens notably viruses and bacteria. In these experimentations, antibodies identify antigens and other molecules particular to the targets [4]. Complete bacterial cells or key cell structures such as lipopolysaccharides or other macro-molecules found in the outer bacterial membrane are being detected using IAs [44]. Flow injection immunoassay, immunochromatography strip test (ICG), enzyme immunoassay (EIA), enzyme-linked immunosorbent assay (ELISA), and immunomagnetic separation are examples of immunological procedures [43]. For example, DynabeadsTM (Dynal Botech, Oslo, Norway) have been found to be effective in isolating a wide range of food microbes, including *L. monocytogenes* and *E. coli* [45–48]. Immunological techniques enable the swift and susceptive identification of a wide spectrum of diseases and toxins, with on-site detection possible. Despite its suitability for quickly preventing the spread of infectious diseases and ensuring food safety, this technology has limitations. Limited sensitivity (susceptibility) and low antibody affinity for the pathogen or interrogating analyte are some of the method's limitations [43]. For example, ELISA has successfully been use to identify complete-cell antigen targets or products (e.g., virulence markers) of microoganisms including *Salmonella* spp., *Escherichia coli O157:H7*, *Campylobacter* spp., *B. cereus*, and *L. monocytogenes* [40].

Armany *et al.* [33] showed that PCR-based approaches may identify foodborne pathogens such as *Staphylococcus aureus*, *Listeria monocytogene*, and *Escherichia coli*. They came to the conclusion that the PCR method of identification was effective, less time consuming, more sensitive, and required less work. The multiplication or amplification and identification of a particular pathogenic organism's unique RNA or DNA sequence is the basis for this pathogen identification method [49]. In-depth PCR method consist of the manipulation of genetic material via denaturation, annealing, and extension/exponential amplification of target DNA. Although amplification enhances detection sensitivity and specificity, it does not provide any information regarding the survivability of identified cells because it cannot differentiate between DNA from dead and alive cells. The PCR viability constraint can be combated by using cell viability dyes in combination with PCR technologies which is now been experimented on [50]. Foodborne pathogens are now identified using PCR-based methods such as simple PCR, multiplex PCR (mPCR), quantitative PCR(qPCR), and real-time PCR (RT-PCR). These approaches are differentiated by their speed and the number of infections detected at a given time. Park and colleagues [51] developed a mPCR technique for detecting *E. coli* O157:H7, *Salmonella spp*, *S. aureus*, and *L. monocytogenes* in a single tube.

12.5 Nanosensors as Pathogen Detectors

Nanotechnology is concerned with substances, structures, and processes that occur on a size of 1 to 100 nm. This technology has the ability to change food safety, particularly in identifying extremely low levels of chemical contaminants and microorganism in food systems [52]. Furthermore, some nanotechnologically manufactured packaging materials can emit antimicrobial agents or hinder water and air interaction with its surroundings, improving the shelf life of food products [53]. Titanium dioxide nanoparticles (NPs) having a size below 100 nm, for example, are commonly used in food processing and preservation containers as a nutritional supplement and an antimicrobial weapon [54]. Biosensors made of platinum and gold nanowires are utilized to improve food analysis [55]. Biosensors built of NPs use cyto-structures like nucleic acids, enzymes, and cells as sensors, which are utilized to monitor biochemical mechanisms or physical alteration [56]. Most of nanotechnology-based sensing techniques use noble metal nanoparticles, like silver and gold. The beneficial optical properties of these nanoparticles enable these applications which can be adjusted by changing their magnitude, morphology, immediate environment, and synthetic technique [57].

Because of their unique features, gold NPs have sparked the most attention as nanobiosensors. These particles, which have a diameter of 1–100 nm and a vast surface area–to-volume ratio and surface energy, can immobilize a significant number of macromolecules while yet allowing them to function. Gold NPs have a unique feature that allows them to be employed as carriers [58], as labels for signal transduction and amplification [59] and as bio-recognition components such as antibodies or aptamers. Futhermore, GNP scan quickly and directly transport electrons between a variety of electroactive species and electrode materials. Due to their light scattering characteristics and unusually substantial augmentation ability of the neighboring electromagnetic field, GNPs can also be used as signal enhancing/amplification tags in a variety of biosensors [56]. Because of these characteristics, gold NPs are great candidates for colorimetric bio-sensing. The colorimetric sensing property of GNPs is founded on the aggregation-induced interparticulate surface plasmon interaction of AuNPs, which culminates in a visible coloration from red to blue (particularly, the absorption-based). This method has resulted in a practical platform for detecting any analyte of interest that triggers AuNP aggregation or re-dispersion [60]. Gold NPs are widely used to functionalize antibodies which does specifically target regions on the membrane of pathogenic organisms. This antibody-antigen interaction causes gold NPs to clump together allover the pathogen of interest, resulting in a colorimetric reaction (Figure 12.4a) [61]. For example, multidrug-resistant *S. Typhimurium* DT104 has been revealed through the aggregation of gold NPs around the bacteria. Because of its long-term survival, the bacterium poses a significant health-care burden. Functionalization of 30 nm popcorn-shaped gold NPs with a monoclonal M3038 antibodies targeting *S. Typhimurium* DT104 allowed for detection. When compared to other Salmonella or *E. coli* strains, the reaction was unique to the drug-resistant *S. Typhimurium* [62]. Transmission electron microscopy measurements of their aggregation suggest that gold NPs associate with lipopolysaccharides on the surface of *E. coli* 055: B5 [63]. Miranda *et al.* [65] developed a hybrid colorimetric enzymatic nanocomposite biosensor for the detection of *E. coli* in aqueous solutions based on enzyme amplification by electrostatic interactions, and an anionic enzyme (-galactosidase) is initially adsorbed

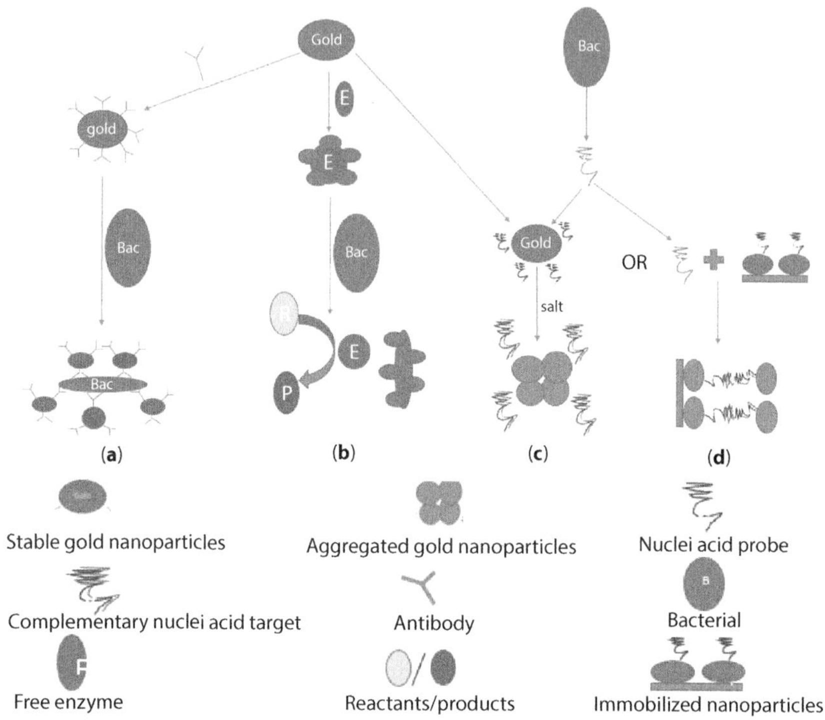

Figure 12.4 The application of gold nanoparticles in pathogen recognition [56, 61, 65].

on the cationic AuNPs with quaternary amine head groups, resulting in enzyme inhibition. In the presence of *E. coli*, gold NPs form around the bacteria and release the enzyme, which aid in the hydrolyzes of chlorophenol red-D-galactopyranoside, thus results in a color shift. This technique was used to measure bacteria at concentrations of 1,102 bacteria/ml in solution and 1,104 bacteria/ml on a field-friendly reagent strip design (Figure 12.4b). Salt-aided aggregation of non-functionalized gold NPs can also be used to identify amplified genetic materials (RNA/DNA) of food pathogens. In the absence of electrostatic repulsion between GNPs, gold NPs will assemble and change its color from red to blue, with the aggregates being stabilized by specific nucleic acids. Impregnation of nucleic acid on the surface or contact with a thiol probe that is hybridized with the target nucleic acids are two basic ways for stabilizing gold NPs (Figure 12.4c) [61]. *Listeria monocytogenes* and *Salmonella enterica* were detected using thiolated probes that targeted the *hly* and *hut* genes, accordingly. Using 13-nm gold NPs, this assay was able to detect bacteria in contaminated milk, and the specificity was validated by the lack of reaction from

E. coli [66]. Jiang and colleagues mounted 17-mer oligonucleotide probes onto GNPs on the gold surface of a QCM (Quartz Crystal Microbalance) in a separate study. Using probes in a hybridization mechanism, the specific DNA was found. The GNPs' vast surface area was considered to aid in the immobilization of additional probes onto the gold surface of the QCM, boosting sensitivity. Jiang's group employed GNPs as amplification tags in a series of DNA QCM bionsensors, and the lowest limit of dection of DNA obtained was 10–16 mol/L (Figure 12.4d) [56].

Due to their optical characteristics, high bio-compatibility, low toxicity, and long-term electrocatalytic activity, silver NPs (AgNps) have also attracted interest [67] and the presence of different types of surface interaction sites on their nanostructure [68]. Basically, AgNps having positively charged surface, ligand replacement ability, and oxidative dissolution capabilities which facilitate their binding to negatively charged surface of bacteria, thus resulting in enhanced bactericidal effects [64, 69] and sensitivity. Using AgNps sensing devices, Gram-negative microbes such as *Escherichia coli*, *Pseudomonas aeruginosa*, and *Yersinia pestis* are efficiently detected [72]. The electrostatic interaction between the positively charged AgNps surface and the negatively charged Lipopolysaccharides on all these Gram-negative bacteria causes aggregate formation, as seen by a color shift. Increased AgNP aggregation due to strong van der Waals forces, on the other hand, could lead to a considerable reduction in electrochemical interaction and detection sensibility [70, 71]. Furthermore, AgNPs functionalization can cause NPs to break down chemically to silver ions and the outer layer of AgNPs to oxidize quickly [72]. To overcome these limitations, Imran *et al.* [73] used positively charged chitosan stabilized AgNPs (Chi-AgNPs) as a sensitive electrochemical nanobiosensor for the detection of LPS, a major endotoxin and component of the outer membrane of numerous Gram-negative bacteria. The identification was done at concentrations ranging from 0.001 to 100 ng/ml and 10–107 CFU/ml.

In sensory tools, silica NPs have also evolved as sensory nanoparticulates. Their distinctive features, including as mechanical stability, optical transparency, ease of manufacturing, and compatibility with biological molecules like proteins and nucleic acids, have led to a variety of uses notably in pathogen detection [74, 75]. Numerous macromolecules and their interaction, including antigen-antibodies, peptides, and DNA, may be linked to silica NPs, giving these nanostructures unique components for bio- and immunosensors [75]. Tuitemwong *et al.* [76] used flourescent dye-doped silica to demonstrate the rapid detection of *E. coli* O157:H7. The dyed NPs binding abilities were improved in the study by conjugating them to anti-*E. coli* and amino groups. In a comparison study to develop a biosensor against viral infections, Enrichi *et al.* [77] looked at the fluorescence activities of QDs, silica NPs, and free dyes separately and found that the dye-doped silica NPs had the best LOD. Free dyes, QDs, and dye-doped SiNPs have limit of detection of 150, 250, and 20 pM, accordingly.

Cerium oxide NPs have also acquired international significance due to their remarkable optical and enzymatic attributes. Their surface interactions with various analytes cause optical changes that may be measured calorimetrically. The optical changes that result from the interaction of cerium oxide nanoparticles with phenolics and hydrogen peroxide, for example, have been used to create handheld colorimetric sensors for the identification of nutritional antioxidant and sugars using nanoparticles as the colorimetric probe [78]. Furthermore, their surface enzymatic capabilities, which include peroxidase, superoxide, and oxidase, suggest that these particles could be used to replace enzymes in the construction of analytical instruments and procedures [79]. Bulbul and colleagues [57] used nanoceria's enzyme mimic properties to develop a mobile enzyme-free electrochemical aptasensor toward detecting dietary contaminants such as mycotoxins.

MNPs have mostly been used in sensing investigations as immobilization aids, as well as in the development of immunoassays separations, magnetically enriched and operated sensing systems [80]. M-NPs come in a variety of forms, including alloys and oxides containing both iron and cobalt (i.e., FePt, $CoPt_3$, Fe_3O_4, Fe_2O_3, and $CoFe_2O_4$). Magnetite (Fe3O4)-based NPs, on the other hand, have been the most commonly

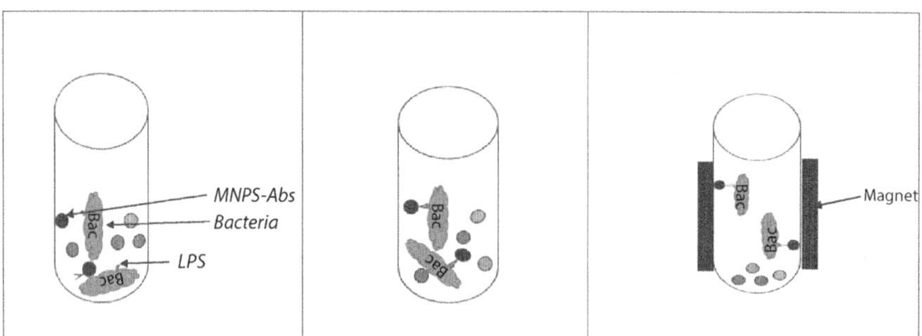

a) Sample matrix containing proteins, bacteria with LPS, lipids, MNPs-Abs etc.

b) MNPs-Abs-Bacteria concentrated, sample matrix that is protein etc.

c) Immuno-magnetic separation. Magnetically marked cells (MNPS-Abs-Bac) separated from sample matrix by magnets.

Figure 12.5 An illustration of pathogenic bacteria identification using immuno-magnetic separation mechanism [82, 83].

used so far. The ease with which these NPs are separated from the matrix in sorbent-based extractions is attributed to their use in pathogen detection. For example, the progress of immunomagnetic separation technologies is based on the simplicity with which these NPs are segregated from the matrix in sorbent-based extractions. These materials are also adjustable, resulting in increased selectivity. Furthermore, their unique electrical, thermal, and optical properties enable the enhancement of sensor signals and the increased sensitivity of analytical equipment [81]. Magnetic NPs are appropriate for direct application in bioassays or as affinity ligands for the capture of target molecules and cells because of their capacity to pair covalently with proteins, enzymes, antibodies, and other ligands [82]. They can be coated with antibodies that target the microorganism's surface proteins or antigens, resulting in magnetically marked cells that can be easily isolated from the specimen matrix. Magnetic NPs have lately been utilized to separate harmful bacteria from food, clinical, and environmental samples, particularly in the detection of pathogens like *Salmonella* and *Staphylococcus aureus* [83].

12.6 Conclusion and Future Perspectives

Packaging has been shown to prevent surface contamination and food disease transmission. Even though packaging and hygiene have been employed to prevent contamination for decades, foodborne pathogens continue to be detected in significant epidemics and pandemics. Early detection using routine traditional approaches such as sensory organs, culturing, immuno assays, and molecular techniques such as polymerase chain reactions remains important avenues for identifying, eliminating, and combating these disease-causing organisms. Even with all of these measures in place to combat food pathogens, global foodborne disease rates continue to rise. The increase suggests that the abovementioned traditional techniques have limitations such as limited sensitivity, low detection limits, not fast enough to detect infections for containment, and inability to distinguish pathogen viability and are complex and labor intensive. Nanomaterials and structures created for diagnosing foodborne pathogens have been found to be less expensive, very specific, and customizable. Noble metallic NPs, such as PtNPs, AgNPs, and AuNPs, for example, have a high stability, are simple to chemically synthesize and have a tuneable surface modification [84]. According to the review, these nanomaterials have also been employed to enrich and increase the specificity, sensitivity, and speed of conventional laboratory techniques without requiring major procedural or infrastructure changes. Giuliana Vinci and Mattia Rapa, for example, used a lateral flow immunoassay in which antibodies were mounted on AuNPs and lateral flow immunoassays were performed with 100% selectivity to identify *E. coli* and *Salmonella* in water and milk [85, 86]. *E. coli* was detected in water and milk samples with concentrations as low as low as 7.8 × 105 and 3 × 106 CFU/ml, respectively, and *Salmonella typhi* was detected in water and milk samples with concentrations as low as 3 × 108 and 3 × 107 CFU/ml using this method. We propose an in-depth study on the toxicity of nanomaterials/nanoformulations, particularly noble NPs, to major organs in the body because few works have cited the effects of these nanomaterials on food samples, especially those incorporated into consumable food commodities like the case of smart packages.

References

1. David, R.J., Jaako, M., Katie, A.M., The importance of food. *Arch. Pediatr. Adolesc. Med.*, 166, 2, 187–188, 2012.
2. Akanele, U., Scholastica Mgbo Otu Chukwu, Ben, Chukwu Mary Ahudie. *Int. J. Sci. Technol. Res.*, 5, 03, 2277–8616, 2016.
3. Hyman, F.N., Wontz, K.C., Tollefson, L., Food and drug administration surveillance of the role of foreign objects in foodborne injuries. *Public Health Rep.*, 108, 54–59, 1993.
4. Xihong, Z., Chii-Wann, L., Jun, W., Deog, H.O., Advances in rapid detection methods for foodborne pathogens. *J. Microbiol. Biotechnol.*, 24, 3, 297–312, 2014.
5. World Health Organization How Safe is your Food? From Farm to Plate, Make Food Safe||, 2015. Available on http://www.afro.who.int/en/rdo/speeches/4599-message-of-the-who-regional-director-for-africa-dr-matshidiso-moeti-for-world-health-day.
6. Kaittanis, C., Santra, S., Perez J, M., Emerging nanotechnology-based strategies for the identification of microbial pathogenesis. *Adv. Drug Deliv. Rev.*, 62, 408–23, 2010.
7. Seema, R., Food spoilage: Microorganisms and their prevention. *Asian J. Plant Sci. Res.*, 5, 4, 47–56, 2015.
8. Tallury, P., Malhotra, A., Byrne, L.M., Santra, S., Nanobioimaging and sensing of infectious diseases. *Adv. Drug Deliv. Rev.*, 62, 424–37, 2010.
9. Nathalie, P. and Thierry, N., Detection of Salmonella in Food Matrices, from Conventional Methods to Recent Aptamer-Sensing Technologies. *Foods*, 8, 371, 2019.
10. Morvarid, S., Abdolmajid, G., Seyyed, K.S.M., Maryam., T., Hossein, R.V., Seyede, A.M., Gold nanoparticle-based colorimetric platform technology

as rapid and efficient bacterial Pathogens detection method from various sources. *Rev. Med. Microbiol.*, 30, 26, 128–132, 2019.
11. Mir, H.J., Hamed, A., Ali, A.P., Hamidreza, P., Bija, S., Various methods of gold nanoparticles (GNPs) conjugation to antibodies. *Sens. Biosens. Res.*, 9, 17–22, 2214–1804, 2016.
12. Chellaram, C., Murugaboopathi, G., John, A.A., Sivakumar, R., Ganesan, S., Krithika, S., Priya, G., Significance of nanotechnology in food industry. *APCBEE Proced*, vol. *8*, pp. 109–113, 2014.
13. Alfadul, S.M. and Elneshwy, A.A., Use of nanotechnology in food processing, packaging and Safety - review. *Afr. J. Food Agric. Nutr. Dev.*, 10, 6, 2010.
14. Baeumner, A., Nanosensors identify pathogens in food. *Food Technol.*, *58*, 51–52, 54–55, 2004.
15. Ravichandran, Nanotechnology applications in food and food processing: Innovative green approaches, opportunities and uncertainties for global market. *Int. J. Green Nanotechnol: Phys. Chem.*, 1:2, P72–P96, 2010.
16. John, D. and Floros, New and emerging applications of nanotechnology in our food supply. *AT IFT International Food Nanoscience Conference*, vol. 17, pp. 25–28, 2010.
17. Food and agriculture organization of the united nations sub-regional office for the Pacific Islands Apia. Manual on food packaging for small and medium-size enterprises in Samoa, 2012.
18. Kenneth, M. and Betty, B., Food packaging—Roles, materials, and environmental issues. *J. Food Sci.*, 72, 3, 2007.
19. R.K. Gupta and Puja Dudeja. *Food packaging*. In book: Food supply safety in India, 547–553, 2017.
20. US Environmental Protection Agency. Reducing wasted food and packaging: A Guide for food services and restaurants, 2015.
21. Nitaigour, P.M., Advances in packaging methods, processes and Systems. *Challenges*, *5*, 374–389, 2014.
22. Fredrik, W., Helén, W., Jakob, T., Zane, R., The importance of packaging functions for food waste of different products in Households. *Sustainability*, 11, 2641, 2019.
23. Maria, L., Giovana, V.B., Carmen, J.C., Microbial deterioration of vacuum-packaged chilled beef cuts and techniques for microbiota detection and characterization: A review. *Braz. J. Microbiol.*, 42, 1–11, 2011.
24. Zuzana, S., Food packaging materials: Comparison of materials used for packaging purposes. Helsinki Metropolia, 37, 2017.
25. Jakobsen, R.A., Heggebø, R., Sunde, E.B., Skjervheim, M., *Staphylococcus aureus* and *Listeria monocytogenes in* Norwegian raw milk cheese production. *Food Microbiology.*, 28, 492–496, 2011.
26. Dalzini, E., Bernini, V., Bertasi, B., Daminelli, P., Losio, M.N., Varisco, G., Survey of prevalence and seasonal variability of Listeria monocytogenes in raw cow milk from Northern Italy. *Food Control.*, 60, 466–470, 2016.

27. Jackson, B.R., Tarr, C., Strain, E., Jackson, K.A., Conrad, A., Carleton, H., Katz, L.S., Stroika, S., Mody, R.K., Silk, B.J., Beal, J., Chen, Y., Timme, R., Doyle, M., Fields, A., Wise, M., Tillman, G., Defibaugh-Chavez, S., Kucerova, Z., Sabol, A., Roache, K., Trees, E., Simmons, M., Wasilenko, J., Kubota, K., Pouseele, H., Klimke, W., Besser, J., Brown, E., Allard, M., Gerner-Smidt, P., Implementation of nationwide real-time whole-genome sequencing to enhance listeriosis outbreak detection and investigation. *Clin. Infect. Dis.*, 63, 3, 380–386, 2016.
28. Wuyts, V., Mattheus, W., De Laminne de Bex, G., Wildemauwe, C., Roosens, N.H., Marchal, K., De Keersmaecker, S.C.J., Bertrand, S., MLVA as a tool for public health surveillance of human salmonella typhimurium: Prospective study in belgium and evaluation of MLVA loci stability. *PloS One*, 8, 12, e84055, 2013.
29. Cancino-Padilla, N., Fellenberg, M.A., Franco, W., Ibañez, R.A., Vargas-Bello- Pérez, E., Foodborne bacteria in dairy products: Detection by molecular techniques. *Cien. Inv. Agr.*, 44, 3, 215–229, 2017.
30. Chaillou, S., Chaulot-Talmon, A., Caekebeke, H., Cardinal, M., Christieans, S., Denis, C., Champomier-Verges, M.C., Origin and ecological selection of core and food-specific bacterial communities associated with meat and seafood spoilage. *ISME J.*, *9*, 5, 1105–1118, 2015.
31. Egan, A.F., Eustace, I.J., Shay, B.J., Meat packaging – maintinaing the quality and prolonging the storage life of chilled beef, pork and lamb. *Proceedings of Meat*, vol. 88, 1988.
32. Chai, S.J., Cole, D., Nisler, A., Mahon, B.E., Poultry: The most common food in outbreaks with known pathogens, United States, 1998–2012. *Epidemiol. Infection*, *145*, 2, 316–325, 2017.
33. George, A.A., Hemmat, M., II, Reham, A.A., Hanaa, A.A., Detection of some foodborne pathogens in meat products by Polymerase. Chain reaction. *BVMJ*, 30, 1, 323–330, 2016.
34. Hui, L., Xiaohong, S., Xiaojun, L., Michael, G., Control of pathogenic and spoilage bacteria in meat and meat products by high pressure: Challenges and future perspectives. *Compr. Rev. Food Sci. Food Saf.*, 19, 6, 3476–3500, 2020. doi: https://doi.org/10.1111/1541-4337.12617.
35. Mandrell, R.E., Gorski, L., Brandl, M.T., In: Microbiology of fresh fruits and vegetables (Eds. G.M. Sapers, J.R. Gorney, and A.E. Yousef). *New York: Taylor Francis Group*, 648, 33–73, 2006.
36. Mahbub, M.K., Tazul, I., Md., Mohammed, M.H.C., Sharmin, R.A., Assessment of microbiological quality of some drinks sold in the streets of Dhaka University Campus in Bangladesh. *Int. J. Food Contamin.*, 2, 4, 2015.
37. Akomolafe, O.M. and Awe, T.V., Microbial contamination and polyethylene packaging of some fruits and vegetables retailed at Akure and Ado Ekiti, South Western Nigeria. *Journal of stored products and post harvest research*, 8, 6, 65–72, 2017.

38. Errampalli, D., Penicillium expansum (Blue mould), in: *Postharvest Decay control strategies*, S. Bautista-Banos (Ed.), pp. 205–206, Academic Press, USA, 2014.
39. Velusamy, V., Arshak, K., Korostynska, O., Oliwa K, A.C., An overview of foodborne pathogen detection: In the perspective of biosensors. *Biotechnol. Adv.*, 28, 232–54, 2010.
40. Kiev, S.G. and John, L.M., A review of conventional detection and enumerationmethods for pathogenic bacteria in food. *Can. J. Microbiol.*, 50, 11, 883–890, 2004. doi:10.1139/w04-080.
41. Ferretti, R., Mannazzu, I., Cocolin, L., Comi, G., Clementi, F., Twelve-Hour PCR-based method for detection of salmonella spp. in food. *Appl. Environ. Microbiol.*, 67, 977–978, 2001.
42. Ayrapetyan, M. and Oliver, J.D., The viable but non-culturable state and its relevance in food safety. *Curr. Opin. Food Sci.*, 8, 127–133, 2016.
43. Nathalie, P. and Thierry, N., Detection of Salmonella in *Food Matrices, from Conventional Methods to Recent Aptamer-Sensing Technologies. Foods*, 8, 371, 2019.
44. Brandão, D., Liébana, S., Pividori, M.I., Multiplexed detection of foodborne pathogens based on magnetic particles. *New Biotechnol.*, 32, 511–520, 2015.
45. Tsen, H.Y., Combination of immunomagnetic separation and polymerase chain reaction for the simultaneous detection of *Listeria monocytogenes* and *Salmonella* spp. in food samples. *J. Food Prot.*, 64, 1744–1750, 2001.
46. Hudson, J.A., Lake, R.J., Savill, M.G., Scholes, P., McCormick, R.E., Rapid detection of *Listeria monocytogenes in* ham samples using immunomagnetic separation followed by polymerase chain reaction. *J. Appl. Microbiol.*, 90, 614–621, 2001.
47. Ogden, I.D., Hepburn, N.F., MacRae, M., The optimization of isolation media used in immunomagnetic separation methods for the detection of *Escherichia coli* O157:H7 in foods. *J. Appl. Microbiol.*, 91, 373–379, 2001.
48. Chandler, D.P., Brown, J., Call, D.R., Wunschel, S., Grate, J.W., Holman, D.A., Olson, L., Stottlemyre, M.S., Bruckner-Lea, C.J., Automated immunomagnetic separation and microarray detection of *E. coli* O157:H7 from poultry carcass rinse. *Int. J. Food Microbiol.*, 70, 143–154, 2001.
49. Priyanka, B., Rajashekhar, K.P., Sulatha, D., A review on detection methods used for foodborne pathogens. *Indian J. Med. Res.*, 144, 327–338, 2016.
50. Pan, Y. and Breidt, F., Jr, Enumeration of viable Listeria monocytogenes cells by real-time PCR with propidium monoazide and ethidium monoazide in the presence of dead cells. *Appl. Environ. Microbiol.*, 73, 8028–8031, 2007.
51. Park, Y.S., Lee, S.R., Kim, Y.G., Detection of Escherichia coli O157:H7, Salmonella spp., Staphylococcus aureus and Listeria monocytogenes in kimchi by multiplex polymerase chain reaction (mPCR). *J. Microbiol.*, 44, 92–97, 2006.
52. Shefer, A. and Shefer, S., Biodegradable bioadhesive controlled release system of nano- particles for food products. *U.S. patent*, 2003.

53. Daniells, S., Thing big, think nano. Food Navigator.com, 2019. Available at: http://www.foodnavigator.com/news/ng.asp?n=8.
54. Long, T.N., Saleh, R., Tilton, R., Lowry, G., Veronesi, B., Titanium dioxide (P25) produces reactive oxygen species in immortalized brain microglia (BV2): Implications for nanoparticle neurotoxicity. *Environ. Sci. Technol.*, 40, 14, 4346–4352, 2006.
55. Wang, B., Feng, W.W., Wang, M., Wang, T., Gu, Y., Zhu, M., Ouyang, H., Shi, J., Zhang, F., Zhao, Y., Chai, Z., Wang, H., Wang, J., A cute toxicological impact of nano- and submicro-scaled zinc oxide powder on healthy adult mice. *J. Nanopart. Res.*, 10, 2, 263–276, 2007.
56. Yuanyuang, L., Hermann, J., Schluesenerb, S.X., Gold nanoparticle-based Biosensors. *Gold Bull.*, 43, 1, 29–49, 2010.
57. Bulbul, G., Hayat, A., Andreescu, S., A generic amplification strategy for electrochemical aptasensors using a non-enzymatic nanoceria tag. *Nanoscale*, *7*, 13230–12338, 2015.
58. Liu, M., Jia, C., Jin, Q., Lou, X., Yao, S., Xiang, J., Zhao, J., Novel colorimetric enzyme immunoassay for the detection of carcinoembryonic antigen. *Talanta*, *81*, 1625–1629, 2010.
59. Ambrosi, A., Airò, F., Merkoçi, A., Enhanced Gold Nanoparticle Based ELISA for a Breast Cancer Biomarker. *Anal. Chem.*, *82*, 1151–1156, 2010.
60. Saha, K., Agasti, S.S., Kim, C., Li, X., Rotello, V.M., Gold Nanoparticles in Chemical and Biological Sensing. *Chem. Rev.*, *112*, 2739–2779, 2012.
61. Mohit, S.V., Jacob, L.R., Lyndon, J., Frank, X.G., Colorimetric biosensing of pathogens using gold nanoparticles. *Biotechnol. Adv.*, xxx, 2015.
62. Khan, S.A., Singh, A.K., Senapati, D., Fan, Z., Ray, P.C., Targeted highly sensitive detection of multidrugresistant Salmonella DT104 using gold nanoparticles. *Chem. Commun. (Camb)*, 1, 33, 6Pt1: 666–680, 2011. doi:10.1016/j.biotechadv.2015.03.003.
63. Sun, J., Ge, J., Liu, W., Wang, X., Fan, Z., Zhao, A facile assay for direct colorimetric visualization of lipopolysaccharides at low nanomolar level. *Nano. Res.*, 5, 486–93, 2012.
64. Su, H., Zhao, H., Qiao, F., Chen, L., Duan, R., Ai, S., Colorimetric detection of Escherichia coli O157: H7 using functionalized Au@Pt nanoparticles as peroxidase mimetics. *Analyst*, 138, 3026–31, 2013.
65. Miranda, O.R., Li, X., Garcia-Gonzalez, L., Zhu, Z.J., Yan, B., Bunz, U.H., Colorimetric bacteria sensing using a supramolecular enzyme-nanoparticle biosensor. *J. Am. Chem. Soc*, 133, 9650–3, 2011.
66. Fu, Z., Zhou, X., Xing, D., Rapid colorimetric gene-sensing of food pathogenic bacteria using biomodification-free gold nanoparticle. *Sensors Actuators B Chem.*, 182, 633–41, 2013.
67. Lei, J. and Ju, H., Signal amplification using functional nanomaterials for biosensing. *Chem. Soc Rev.*, 41, 2122–2134, 2012.
68. Roy, K.J., Rahman, A., Hossain, K.S., Rahman, M., Kafi, M.A., Antibacterial *Investigation of Silver Nanoparticle* against *Staphylococcus, E. Coli* and

Salmonella isolated from selected live bird markets. *Appl. Microbiol.*, 6, 170, 2020.
69. Ouay, B.L. and Stellacci, F., Antibacterial activity of silver nanoparticles: *A surface science insight. Nano Today*, 10, 339–354, 2015.
70. Tran, H.V., Huynh, C.D., Tran, H.V., Piro, B., Cyclic voltammetry, square wave voltammetry, electrochemical impedance spectroscopy and colorimetric method for hydrogen peroxide detection based on chitosan/silver nanocomposite. *Arab. J. Chem.*, 11, 453–459, 2018.
71. Zahed, F.M., Hatamluyi, B., Lorestani, F., Es'haghi, Z., Silver nanoparticles decorated polyaniline nanocomposite based electrochemical sensor for the determination of anticancer drug 5-fluorouracil. *J. Pharm. Biomed. Anal.*, 161, 12–19, 2018.
72. Wei, H., Chen, C., Han, B., Wang, E., Enzyme colorimetric assay using unmodified silver nanoparticles. *Anal. Chem.*, *80*, 7051–7055, 2008.
73. Muhammad, I., Christopher, J.E., Massimo, F.B., Muhammad, R.S., Vamsi, K.Y., Chitosan stabilized silver nanoparticles for the electrochemical detection of lipopolysaccharide. *A Facile Biosensing Approach Gram-Negative Bacteria. Micromachines*, 11, 413, 2020.
74. PRIMEC.Nanotechnology in food safety and quality assessment: Potentiality of nanoparticles in diagnosis of foodborne pathogens. *Agricultura*, 13, 1-2, 21–32, 2017.
75. Ahad, M., Reza, E.-K., Paria, P., Maryam, H., Nasrin, G., Mohammad, H., Behzad, B., de la Guardia, M., Nanomaterial-based biosensors for detection of pathogenic virus. *Trends Analytic. Chem.*, 97, 445e457, 2017.
76. Tuitemwong, P., Songvorawit, N., Tuitemwong, K., Facile and Sensitive Epifluorescent Silica Nanoparticles for the Rapid Screening of EHEC. *J. Nanomater*, 2013, Article ID 706354, 1–8, 2013.
77. Ricco., R., Meneghello, A., Enrichi, F., Signal enhancement in DNA microarray using dye doped silica nanoparticles: application to human papilloma virus (HPV) detection, Biosens. *Bioelectron.*, 26, 2761e2765, 2011.
78. Gonca, B., Akhtar, H., Silvana, A., Portable Nanoparticle-Based Sensors for Food Safety Assessment. *Sensors, 15*, 30736–30758, 2015.
79. Asati, A., Santra, S., Kaittanis, C., Nath, S., Perez, J.M., Oxidase-like activity of polymer-coated cerium oxide nanoparticles. *Angew. Chem. Int. Ed.*, *48*, 2308–2312, 2009.
80. Hayat, A., Yang, C., Rhouati, A., Marty, J.L., Recent advances and achievements in nanomaterial-based, and structure switchable aptasensing platforms for ochratoxin A detection. *Sensors, 13*, 15187–15208, 2013.
81. Bárbara, S., Antonio, V.H., María, A., Miguel, Á. R., Review Recent applications of magnetic nanoparticles in food analysis. *Processes*, 8, 1140, 2020.
82. Varshney, M., Li, Y., Nanapanneni, R., Johnson, M.G., Griffis, C.A., A chemiluminescence biosensor coupled with imunomagnetic separation for rapid detection of *Salmonella typhimorium*. *J. Rap. Meth. Automat. Microbiol.*, 11, 111–131, 2003.

83. Dimitra, H., Joseph, P., Katerina, Z., Nikolaos, N., Hanan, I.M., Marie-Paule, M., Spyros, K., Stamatis, K., Athanassios, T., Ekatherina, C., Magnetic nanoparticle-enhanced PCR for the detection and identification of *Staphylococcus aureus* and *Salmonella enteritidis*. *New Microbiologica*, 40, 3, 165–1691121–7138, 2017.
84. Neuschmelting, V., Harmsen, S., Beziere, N., Lockau, H., Hsu, H.T., Huang, R., Razansky, D., Ntziachristos, V., Kircher, M.F. Dual-modality surface-enhanced resonance raman scattering and multispectral optoacoustic tomography nanoparticle approach for brain tumor delineation. *Small*, 14, 1800740, 2018.
85. Giuliana, V. and Mattia, R., Noble metal nanoparticles applications: Recent trends in food control. *Bioengineering*, 6, 1, 10, 2019.
86. Mohammad Lukman, Y., Nor Dyana, Z., Rahmah, N., Khairunisak, A.R., Development and evaluation of colloidal gold lateral flow immunoassays for detection of *Escherichia Coli* O157 and *Salmonella Typhi*. *J. Phys. Conf. Ser.*, 1082, 012049, 2018.

13

Nanotechnology in Packaging for Food Preservation

Ravish Choudhary[1], Varun Kumar[2] and Reena Yadav[3]*

[1]Division of Seed Science and Technology ICAR Indian Agricultural Research Institute, New Delhi, India
[2]AcSIR - Academy of Scientific and Innovative Research, CSIR, Ghaziabad, UP, India
[3]Department of Biotechnology, Kumaun University, Nainital, Uttarakhand, India

Abstract
Food packaging is a very important and essential part to store the food for longer period and deliver the food product to the end user without spoilage. A good packaging protects the food from outer environmental contaminations like moister, dust, microbes and pollutants, and chemical and physical damages. Nowadays, a very advanced technology has been used to improve the quality of packaging that is nanotechnology. Nanotechnology-based packaging is a smart and active packaging technique in which nano-sized packaging materials are used that contains nanosensor to detect the biological and chemical contaminations at very low level, antimicrobial properties, and high environmental barriers protection. These properties of nanomaterials improve the quality of packaging in terms of enhancing the shelf life of food and longer storage without the spoilage of food.

***Keywords*:** Nanotechnology, nanomaterial, food packaging, antimicrobial, nanosensor

**Corresponding author*: reenayadavbt@gmail.com

Annu, Tanima Bhattacharya and Shakeel Ahmed (eds.) Nanotechnology in Intelligent Food Packaging, (313–342) © 2022 Scrivener Publishing LLC

13.1 Introduction

Global food industry is facing immerse consumers' pressure to fulfill their concerns about food quality, safety, and hygiene. In the process of betterment of food preservation, humans have continuously improved several techniques in food processing to preserve food for long-term uses [1]. As we know, preserving food is an extremely old phenomenon since the prehistorical era. From the Palaeolithic period, humans start preserving food by storing freshly hunted animal meat in the cave for days and consumed it when needed. Roasting, drying, and fermentation practices adapted from the early Palaeolithic periods, and nowadays, we use the modified versions of these practices [2]. Roasting, sun drying, steaming, salting, jellying, food fermenting, and artificial preservatives/chemicals have been normally used by people in regular mode. Due to these applications, all the pathogens that can cause the any kind of disease were inactive or may be killed. Preservation techniques have own nature so that food can be keep longer. To fulfill such kind of demands, very cheap, effective, speedy, and steady tools were developed by some companies those produce food, dealers, consumers, and food authoritarian bodies to scrutinize the quality of food effectively on the conventional inactive fence model because the main important part for any food business is food packaging [3].

Nanotechnology emerges as an influential technique that connects several disciplines to develop the pioneering goods and also known as emerging branch of science and engineering with its wide range of applications in science and technology to synthesize new materials at different nanoscale level ranges from 1 to 100 nm [4, 5]. It includes manipulation and characterization of nano range molecules and their exceptional properties that were not found in the original form of stuff. Throughout the world, all the scientists who belong to nanotechnology are going to focus mainly to develop unique, emerging, and advance products using green method due to considerate these unique properties of nanomaterial [6]. European countries have significantly increased the use of nanomaterial in food industry to improve food quality, shelf life, and safety. In 2015, more than \$2.5 million revenue was generated by nanomaterial market, and around \$9 million is expected by 2022 in Europe [7]. Nanoparticles (NPs) have antimicrobial activity so it has been used as a disinfectant agent in textile fabrics, water disinfection, medicine, and food packaging [8, 9].

As broad level of beneficial applications related to food present in nanotechnology so a unique and a particular nanomaterial is used for packaging the food product to adapt convinced and much needed properties [10].

Recent studies confirmed that nanomaterials can be used for improvement the food safety issues and their improve food safety and their superiority by enhancement of the efficacy of quality of food packaging, improving the shelf life, nutrient rate without its varying of test, and substantial properties [11, 12]. Thus, nanotechnology improves the mechanical and biological barrier of food packaging to offer active and intelligent packaging systems [9]. NPs inside the food packages were used for inhibition the spoilage and microbe/pathogen growth or may be killed, which caused food contamination. Recently, there are so many naturally derived antimicrobial agents, *viz.*, essential oils, antibodies, and enzymes; some synthetic and artificially produced inorganic or organic materials, namely, ZnO, TiO_2, Ag, and graphene, are using for the improvement of food safety issues with lengthened shelf life during food processing [13]. These antimicrobial properties of NPs are imitative from a number of methods, *viz.*, harmful integrity of cell membrane [14], spoiling the biomolecules inside the cells [15], regulating metabolic systems, and exerting oxidative stress [16]. Silver NPs have unique substantial properties, physically powerful facade plasma quality, a high surface-to-size proportion, well-organized catalytic action, and extraordinary wide-ranging antimicrobial activity, which are used next to several microbial pathogens [17]. TiO_2 NPs have antimicrobial property, low cytotoxicity, UV blocking, and modification of color in food packaging material [18]. In food packaging materials, the antimicrobial activities and application of these NPs have been studied intensively. The composition, shape, size, attentiveness, constancy, allocation, and scattering ability of these NPs determine their antimicrobial activities, which is used alone or jointly according to utility, mechanisms, target-specific pathogen, and food system. The application of nanotechnology has been explored in a diverse area of food industries, *viz.*, food security, food covering, food dispensation, food strengthening, encapsulation, and pathogen recognition [19]. The discovery of nanotechnology for food packaging has several potential to straightforward integration of food commodities, wrapping material, and food dispensation [20]. Nanomaterials offer several advantages over traditional food packaging such as antimicrobial properties, challenging to temperature or fire, improved sturdiness, obstruction, salvage, and optical belongings [21]. These properties of nanomaterials make it an ideal for food packaging and processing of several products like meat, cheese, cereals, ready-to-eat foods, beer bottles, and fizzy drinks and also have application for the fruit juice and dairy products [22]. The advance technology improves the properties of nanomaterial such as its mechanical strength, durability, diffusivity, and solubility that provide a degradable food packaging with environmental safety [23]. In this book chapter, we discussed

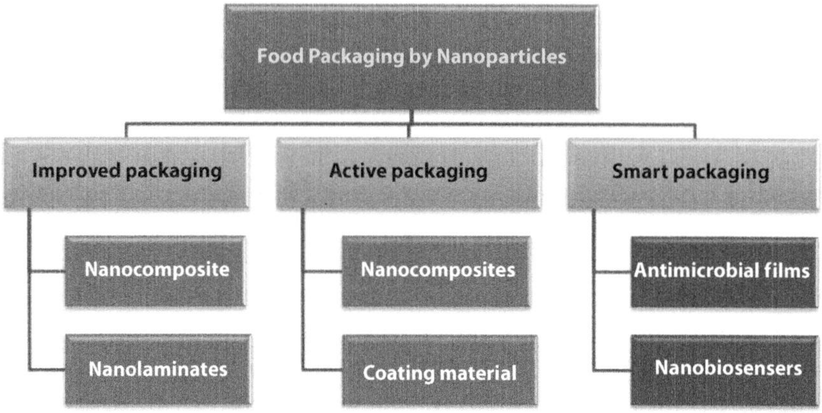

Figure 13.1 Application of NPs in food packaging.

about the different types of nanopackaging, different types of nanomaterial used in nanopackaging, and their properties (Figure 13.1).

13.2 Types of Packaging

These packaging resources become smart and more active from conventional packaging methods due to nanotechnology.

1. *Advanced packaging by nanoparticles*
2. *Active packaging by nanoparticles*
3. *Smart packaging using nanoparticles*

13.2.1 Advanced Packaging by Nanoparticles

Nanocomposite and nanolaminates used for food packaging actively and improve food shelf life. NPs have antimicrobial property for food packaging and provide a strong barrier against adverse thermo and mechanical shock. Quality food with extremely long shelf life can be achieved using NPs into food packaging materials. Food packaging material can be thermal and mechanically stable using improved polymer composites (inorganic or organic fillers). Polymers can be more cost effective using NPs and can develop more resist packaging material [24]. Polymer matrices that have silicate nanoplatelets, clay, SiO_2 NPs, chitin, or chitosan (inert nanoscale fillers) make it light, strong, and resistant against to fire and improve thermal properties [25]. Antimicrobial nanocomposite films into the polymers have structural integrity and barrier properties that provide more advantage in food packaging A material for food packaging, having strength and biodegradable property need sufficient amount of suitable gas and optimum moisture permeability [26]. Nanopackaging materials deliberate more benefits, provide food safety status with high mechanical strength, and improve barrier activities and detection of pathogen by anti-microbial films using nanosensing [9]. To reduce the negative environmental effects, we need to adapt green packaging that includes plant extracts, biodegradable edible materials, and nanocomposite materials. In this system, some food products used for the power build up are more horizontal in the direction of microbial infection have been augmented exponentially. These power products protect the processing muscles from microbial spoilage, bypass contamination, and retain their tenderness and freshness [27]. In smart packaging, nanosensors in the food products provide information about food spoilage to patrons on the basis of flavor and color creation by toxins, pesticides detections, and microbial infection [28]. NPs, nanoemulsions (NEs), and nanoclays (NCs) are commonly used nanomaterials that provide several applications in the food sector [29]. Besides, these some metals, alone or with oxide NPs, *viz.*, ZnO, MgO, TiO_2, SiO_2, are often used like anti-microbial and photo-catalyst agents for the degradation of organic molecules [30]. ROS production by the photolytic reaction of zinc oxide (ZnO) and titanium oxides produces ROS, which leads bacterial cells death due to cytoplasmic oxidation [31]. The efficiency of ZnO is relatively more attractive than silver NPs because of their minimum toxicity and economical [32]. Nanocomposites were used in food packaging and coating purposes [9]. Many NPs such as silica oxide, carbon nanotubes (CNTs), clay, starch nanocrystals, chitosan NPs, and cellulose-based nanofibers are used in a polymeric matrix, which improves thermal properties and lower permeability to gases [25]. Nanomaterials have been associated with so many other properties like mechanical strength and heat resistance along with barrier against oxygen, carbondioxide, UV rays, moisture, and volatiles.

13.2.2 Active Packaging by Nanoparticles

In food package material, the food quality, safety, and their shelf life can be increased by the active packaging using nanotechnology in which active carrier component interact with internal and external factors followed by stimulate their positive actions [7]. Food packaging can be improved by application of nanocomposites and coating material. Active packaging can be generated using different various compounds, which may be added into food packaging materials to control the microbial proliferation. As organic compounds, *viz.*, essential oils, organic acids, and bacteriocins have antimicrobial properties those can be used for antimicrobial packaging [33], but organic compounds need high temperatures and pressures for food processing, which make these for non-ideal for this process. However, inorganic NPs used in antimicrobial food packaging showed more stable in extreme situations and showed highly antibacterial activity in low concentrations. An antimicrobial packaging also known as active packaging showed retard microbial growth present on the surface of food products [34]. Shelf life and microbial quality of food products can be improve using embedded film surface of various active compounds like metal NPs that showed the properties against various food spoilage strains and make capable them for ethylene oxidation or oxygen scavenging. Some most commonly used NPs like chitosan, copper, gold, silver, and metal oxide NPs (TiO and ZnO) have antibacterial property and found more effective in several food applications [35, 36].

13.2.3 Smart Packaging Using Nanoparticles

Smart packaging using NPs is one more additional rising packaging tools that can provide information at the same time on novelty, observe the microbial growth of food, and exchange a few words with consumers. Such kind of smart packaging can be attained by using sensing technologies or smart labels that give information on food spoilage, gas sensing, microbial growth, and measure changes in the internal environmental condition of packaging materials [7]. As compared to traditional food packaging that only interact with elements, which are present inside the food packages, but in the case of smart packaging, it also offers signal information to customers and moves forward to food safety and quality that were controlled during storage. The conception of keenly intellective aliment packaging is widely apperceived and much research being carried out to developed antigen-categorical biomarkers utilized in victuals packaging and additionally the incorporation of active NPs to prepare polymer nanocomposite films [37]. Various food and beverage products like meat and seafood degradation could change the pH inside packaging due to their nature of microbial growth, biogenic amines [38, 39], and volatile organic amines [40]. In the case of fruit, an excessive amount of ethylene secretion causes their ripening [41]. These parameters are uses as filth markers to assess the vicissitudes that come in the superiority and originality of packaged aliment. The transmuting environment within the packaged food can withal be detected by quantifying the level of O_2/CO_2, moisture content, and microbial magnification, all these parameters designate the vicissitudes in freshness conditions that lead to their degradation. The sensors that are going to develop and tags are utilized in smart packaging for identifying the alteration or modification in the pH, O_2, CO_2, RH, and CH_2CH_2. New technology with degradation compounds makes available information to the customer/trader for the food excellence. A zeolite-molybdate nanopores tablet used in avocados packages to scavenge and measure ethylene production. It undergoes a color change like yellow to blue within 10 days of storage by the reduction activity of Mo (VI) to Mo (V), and it gives information of packaging materials [42]. In smart packaging material, there are two common elements used. First one who detect the targets called as receptors and second one is reporter by which the changes or alteration in signals can be detected. There are so many receptors to detect the targets including antibodies, DNAzymes, RNAs, aptamers, and enzymes along with some small other molecules like oligothiophenebenzothiazole-based cyanide sensor [43]. Receptors also include many NPs and enzymes that convert the molecular indicators to measurable information in the form of color, fluorescence, and electronic indications that can be measured by the eye and equipment. Some microbes responsible for the food spoilage can be detected by using antigen-specific biomarkers that detect their presence in food materials. The nanopore-based material, BioSilicon, designed by pSivida, Australia, is used in the food packaging industry [44]. The specialty lies in BioSilicon composition, and it is made up of nanostructured silicon. Newly engineered device, namely, Radio Frequency Identification Display (RFID), is useful for the quick food products supply those have a shorter shelf period. The RFID-based transistors used in packaging to detect extremely miniature alteration of pH, pressure, and temperature. The use of nanotechnology in food packaging systems introduces a new function that greatly enhances the performance of packaging materials by providing thermal stability and barrier properties like permeability of gas and water vapor, along with mechanical properties, *viz.*, strength of tensile and stretchy modules.

13.3 Types of Nanomaterial and Their Advantages in Food Packaging

In food processing, packaging is important in food trade with some limitation, which causes the foremost imperfection or fault in conventional food packaging. Due to lack of such packaging materials, food items cannot be

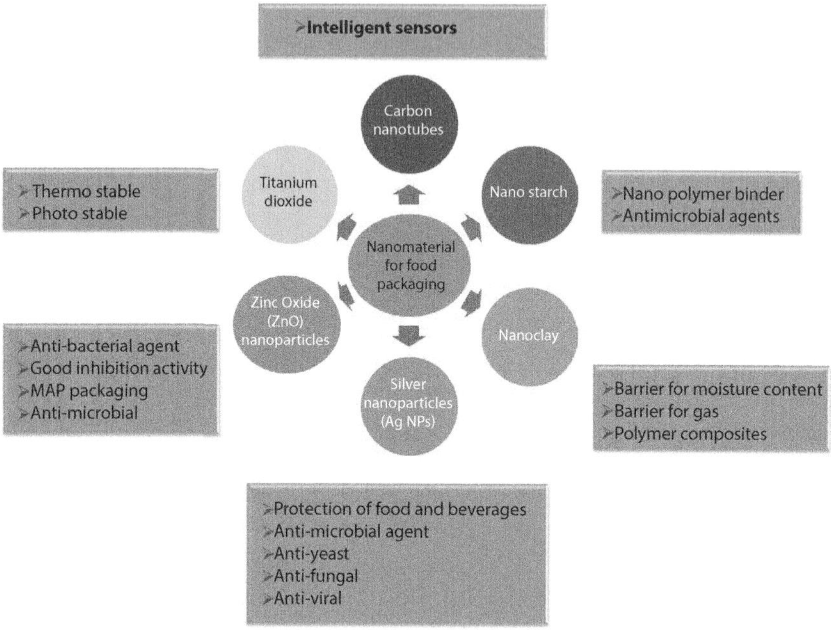

Figure 13.2 Nanomaterial used in food packaging and their properties.

completely protect packaging from atmospheric and water vapor resistance [45]. As a result, there is a need for the continuous discovery of innovative and eco-friendly food packaging materials [46, 47], which not only provide protection from environmental contamination and facilitate the handling and transportation but also provide high-quality nutritious food products [48, 49]. In traditional packaging, maximum use of non-biodegradable such as plastics, glass, and metals for packaging materials raises atmospheric pollution that cause a serious problem for environment [50]. We mentioned here few nanomaterials which are being used for food packing and their advantages (Figure 13.2).

13.3.1 Silver Nanoparticles (Ag NPs)

For many years, silver particles are being used to protect food and beverages as an antimicrobial agent. From ancient times, silver used as a container for water, wine, and silver serves were mounted at the bottle underneath to preserve liquids such as milk and water for travellers [25]. It has been known to be quite a stable metal element with no major threat to the biological system if used in less quantity as assigned by FDA (Food and Drug

Association) standards [51]. Silver NPs have extensive biocide and low volatility characteristics containing broad level of anti-microbial, anti-yeasts, anti-fungal, and antiviral properties as they have more immensely colossal facade vicinity for every mass against to micro-scale Ag NPs along with toxic in smallest amount for eukaryotic cells [52–54]. Silver NPs are most efficient in direction for Gram-negative as compared to Gram-positive bacteria as it is easily go through to Gram-negative bacterial cell wall [51]. Its NPs bind to the surface of microbes which leads to degrade lipopolysaccharides and form a pit within the cell membrane. Silver NPs also disrupt the ribosomal activity of many microbes that inhibit several important enzymes production [46]. These days use of silver NPs into plastic polymers for making packaging material is mainly utilized due to its smaller tendency to make resistant against microbes which is implemented in several different methods [55]. For example, reports say that silver ion, those were placed into the porous zeolite, can be used in plastic polymers. It was observed previously that NPs like silver and ZnO with very low density of PE (LDPE) are being used to preserve and enhance the shelf life of vegetables and fruits by absorbing and decomposing ethylene hormone [51]. This type of active silver nanocomposite has antimicrobial nanomaterial in combination with heat treatment property and was found very helpful and more efficient [56]. In the last few years, silver has antibiotic-resistant against the strains of *Staphylococcus aureus*, which become most popular itself and being sporadic with a low clinical incidence [57].

13.3.2 Nanoclay

The appearance of nanoclay among the polymers of nanocomposites was reported first in the market, and it has been the most commonly used nanomaterial in food packaging materials [58]. Nanoclay mainly plays an important role to get better plastic material properties physically and builds up obstruction for moisture content and gas. It has been broadly used in polymer composites because of its magnificent cation exchange efficiency, wide surface area, and good swelling properties [59]. Montmorillonite (MMT) is an example of nanoclay, which consists of an octahedral wedge-shaped sheet of aluminum hydroxide in between two-fold tetrahedral silica. MMT is naturally obtained from volcanic ash and rocks; therefore, it is fairly inexpensive and easily available in nature [60]. The scattering pattern of nanoclay in the polymer matrix has been classified into three main classes, namely, tactoid, intercalated, and exfoliated. The tactoid structure of nanoclay is remarkable to come across within the nanocomposite and the clay platelet cluster shows a high level of fence like arrangement due to its less compatibility to polymers [61]. The NPs, those having minimum amount of clay with same biodegradability, are generally used as essential filler in bio-based polymer, which normally boosts the blockade properties, is resistant to move stealthily, and provides more potential to biopolymer mechanically [62]. Bio-composite films prepared with Agar and Cloisite-Na+ that resulted in nanoclay greatly increase the tensile strength, water vapor permeability, and hydrophobic capacity of biocomposite. The uses of nanoclay in food packaging and medical started in the 1990s. Bayer developed a nanoclay-engineered polyamide film, which is known in the market as Durethan® KU2-2601 [63] and get commercialized by Lanxess Deutschland [64]. Durethan is commonly used in beverage packaging because of its barrier characteristics [65]. From the last few years, uses of plastic bottles in the beer and soda industries extensively decrease due to their poor gas barrier property. However, the use of a polymer matrix of nanoclay overcomes these challenges.

13.3.3 Nanostarch

Over the decades, starch has commonly known material for packaging the food material because of several associated benefits like easy availability, renewable, biodegradable, biocompatibility, non-toxic, stable in air, eco-friendly, abundant supply, and also due to its low cost [66]. The inert inorganic material and artificial polymers upgrades itself the properties against water resistance [67]. In normal condition, starch present in separate and crystalline granules form which mostly composed of two main glucosidic macromolecules like linear amylose and amylopectin which has branched pattern. In the process of producing the nanostarch, it is necessary to split its native form of starch into crystalline starch. Enzyme hydrolysis, acid hydrolysis, mechanical micro-fluidizer treatment, and precipitation are common methods are being used for amylose extraction [68, 69]. In the process of acid hydrolysis, the starch is incubated for an extended time of hydrolysis at lower the gelatinization temperature point of diluted acid. The amorphous regions have been targeted by acid molecules during the hydrolysis and separation of crystalline lamellae. Due to its natural and low-cost binder properties, nanocrystals of starch are found to be a promising alternative option to artificial emulsion latex. The artificial and dehydrated potential resins (e.g., glyoxalated poly acrylamide), which are processed using cationic starch can also used for the preparing nanopolymer binder. The crystalline particles of starch display platelet morphology with 6–8 nm in diameters can be used for the improvement of tensile potential and modulus of pullulan films, along with reduced the elongation property [61]. The antimicrobial agents have positive charge, which shows antimicrobial activity and utilized for adsorption of antimicrobial spectrum of metals onto polysaccharides surface leads to enhancement of facade region [70]. Starch nanocrystals are found to be favorable filler for packaging the food material due to improved its mechanical and obstacle properties as they are exceedingly vulnerable to moisture [71].

13.3.4 Carbon Nanotubes

CNTs are most practiced nanomaterial to developed intelligent sensors and antimicrobial polymers like polypropylene, nylon, and polylactic acid. CNTs are found in two forms: either single-walled nanotube (SWNT), generally, one atom thick; or multiwalled nanotubes (MWNT) where it comprises manifold concentric tubes along with elastic modulus at high ratios [72]. Both types of CNTs provide remarkable elevated elastic modulus (upto 1 TPa) and tensile power (upto 200 GPa) during the addition to the polymer surrounding substance. These antimicrobial properties directly assign in the direction of damage the cell membrane of microbes [50]. In the case of *E. coli* bacteria having negligible metal filtrate and very much toxic toward pure SWCNTs, and in other cases, when CNTs are combined with allylisothiocyanate, then they could inhibit the growth of Salmonella choleraesuis over 40 days of storage [73]. In another study, it was found that fungal attack can be stopped for up to 3 months when CNTs combined with polyethylene films that are used for the packaging of Mazafati dates [74]. A complex of SWNTs and cobalt mesoarylporphyrin are prepared for the development of a chemiresistive detector for amines detection, which are produced during the meat spoilage [75].

13.3.5 Zinc Oxide Nanoparticles

ZnO NPs exhibit good inhibition activity against the growth of broad-spectrum bacterial species and further enhances with a decrease in particle size [76]. Its NPs require visible light for activation. ZnO NPs are extensively used in atmospheric processed (MAP) packaging, cosmetics, medication delivery, and medical devices. As compared to other NPs like AgNPs, ZnO NPs have frequently used for food packaging purposes due to its extra affordable with mild toxic characteristics for eukaryotic organisms [32]. ZnO produces a decent amount of H_2O_2 under ultraviolet light resulted for oxidative stress production. ZnO NPs also showed antimicrobial activity for several pathogenic microbes, bacteria, and germs with the effects of UV rays [77]. Earlier in the study, it was also suggested that Zn^+

ions are suitable for inhibition of bacterial growth in strains like bacteriostatic [8]. It has been reported that ZnO can be directly in contact with cell wall of microorganisms and destroy the reliability of bacterial cell by which antimicrobial ions like Zn^{2+} can be released and generated reactive oxygen species (ROS) [78]. It was observed that ZnO perform better against E. coli, *B. atrophaeus*, and *S. aureus* growth inhibition when compared to magnesium oxide (MgO) and calcium oxide (CaO) [79, 80]. In another study, it was reported that ZnO inhibited the growth of *S. sobrinus* biofilms nearly 80% in dental materials, and it displayed antibacterial effects for *S. enteritidis* monocyte genes [81]. The essential oils as a promising antimicrobial agent have nasty smell used in PBS film may be removed by ZnO and performed better antibacterial effects. Also, decreased fresh-cut apples decay, many companies use ethylene, malondialdehyde, polyphenol oxidase, and pyrogallol peroxidase activities. The addition of ZnO NPs can oxidize ethylene into H_2O and CO_2 followed by reduction and build up malondialdehyde, polyphenol oxidase, and activities under UV radiation. From this technique, shelf life of commodities can be extended [82, 83].

13.3.6 Titanium Dioxide

Titanium dioxide is among the very discover materials to be considered as very significant metal oxide nanomaterials that have thermo-stability properties. Naturally, on the basis of its crystal size, it is classified in three main phases, *viz.*, anatase, rutile, and brookite [84]. Titanium dioxide is cheaper, nontoxic, and photostable in nature along with the capability of property to change biodegradable films. The USFDA approved NMs for food supplement and material for contact to food in 1996 [85]. In food packaging, titanium dioxide is typically used as a pigment and coating additive, and the white NPs of titanium dioxide successfully dispersed visible light. By the use of this nanoparticle, more brightness, white light, and power can be obtained using coated object. In dairy industries, titanium dioxide NPs along with milk after pasteurization, followed by packaging in PET and HDPE bottles after co-extruded, effectively reduce the adverse effects of light on milk quality [86]. A thin coating of titanium dioxide in food packaging materials is totally non-volatile, non-flammable, insoluble, and completely inert in all foods and food substances [35]. In 2011, the European Food Safety Authority (EFSA) approved the titanium nitride (TiN), which is best example of popular titanium-based products used to made plastics for food packaging. For economic and environmental fields, it acts as a whitener or photocatalyst and used for food packaging materials. It has antimicrobial, self-cleaning structures, with water filtration activity,

Table 13.1 Nanotechnology in packaging for food preservation.

Nanomaterial	Packaging materials	Nanomaterial function	Reference
AgNPs-organoclay	Cellulose acetate	Enhanced UV and O_2 blockade, tensile, antioxidant, and antimicrobial activity with reduced optical clarity and thermal stability	[103]
Titanium dioxide	Zein or chitosan	Improved antibacterial, hydrophobicity, and mechanical properties with thermal stability	[104]
ZnO	Soya protein	Improved antimicrobial activity and tensile strength (16% and 234%, respectively) with thermal stability and mechanical properties	[105]
Essential oil of Cinnamon	Chitosan	Improved the tensile power with antimicrobial activity with reduced WVP and EAB	[106]

(*Continued*)

Table 13.1 Nanotechnology in packaging for food preservation. (*Continued*)

Nanomaterial	Packaging materials	Nanomaterial function	Reference
AgNPs and CNC	Cellulose	Enhanced mechanical properties Increase antibacterial and antioxidant activities	[107]
Halloysite nanotubes	Lysozyme	Antimicrobial and mechanical properties are increased with more elastic modulus	[108]
TiO_2	Starch/kefiran	Improved miscibility and compatibility	[109]
Halloysite nanotubes	Lysozyme	Reinforced WVP and mechanical properties decreased elongation at break	[110]
TiO_2	Starch	Increased the UV-barrier hydrophobicity and thermal stability with reduction	[111]
Nanoclay/zein Essential oil	Sodium bentonite	Enhanced antimicrobial, water vapor permeability, WS, and mechanical properties	[112]

(*Continued*)

Table 13.1 Nanotechnology in packaging for food preservation. (*Continued*)

Nanomaterial	Packaging materials	Nanomaterial function	Reference
ZnO	Starch	Enhanced tensile, mechanical, and WVP properties	[113]
ZnO/CuO	Carrageenan hydrogel	Enhanced and improved antibacterial, thermal stability, and water-holding capability	[114]
MgO	Chitosan	Improved UV and moisture obstacle, mechanical activity, thermal stability, and flame retardant	[115]
TiO_2	Chitosan/PVA	Increased antimicrobial, mechanical, tensile strength, film thickness, transparency, and barrier activities	[116]
Nano-ZnO	Chitosan coated on PE film	Enhanced and improved solubility, and antimicrobial activity	[117]
Zein NPs	Whey protein isolate (WPI)	Enhanced mechanical properties and water vapor barrier	[118]

(*Continued*)

Table 13.1 Nanotechnology in packaging for food preservation. (*Continued*)

Nanomaterial	Packaging materials	Nanomaterial function	Reference
TiO_2	Amylose starch/PVA	Improved water resistance, antibacterial, and mechanical properties	[75]
ZnO NPS	Chitosan	Increased antimicrobial, tensile strength, film thickness, transparency, and barrier like mechanical properties	[119]
Nano-graphene	Chitosan–tapioca starch	Enhanced thermal and mechanical properties reduction of water vapor permeability	[120]
ZnO NPs	Bacterial polyester poly(3hydroxybutyrate-co-3hydroxyvalerate) (PHBV)	Enhanced thermal and mechanical properties with antimicrobial and tensile strength (16% and 234%, respectively)	[121]
MgO	Chitosan	Enhanced antimicrobial activity with tensile strength, film thickness, transparency, and barrier like mechanical properties	[122]

(*Continued*)

Table 13.1 Nanotechnology in packaging for food preservation. (*Continued*)

Nanomaterial	Packaging materials	Nanomaterial function	Reference
CNC	Cellulose nanocrystals	Increased thermal stability and mechanical properties with obstacle movement	[123]
Nanoclay(NC)	Polylactide (PLA)	Significant oxygen barrier improvement	[124]
PMMA-coAAm (Poly(methyl methacrylate-coacrylamide)	Corn starch–PVA	Increased water resistance and mechanical properties with degradability (45%–65%)	[125]
Nanoclay	WPI	Improved tensile strength and water vapor barrier like mechanical properties	[126]

and also utilized for purification of air and water [87]. Titanium dioxide is non-toxic to animals and humans; it oxidized the unsaturated poly phospholipid component of a microbial cell membrane, and therefore, it acts as antimicrobial or a biocidal for pathogens. Due to photocatalytic activity of titanium dioxide, become more reactive to attach the biological molecules, namely, phosphorylated proteins, peptides, and DNA to its surface at the nanoscale [88–90]. Its NP facade force increases with size, which has a significant factor in polymer/filler interaction. Titanium dioxide mostly used for preparing several nanomaterials like NPs, nanorods, nanotubes, nanowires, nanoporous, and mesoporous, those have antibacterial properties are confined by UV irradiation exposure [91–94]. Several studies also suggested that ROS like superoxide anions, hydrogen peroxide, and hydroxyl radicals are generated through antimicrobial effects of titanium dioxide under UV light or sunlight may damaged the cell membrane of microbes directly [95, 96]. Moreover, its exact mechanism of biocidal activity is not well known; it may initially attack to outer/inner bacterial cell membrane by oxidative stress and alter and modify the specific Coenzyme A activity and finally damage the DNA using hydroxyl radicals [97]. Earlier, it was studied the effect of titanium dioxide NP on different nine microorganisms, *viz.*, *B. stearothermophilus*, *S. aureus*, *E. coli*, *P. fluorescens*, *Bacillus* sp., *L. plantarum*, *E. caratovora*, *P. jadinii*, and *Z. rouxii* and suggested that biocidal properties of this nanoparticle after photoactivtion of EVOH films against these nine microorganism [98]. It was found that, in the presence of the TiO_2/EVOH materials, *B. stearothermophilus*, *Bacillus sp.*, *L. plantarum*, and *P. jadinii* get over 5-log reduction after 30 min of irradiation. In the case of *E. coli*, the fresh cut of lettuce showed 3-log reduction after the TiO_2 NPs coated oriented-polypropylene (OPP) film treatment from 3 hours of illumination [99]. In another study, it was found that titanium dioxide NP-coated plastic films reduced the spoilage in apple, tomato, and lemons against *Penicillium expansum*, which indicates the suppression growth using titanium dioxide NPs photocatalytic activity with light exposure [100]. It was also found that titanium dioxide along with silver NPs enhanced its antimicrobial properties significantly [101, 102]. Types of NP used in food packaging and their function were described in Table 13.1.

13.4 Advantages of Nanopackaging

A nutritional meal is an important element of a balanced diet and lifestyle. It becomes a very challenging task for food industries to avoid food spoilage by air, oxygen, and microbes. It is believed that nanotechnology persuades diverse sector in food industries, enhancing the shelf span of packaging food materials. Nanotechnology plays an important role for many industries related to food through the improvement of quality, processing and packaging of food, efficiently increased the shelf life of packaging food, and long-term storage capacity, which provides huge growth in food industries through enhancing food quality by improving physicochemical properties of food and minimizing microbial loads. The hydrophobic nature of many bioactive against several diseases is showing least bioavailability and stability to increase bioavailability, and their targeted delivery of food bioactive compounds is mainly achieved by nanotechnology-based delivery systems. Using nanomaterial and nanosensors helps the consumers in providing information about food nutritional status, flavor or color, microbes, and contamination status of food packaging. Traditionally, the use of non-renewable fossil materials like petroleum-based plastic stuffs in food packaging materials conventionally creates problem in disposing and recycling. On the other hand, NPs not only are a good source of their substitution due to their biodegradability but also have the potential to become alternate for artificial plastic, easy accessibility, economically, and can minimize wide uses of harmful and non-renewable fossil fuels. Progressive uses of NPs in several branches of food industries for improving their nutritional value, quality, processing, and packaging have been globally reported. The uses of NPs, properties, and behavior in food industries seem important to design foodstuffs for their better quality, sustainability, safer, and healthier. The government, as well as other food industries, faced significant challenges for supply non-nanotechnology-based foodstuffs, besides this nanotechnology-based foods not only ensuring consumer confidence but also gain acceptance of nanofoods available in the market. The use of nanomaterial is much beneficial for packaging the food than food additive directly. To minimize its toxicity, NPs use with biopolymers in food packaging materials which also make it more sustainable and eco-friendly. Different types of NPs such as nanometals [copper(Cu) and silver (Ag)], metal oxides (MgO, TiO2, and ZnO), and many organic materials such these NP considerably attract food packaging industries because of their diversified application in packaging such as lox toxicity, antimicrobial activity, and its thermal stability [4, 7]. However, there is a proper guideline for the uses of NPs in foodstuffs, for example, the EFSA set the upper limitation of silver NPs uses in water and food (0.05 and 0.05 mg kg^{-1}, respectively). Although a very few studies have been performed on the effect of nano foods on human and animal health. Therefore, more observation needed on managing and regulating the use of NPs in packaging materials for improving food processing and their quality, which will make it beneficial to human health and eco-friendly.

13.5 Conclusion and Future Aspect

The nanotechnology is applicable in terms of preservation of food with enriched nutrition and also delivers micronutrients and bioactive components safely. In current chapter, it deeply discussed food packaging using advanced nanotechnology along with promising prospects and confronts and found that it is a key technology for food industry. In food sectors, it is very much important that nanomaterial production should be free from risk and health issues. Even though the application of promising nanotechnology are using in food industry, many more unmoving efforts are necessary for nano food structure using meticulous investigation so that public consumer attentiveness may be build up. In food preservation, there is less information is available for the use of nanomaterials and their mechanism. The nanomaterial are using for food preservation with minimum food spoilage and also provide an alternative to enhance quality and produce with more protection of food products preservation. On the other hand, the toxic effects may be raised on human health and environment using NMs, but a huge covenant of investigation has established and revealed the benefits of these different new nanomaterials in terms of food monitoring improvement, high quality and freshness with low toxicity, and easy degradation. Hence, more studies is a must for exact and deeply analysis of nanomaterials on food packaging and environment for having their full potential. The characterization method of nanomaterials with its properties along with their biotransformation in food must be developed to know about its risks. Recently, more opportunities have been build up using advancement of new nanotechnology method and solved many challenges in food sector by new sensing and food packaging. The shelf life of food has been extended, waste has been reduced, more protection has been assessed, and quality has been improved using nanotechnology. The several applications of NMs have been discussed in this chapter and revealed that mechanical strength can be emphasized, the properties for gas barrier may be enhanced with increasing the water repellence, and antimicrobial and scavenging activity may be provided for food packaging. The chemical and biological sensors facilitated the invention a fast and susceptible devices to assess novelty and also used against some pathogenic contaminants, allergens, and toxins. This chapter concludes the use of different nanomaterials in food packaging along with the active food packaging and environment safety and health issues of nanomaterials.

References

1. Sridhar, A., Ponnuchamy, M., Kumar, P.S., Kapoor, A., Food preservation techniques and nanotechnology for increased shelf life of fruits, vegetables, beverages and spices: a review. *Environ. Chem. Lett.*, 19, 1715, 2021.
2. Chellaram, C., Murugaboopathi, G., John, A.A., Sivakumar, R., Ganesan, S., Krithika, S., Priya, G., Significance of nanotechnology in food industry. *APCBEE Proc.*, 8, 109, 2014.
3. Janjarasskul, T. and Suppakul, P., Active and intelligent packaging: The indication of quality and safety. *Crit. Rev. Food Sci. Nutr.*, 58, 5, 808, 2018.

4. Agriopoulou, S., Stamatelopoulou, E., Skiada, V., Tsarouhas, P., Varzakas, T., Emerging Nanomaterial Applications for Food Packaging and Preservation: Safety Issues and Risk Assessment. *Proc.*, 70, 7, 2020.
5. Chandra, P., *Nanobiosensors for personalized and onsite biomedical diagnosis*, *Inst. Eng. Tech.*, London, 2016.
6. Momin, J.K. and Joshi, B.H., Nanotechnology in foods, in: *Nanotechnologies in Food and Agriculture*, Springer, Basel, pp. 3–24, 2015.
7. Primožic, M., Knez, Ž., Leitgeb, M., (Bio)Nanotechnology in Food Science-Food Packaging. *Nanomaterials*, 11, 292, 2021.
8. Seil, J.T. and Webster, T.J., Antimicrobial applications of nanotechnology: methods and literature. *Int. J. Nanomedicine*, 7, 2767, 2012.
9. Mihindukulasuriya, S.D.F. and Lim, L.T., Nanotechnology development in food packaging: A review. *Trends Food Sci. Technol.*, 40, 2, 149, 2014.
10. Sandoval, B., Perspectives on FDA's regulation of nanotechnology: emerging challenges and potential solutions. *Compr. Rev. Food Sci. F.*, 8, 4, 375, 2009.
11. Das, M.P., Livingstone, J.R., Veluswamy, P., Das, J., Exploration of Wedelia chinensis leaf-assisted silver nanoparticles for antioxidant, antibacterial and *in vitro* cytotoxic applications. *J. Food Drug Anal.*, 26, 2, 917, 2018.
12. He, X. and Hwang, H.M., Nanotechnology in food science: Functionality, applicability, and safety assessment. *J. Food Drug Anal.*, 24, 4, 671, 2016.
13. Khaneghah, A.M., Hashemi, S.M.B., Limbo, S., Antimicrobial agents and packaging systems in antimicrobial active food packaging: An overview of approaches and interactions. *Food Bioprod. Process.*, 111, 1, 2018.
14. Bondarenko, O.M., Sihtmäe, M., Kuzmičiova, J., Ragelienė, L., Kahru, A., Daugelavičius, R., Plasma membrane is the target of rapid antibacterial action of silver nanoparticles in Escherichia coli and Pseudomonas aeruginosa. *Int. J. Nanomedicine.*, 13, 6779, 2018.
15. Biswas, M.C., Tiimob, B.J., Abdela, W., Jeelani, S., Rangari, V.K., Nano silica-carbon-silver ternary hybrid induced antimicrobial composite films for food packaging application. *Food Packag. Shelf Life*, 19, 104, 2019.
16. Hoseinnejad, M., Jafari, S.M., Katouzian, I., Inorganic and metal nanoparticles and their antimicrobial activity in food packaging applications. *Crit. Rev. Microbiol.*, 44, 2, 161, 2018.
17. Amirsoleimani, M., Khalilzadeh, M.A., Sadeghifar, F., Sadeghifar, H., Surface modification of nanosatrch using nano silver: a potential antibacterial for food package coating. *J. Food Sci. Technol.*, 55, 3, 899, 2018.
18. Garcia, C.V., Shin, G.H., Kim, J.T., Metal oxide-based nanocomposites in food packaging: Applications, migration, and regulations. *Trends Food Sci. Technol.*, 82, 21, 20182018.
19. Weiss, J., Takhistov, P., Mcclements, D.J., Functional materials in food nanotechnology. *J. Food Sci.*, 71, 9, 107, 2006.

20. Ravichandran, R., Nanotechnology applications in food and food processing: innovative green approaches, opportunities and uncertainties for global market. *Int. J. Green Nanotechnol.: Phys. Chem.*, 1, 2, 72, 2010.
21. Bumbudsanpharoke, N. and Ko, S., Nano-food packaging: an overview of market, migration research, and safety regulations. *J. Food Sci.*, 80, 5, 910, 2015.
22. Trujillo, L.E., Ávalos, R., Granda, S., Guerra, L.S., País-Chanfrau, J.M., Nanotechnology applications for food and bioprocessing industries. *Med. Biol.*, 8, 3, 1, 2016.
23. Mustafa, F. and Andreescu, S., Nanotechnology-based approaches for food sensing and packaging applications. *RSC Adv.*, 10, 19309, 2020.
24. Sorrentino, A., Gorrasi, G., Vittoria, V., Potential perspectives of bio-nanocomposites for food packaging applications. *Trends Food Sci. Technol.*, 18, 2, 84, 2007.
25. Duncan, T.V., Applications of nanotechnology in food packaging and food safety: barrier materials, antimicrobials and sensors. *J. Colloid Interface Sci.*, 363, 1, 1, 2011.
26. Sahoo, M., Vishwakarma, S., Panigrahi, C., Kumar, J., Nanotechnology: Current applications and future scope in food. *Food Front.*, 2, 3, 2021.
27. Kerry, J.P., O'grady, M.N., Hogan, S.A., Past, current and potential utilisation of active and intelligent packaging systems for meat and muscle-based products: A review. *Meat Sci.*, 74, 1, 113, 2006.
28. He, X., Deng, H., Hwang, H.M., The current application of nanotechnology in food and agriculture. *J. Food Drug Anal.*, *27*, 1, 1, 2019.
29. Ranjan, S., Dasgupta, N., Chakraborty, A.R., Samuel, S.M., Ramalingam, C., Shanker, R., Kumar, A., Nanoscience and nanotechnologies in food industries: opportunities and research trends. *J. Nanopart. Res.*, 16, 6, 1, 20142014.
30. Majeed, K., Jawaid, M., Hassan, A.A.B.A.A., Bakar, A.A., Khalil, H.A., Salema, A.A., Inuwa, I., Potential materials for food packaging from nanoclay/natural fibres filled hybrid composites. *Mater. Des.*, 46, 391, 2013.
31. Bodaghi, H., Mostofi, Y., Oromiehie, A., Zamani, Z., Ghanbarzadeh, B., Costa, C., Conte, A., Del Nobile, M.A., Evaluation of the photocatalytic antimicrobial effects of a TiO2 nanocomposite food packaging film by *in vitro* and *in vivo* tests. *LWT-FOOD Sci. Technol.*, 50, 2, 702, 2013.
32. Silvestre, C., Duraccio, D., Cimmino, S., Food packaging based on polymer nanomaterials. *Prog. Polym. Sci.*, 36, 12, 1766, 2011.
33. Schirmer, B.C., Heiberg, R., Eie, T., Møretrø, T., Maugesten, T., Carlehøg, M., Langsrud, S., A novel packaging method with a dissolving CO2 headspace combined with organic acids prolongs the shelf life of fresh salmon. *Int. J. Food Microbiol.*, 133, 1–2, 154, 2009.
34. Soares, N.D.F., Pires, A.C.S., Camilloto, G.P., Santiago-Silva, P., Espitia, P.J.P., Silva, W.A., Recent patents on active packaging for food application. *Recent Pat. Food Nutr. Agric.*, 1, 2, 171, 2009.

35. Bradley, E.L., Castle, L., Chaudhry, Q., Applications of nanomaterials in food packaging with a consideration of opportunities for developing countries. *Trends Food Sci. Technol.*, 22, 11, 604, 2011.
36. Tan, A., Farhatnia, Y., de Mel, A., Rajadas, J., Alavijeh, M.S., Seifalian, A.M., Inception to actualization: next generation coronary stent coatings incorporating nanotechnology. *J. Biotechnol.*, 164, 1, 151, 2013.
37. Cho, Y.J., Kim, C.J., Kim, N., Kim, C.T., Park, B., Some cases in applications of nanotechnology to food and agricultural systems. *Biochip J.*, 2, 3, 183, 20082008.
38. Schaude, C., Meindl, C., Fröhlich, E., Attard, J., Mohr, G.J., Developing a sensor layer for the optical detection of amines during food spoilage. *Talanta*, 170, 481, 2017.
39. El-Nour, K.M.A., Salam, E.T.A., Soliman, H.M., Orabi, A.S., Gold nanoparticles as a direct and rapid sensor for sensitive analytical detection of biogenic amines. *Nanoscale Res. Lett.*, 12, 1, 1, 2017.
40. Kuswandi, B. and Nurfawaidi, A., On-package dual sensors label based on pH indicators for real-time monitoring of beef freshness. *Food Control*, 82, 91, 2017.
41. De Chiara, M.L.V., Pal, S., Licciulli, A., Amodio, M.L., Colelli, G., Photocatalytic degradation of ethylene on mesoporous TiO2/SiO2 nanocomposites: Effects on the ripening of mature green tomatoes. *Biosyst. Eng.*, 132, 61, 2015.
42. Lang, C. and Hübert, T., A colour ripeness indicator for apples. *Food Bioprocess Tech.*, 5, 8, 3244, 2012.
43. Niu, G. and Chen, X., When radionuclides meet nanoparticles. *Nat. Nanotechnol.*, 13, 5, 359, 2018.
44. Coffer, J.L. and Canham, L.T., Nanoporous Silicon as a Green, High-Tech Educational Tool. *Nanomaterials*, 11, 2, 553, 2021.
45. Sharma, C., Dhiman, R., Rokana, N., Panwar, H., Nanotechnology: an untapped resource for food packaging. *Front. Microbial.*, 8, 1735, 2017.
46. Wang, L., Hu, C., Shao, L., The antimicrobial activity of nanoparticles: present situation and prospects for the future. *Int. J. Nanomedicine.*, 12, 1227, 2017.
47. Suh, S., Meng, X., Ko, S., Proof of concept study for different-sized chitosan nanoparticles as carbon dioxide (CO2) indicators in food quality monitoring. *Talanta*, 161, 265, 2016.
48. Han, J.W., Ruiz-Garcia, L., Qian, J.P., Yang, X.T., Food packaging: A comprehensive review and future trends. *Compr. Rev. Food Sci. F.*, 17, 4, 860, 2018.
49. Youssef, A.M. and El-Sayed, S.M., Bionanocomposites materials for food packaging applications: Concepts and future outlook. *Carbohydr. Polym.*, *193*, 19, 2018.
50. Kuswandi, B., Environmental friendly food nano-packaging. *Environ. Chem. Lett.*, 15, 2, 205, 2017.

51. Zhao, R., Torley, P., Halley, P.J., Emerging biodegradable materials: starch- and protein-based bio-nanocomposites. *J. Mater. Sci.*, 43, 9, 3058, 2008.
52. Sondi, I. and Salopek-Sondi, B., Silver nanoparticles as antimicrobial agent: a case study on E. coli as a model for Gram-negative bacteria. *J. Colloid Interface Sci.*, 275, 1, 177, 2004.
53. Marambio-Jones, C. and Hoek, E.M., A review of the antibacterial effects of silver nanomaterials and potential implications for human health and the environment. *J. Nanopart. Res.*, 12, 5, 1531, 2010.
54. Nowack, B., Krug, H.F., Height, M., 120 years of nanosilver history: implications for policy makers. *Environ. Sci. Technol.*, 45, 1177, 2011.
55. Simbine, E.O., Rodrigues, L.D.C., Lapa-Guimaraes, J., Kamimura, E.S., Corassin, C.H., OLIVEIRA, C.A.F.D., Application of silver nanoparticles in food packages: a review. *Food Sci. Technol. Res.*, 39, 4, 793, 2019.
56. Emamifar, A., Kadivar, M., Shahedi, M., Soleimanian-Zad, S., Evaluation of nanocomposite packaging containing Ag and ZnO on shelf life of fresh orange juice. *INNOV. Food Sci. Emerg. Technol.*, 11, 4, 742, 2010.
57. Ottman, N., Ruokolainen, L., Suomalainen, A., Sinkko, H., Karisola, P., Lehtimäki, J., Lehto, M., Hanski, I., Alenius, H., Fyhrquist, N., Soil exposure modifies the gut microbiota and supports immune tolerance in a mouse model. *J. Allergy Clin. Immunol.*, 143, 3, 1198, 2019.
58. Bumbudsanpharoke, N. and Ko, S., Nanoclays in food and beverage packaging. *J. Nanomater.*, 8927167, 1, 2019. https://doi.org/10.1155/2019/8927167.".
59. De Azeredo, H.M., Nanocomposites for food packaging applications. *Food Res. Int.*, 42, 1240, 2009.
60. De Paiva, L.B., Morales, A.R., Díaz, F.R.V., Organoclays: properties, preparation and applications. *Appl. Clay Sci.*, 42, 1–2, 8, 2008.
61. Arora, A. and Padua, G.W., Review: nanocomposites in food packaging. *J. Food Sci.*, 75, 43, 2010.
62. Mallakpour, S. and Dinari, M., Surface treated montmorillonite: structural and thermal properties of chiral poly (amide-imide)/organoclay bionanocomposites containing natural amino acids. *J. Inorg. Organomet. Polym. Mater.*, 22, 5, 929, 2012.
63. Cushen, M., Kerry, J., Morris, M., Cruz-Romero, M., Cummins, E., Nanotechnologies in the food industry–Recent developments, risks and regulation. *Trends Food Sci. Technol.*, 24, 30, 2012.
64. Hatzigrigoriou, N.B. and Papaspyrides, C.D., Nanotechnology in plastic food-contact materials. *J. Appl. Polym. Sci.*, 122, 6, 3719, 2011.
65. Sekhon, B.S., Food nanotechnology–an overview. *Nanotechnol. Sci. Appl.*, 3, 1, 2010.
66. Wesley, S.J., Raja, P., Raj, A.A., Tiroutchelvamae, D., Review on - nanotechnology applications in food packaging and safety. *Int. J. Eng. Res.*, 3, 645, 2014.

67. Cyras, V.P., Manfredi, L.B., Ton-That, M.T., Vázquez, A., Physical and mechanical properties of thermoplastic starch/montmorillonite nanocomposite films. *Carbohydr. Polym.*, 73, 1, 55, 20082008.
68. Pérez-Pacheco, E., Canto-Pinto, J.C., Moo-Huchin, V.M., Estrada-Mota, I.A., Estrada-León, R.J., Chel-Guerrero, L., Thermoplastic starch (TPS)-cellulosic fibers composites: mechanical properties and water vapor barrier: a review, in: *Composites from Renewable and Sustainable Materials*, pp. 85–105, 2016.
69. Da Silva, N.M., Lima, F., Fialho, R.L., Albuquerque, E.C.M.C., Velasco, J.I., Fakhouri, F.M., Production and Characterization of Starch Nanoparticles, in: *Applications of Modified Starches*, p. 41, 20182018.
70. Arora, D., Sharma, N., Sharma, V., Abrol, V., Shankar, R., Jaglan, S., An update on polysaccharide-based nanomaterials for antimicrobial applications. *Appl. Microbiol. Biotechnol.*, 100, 6, 2603, 2016.
71. Salama, A., Diab, M.A., Abou-Zeid, R.E., Aljohani, H.A., Shoueir, K.R., Crosslinked alginate/silica/zinc oxide nanocomposite: a sustainable material with antibacterial properties. *Compos. Commun.*, 7, 7, 2018.
72. Moghadam, A.D., Omrani, E., Menezes, P.L., Rohatgi, P.K., Mechanical and tribological properties of self-lubricating metal matrix nanocomposites reinforced by carbon nanotubes (CNTs) and graphene–a review. *Compos. B. Eng.*, 77, 402, 2015.
73. Dias, M.V., Nilda de Fátima, F.S., Borges, S.V., de Sousa, M.M., Nunes, C.A., de Oliveira, I.R.N., Medeiros, E.A.A., Use of allyl isothiocyanate and carbon nanotubes in an antimicrobial film to package shredded, cooked chicken meat. *Food Chem.*, *141*, 3, 3160, 2013.
74. Asgari, P., Moradi, O., Tajeddin, B., The effect of nanocomposite packaging carbon nanotube base on organoleptic and fungal growth of Mazafati brand dates. *Int. Nano Lett.*, *4*, 1, 98, 2014.
75. Liu, C., Xiong, H., Chen, X., Lin, S., Tu, Y., Effects of nano-tio2 on the performance of high-amylose starch based antibacterial films. *J. Appl. Polym. Sci.*, 132, 32, 42339, 2015.
76. Azam, A., Ahmed, A.S., Oves, M., Khan, M.S., Habib, S.S., Memic, A., Antimicrobial activity of metal oxide nanoparticles against Gram-positive and Gram-negative bacteria: a comparative study. *Int. J. Nanomedicine*, 7, 6003, 2012.
77. Mizielińska, M., Kowalska, U., Jarosz, M., Sumińska, P., Landercy, N., Duquesne, E., The effect of UV aging on antimicrobial and mechanical properties of pla films with incorporated zinc oxide nanoparticles. *Int. J. Environ. Res. Public Health*, 15, 4, 794, 2018.
78. Sirelkhatim, A., Mahmud, S., Seeni, A., Kaus, N.H.M., Ann, L.C., Bakhori, S.K.M., Hasan, H., Mohamad, D., Review on zinc oxide nanoparticles: antibacterial activity and toxicity mechanism. *Nano-micro Lett.*, 7, 3, 219, 2015.
79. Sawai, J., Quantitative evaluation of antibacterial activities of metallic oxide powders (ZnO, MgO and CaO) by conductimetric assay. *J. Microbiol. Methods*, 54, 2, 177, 2003.

80. Shi, L.E., Li, Z.H., Zheng, W., Zhao, Y.F., Jin, Y.F., Tang, Z.X., Synthesis, antibacterial activity, antibacterial mechanism and food applications of ZnO nanoparticles: a review. *Food Addit. Contam: Part A*, 31, 2, 173, 2014.
81. Jin, T., Sun, D., Su, J.Y., Zhang, H., Sue, H.J., Antimicrobial efficacy of zinc oxide quantum dots against Listeria monocytogenes, Salmonella enteritidis, and Escherichia coli O157: H7. *J. Food Sci.*, 74, 1, 46, 2009.
82. Murariu, M., Doumbia, A., Bonnaud, L., Dechief, A.L., Paint, Y., Ferreira, M., Campagne, C., Devaux, E., Dubois, P., High-performance polylactide/ZnO nanocomposites designed for films and fibers with special end-use properties. *Biomacromolecules*, 12, 5, 1762, 2011.
83. Zaman, H.U., Hun, P.D., Khan, R.A., Yoon, K.B., Morphology, mechanical, and crystallization behaviors of micro-and nano-ZnO filled polypropylene composites. *J. REINF. Plast. COMP.*, 31, 5, 323, 2012.
84. Naicker, P.K., Cummings, P.T., Zhang, H., Banfield, J.F., Characterization of titanium dioxide nanoparticles using molecular dynamics simulations. *J. Phys. Chem. B*, 109, 32, 15243, 2005.
85. Kamat, P.V., Graphene-based nanoassemblies for energy conversion. *J. Phys. Chem. Lett.*, 2, 3, 242, 2011.
86. Moyssiadi, T., Badeka, A., Kondyli, E., Vakirtzi, T., Savvaidis, I., Kontominas, M.G., Effect of light transmittance and oxygen permeability of various packaging materials on keeping quality of low fat pasteurized milk: chemical and sensorial aspects. *Int. Dairy J.*, 14, 5, 429, 2004.
87. Yadav, H.M., Kim, J.S., Pawar, S.H., Developments in photocatalytic antibacterial activity of nano TiO 2: A review. *Korean J. Chem. Eng.*, 33, 7, 1989, 2016.
88. Liang, S.S., Makamba, H., Huang, S.Y., Chen, S.H., Nano-titanium dioxide composites for the enrichment of phosphopeptides. *J. Chromatogr. A*, 1116, 1–2, 38, 2006.
89. Brown, E.M., Paunesku, T., Wu, A., Thurn, K.T., Haley, B., Clark, J., Priester, T., Woloschak, G.E., Methods for assessing DNA hybridization of peptide nucleic acid–titanium dioxide nanoconjugates. *Anal. Biochem.*, 383, 2, 226, 2008.
90. Ziental, D., Czarczynska-Goslinska, B., Mlynarczyk, D.T., Glowacka-Sobotta, A., Stanisz, B., Goslinski, T., Sobotta, L., Titanium Dioxide Nanoparticles: Prospects and Applications in Medicine. *Nanomaterials*, 10, 387, 2020.
91. Chen, X. and Mao, S.S., Synthesis of titanium dioxide (TiO2) nanomaterials. *J. Nanosci. Nanotechnol.*, 6, 4, 906, 2006.
92. Shi, L., Zhao, Y., Zhang, X., Su, H., Tan, T., Antibacterial and anti-mildew behavior of chitosan/nano-TiO 2 composite emulsion. *Korean J. Chem. Eng.*, 25, 6, 1434, 2008.
93. Macwan, D.P., Dave, P.N., Chaturvedi, S., A review on nano-TiO 2 sol–gel type syntheses and its applications. *J. Mater. Sci.*, 46, 11, 3669, 2011.

94. Montazer, M. and Seifollahzadeh, S., Enhanced self-cleaning, antibacterial and UV protection properties of nano TiO2 treated textile through enzymatic pretreatment. *Photochem. Photobiol.*, 87, 877, 2011.
95. Xing, Y., Li, X., Zhang, L., Xu, Q., Che, Z., Li, W., Bai, Y., Li, K., Effect of TiO2 nanoparticles on the antibacterial and physical properties of polyethylene-based film. *Prog. Org. Coat.*, 73, 2–3, 219, 20122012.
96. Azizi-Lalabadi, M., Ehsani, A., Divband, B., Alizadeh-Sani, M., Antimicrobial activity of Titanium dioxide and Zinc oxide nanoparticles supported in 4A zeolite and evaluation the morphological characteristic. *Sci. Rep.*, 9, 1, 1, 2019.
97. Kubacka, A., Diez, M.S., Rojo, D., Bargiela, R., Ciordia, S., Zapico, I., Understanding the antimicrobial mechanism of TiO2-based nanocomposite films in a pathogenic bacterium. *Sci. Rep.*, 4, 4134, 2014.
98. Cerrada, M.L., Serrano, C., Sánchez-Chaves, M., Fernández-García, M., Fernández-Martín, F., de Andres, A., Rioboo, R.J.J., Kubacka, A., Ferrer, M., Fernández-García, M., Self-sterilized EVOH-TiO2 nanocomposites: interface effects on biocidal properties. *Adv. Funct. Mater.*, 18, 13, 1949, 2008.
99. Chawengkijwanich, C. and Hayata, Y., Development of TiO2 powder-coated food packaging film and its ability to inactivate Escherichia coli *in vitro* and in actual tests. *Int. J. Food Microbiol.*, 123, 3, 288, 2008.
100. Maneerat, C. and Hayata, Y., Antifungal activity of TiO2 photocatalysis against Penicillium expansum *in vitro* and in fruit tests. *Int. J. Food Microbiol.*, 107, 2, 99, 2006.
101. Li, X., Xing, Y., Jiang, Y., Ding, Y., Li, W., Antimicrobial activities of ZnO powder-coated PVC film to inactivate food pathogens. *Int. J. Food Sci. Technol.*, 44, 11, 2161, 2009.
102. Wu, T.S., Wang, K.X., Li, G.D., Sun, S.Y., Sun, J., Chen, J.S., Montmorillonite-supported Ag/TiO2 nanoparticles: an efficient visible-light bacteria photodegradation material. *ACS Appl. Mater. Interfaces*, 2, 2, 544, 2010.
103. Dairi, N., Ferfera-Harrar, H., Ramos, M., Garrigós, M.C., Cellulose acetate/AgNPs-organoclay and/or thymol nano-biocomposite films with combined antimicrobial/antioxidant properties for active food packaging use. *Int. J. Biol. Macromol.*, 121, 508, 2019.
104. Qu, L., Chen, G., Dong, S., Huo, Y., Yin, Z., Li, S., Chen, Y., Improved mechanical and antimicrobial properties of zein/chitosan films by adding highly dispersed nano-TiO2. *Ind. Crops Prod.*, 130, 450, 2019.
105. Tang, S., Wang, Z., Li, W., Li, M., Deng, Q., Wang, Y., Li, C., Chu, P.K., Ecofriendly and biodegradable soybean protein isolate films incorporated with ZnO nanoparticles for food packaging. *ACS Appl. Bio Mater.*, 2, 5, 2202, 2019.
106. Vahedikia, N., Garavand, F., Tajeddin, B., Cacciotti, I., Jafari, S.M., Omidi, T., Zahedi, Z., Biodegradable zein film composites reinforced with chitosan nanoparticles and cinnamon essential oil: Physical, mechanical, structural and antimicrobial attributes. *Colloids Surf. B Biointerfaces*, 177, 25, 2019.

107. Wu, Z., Deng, W., Luo, J., Deng, D., Multifunctional nano-cellulose composite films with grape seed extracts and immobilized silver nanoparticles. *Carbohydr. Polym.*, 205, 447, 2019.
108. Bugatti, V., Vertuccio, L., Viscusi, G., Gorrasi, G., Antimicrobial membranes of bio-based PA 11 and HNTs filled with lysozyme obtained by an electrospinning process. *Nanomaterials*, 8, 3, 139, 2018.
109. Goudarzi, V. and Shahabi-Ghahfarrokhi, I., Development of photo-modified starch/kefiran/TiO2 bio-nanocomposite as an environmentally-friendly food packaging material. *Int. J. Biol. Macromol.*, 116, 1082, 20182018.
110. Bugatti, V., Sorrentino, A., Gorrasi, G., Encapsulation of Lysozyme into halloysite nanotubes and dispersion in PLA: Structural and physical properties and controlled release analysis. *Eur. Polym. J.*, 93, 495, 2017.
111. Goudarzi, V., Shahabi-Ghahfarrokhi, I., Babaei-Ghazvini, A., Preparation of eco-friendly UV-protective food packaging material by starch/TiO2 bio-nanocomposite: Characterization. *Int. J. Biol. Macromol.*, 95, 306, 2017.
112. Kashiri, M., Maghsoudlo, Y., Khomeiri, M., Incorporating Zataria multiflora Boiss essential oil and sodium bentonite nano-clay open a new perspective to use zein films as bioactive packaging materials. *Food Sci. Technol. Res.*, 23, 7, 582, 2017.
113. Mirjalili, F. and Yassini Ardekani, A., Preparation and characterization of starch film accompanied with ZnO nanoparticles. *J. Food Process Eng.*, 40, 6, 12561, 2017.
114. Oun, A.A. and Rhim, J.W., Carrageenan-based hydrogels and films: Effect of ZnO and CuO nanoparticles on the physical, mechanical, and antimicrobial properties. *Food Hydrocoll.*, 67, 45, 2017.
115. De Silva, R.T., Mantilaka, M.M.M.G.P.G., Ratnayake, S.P., Amaratunga, G.A.J., de Silva, K.N., Nano-MgO reinforced chitosan nanocomposites for high performance packaging applications with improved mechanical, thermal and barrier properties. *Carbohydr. Polym.*, 157, 739, 2017.
116. Lian, Z., Zhang, Y., Zhao, Y., Nano-TiO2 particles and high hydrostatic pressure treatment for improving functionality of polyvinyl alcohol and chitosan composite films and nano-TiO2 migration from film matrix in food simulants. *INNOV. Food Sci. Emerg. Technol.*, 33, 145, 2016.
117. Al-Naamani, L., Dobretsov, S., Dutta, J., Chitosan-zinc oxide nanoparticle composite coating for active food packaging applications. *INNOV. Food Sci. Emerg. Technol.*, 38, 231, 2016.
118. Oymaci, P. and Altinkaya, S.A., Improvement of barrier and mechanical properties of whey protein isolate based food packaging films by incorporation of zein nanoparticles as a novel bionanocomposite. *Food Hydrocoll.*, 54, 1, 2016.
119. Sanuja, S., Agalya, A., Umapathy, M.J., Synthesis and characterization of zinc oxide–neem oil–chitosan bionanocomposite for food packaging application. *Int. J. Biol. Macromol.*, 74, 76, 2015.

120. Ashori, A. and Bahrami, R., Modification of physico-mechanical properties of chitosan-tapioca starch blend films using nano graphene. *Polym. Plast. Technol. Eng.*, 53, 3, 312, 2014.
121. Díez-Pascual, A.M. and Diez-Vicente, A.L., ZnO-reinforced poly (3-hydro xybutyrate-co-3-hydroxyvalerate) bionanocomposites with antimicrobial function for food packaging. *ACS Appl. Mater. Interfaces*, 6, 12, 9822, 2014.
122. Sanuja, S., Agalya, A., Umapathy, M.J., Studies on magnesium oxide reinforced chitosan bionanocomposite incorporated with clove oil for active food packaging application. *Int. J. Polym. Mater. Polym. Biomater.*, 63, 14, 733, 2014.
123. Fortunati, E., Peltzer, M., Armentano, I., Torre, L., Jiménez, A., Kenny, J.M., Effects of modified cellulose nanocrystals on the barrier and migration properties of PLA nano-biocomposites. *Carbohydr. Polym.*, 90, 2, 948, 20122012.
124. Svagan, A.J., Åkesson, A., Cárdenas, M., Bulut, S., Knudsen, J.C., Risbo, J., Plackett, D., Transparent films based on PLA and montmorillonite with tunable oxygen barrier properties. *Biomacromolecules*, 13, 2, 397, 2012.
125. Yoon, S.D., Park, M.H., Byun, H.S., Mechanical and water barrier properties of starch/PVA composite films by adding nano-sized poly (methyl methacrylate-co-acrylamide) particles. *Carbohydr. Polym.*, 87, 1, 676, 2012.
126. Sothornvit, R., Rhim, J.W., Hong, S.I., Effect of nano-clay type on the physical and antimicrobial properties of whey protein isolate/clay composite films. *J. Food Eng.*, 91, 3, 468, 2009.

14

Food Science Nanotechnologies: Implementations, Recent Developments, and Prospects

Rokeya Akter[1,2], Tanima Bhattacharya[3] and Md. Habibur Rahman[2,4*]

[1]Department of Pharmacy, Jagannath University, Sadarghat, Dhaka, Bangladesh
[2]Department of Global Medical Science, Yonsei University Wonju College of Medicine, Yonsei University, Wonju, Gangwon-do, South Korea
[3]Hubei Collaborative Innovation Center for Advanced Organic Chemical Materials, Ministry of Education Key Laboratory for the Synthesis and Application of Organic Functional Molecules, College of Chemistry and Chemical Engineering, Hubei University, Wuhan, P. R. China
[4]Department of Pharmacy, Southeast University, Banani, Dhaka, Bangladesh

Abstract
Smart packaging, on-demand preservatives, and digital foods are now used in nano applications. Based on an on-demand food model, the idea of interactive foods was to allow consumers to adjust food in accordance with their individual nutritional needs or tastes. Nanomaterials have the ability to produce healthier, safer, and high-quality functionally consumable or semi-perishable foods in quality and quantity. In the case of food products with enhanced food safety and prevention of contamination and production, nanoservices are superior to conventional processing techniques. This extensive nanotechnology chapter addresses current developments and potential prospects in the food processing, packaging, protection, and storage of advanced nanomaterials. Nanotechnology applications increase the nutritional bioavailability, taste, texture, and quality of products obtained by altering particle size, clustering, and the surface charge of nanomaterials. Additionally, nutraceuticals mediated by nanodelivery, nanomaterials synergistic action in food security, and the use of nanosensors in intelligent food packaging to track food quality and common methods used to assess the impact of nanomaterials in biological systems are being addressed.

**Corresponding author*: pharmacisthabib@gmail.com
All authors contributed equally to this work

Annu, Tanima Bhattacharya and Shakeel Ahmed (eds.) Nanotechnology in Intelligent Food Packaging, (343–360) © 2022 Scrivener Publishing LLC

Keywords: Nanomaterials, bioavailability, nano-delivery, food science, food processing

14.1 Introduction

Nanotechnology is the manipulation of nanomaterials for various purposes. By using novel and creative approaches, it plays an important role in the food and agriculture industries [1, 2]. As a result of their high surface-to-volume ratios, modified solubility, and toxicity, engineered nanoparticles have attracted greater attention in medicine, agro-food industries, wastewater management, and other industries [3, 4]. Due to their potential antibiologically based properties, the volumes of other nanomaterials are 10-fold, and they are used in air filters, food storage containers, deodorizers, bandages, toothpastes, and paints [5, 6]. Nanoparticles of various sizes are used in food science nanotechnologies for the future development and processing of healthier, safer, and high-quality foods (Figure 14.1). Furthermore, nanosized copper oxides (nCuO) have found widespread use in commercial nano-biocide products due to their potent antibiotic activity [7]. Nanomaterials are small and small, insoluble or bio-persistent particles ranging from 1 to 100 nm in size, synthesized along various roads, and

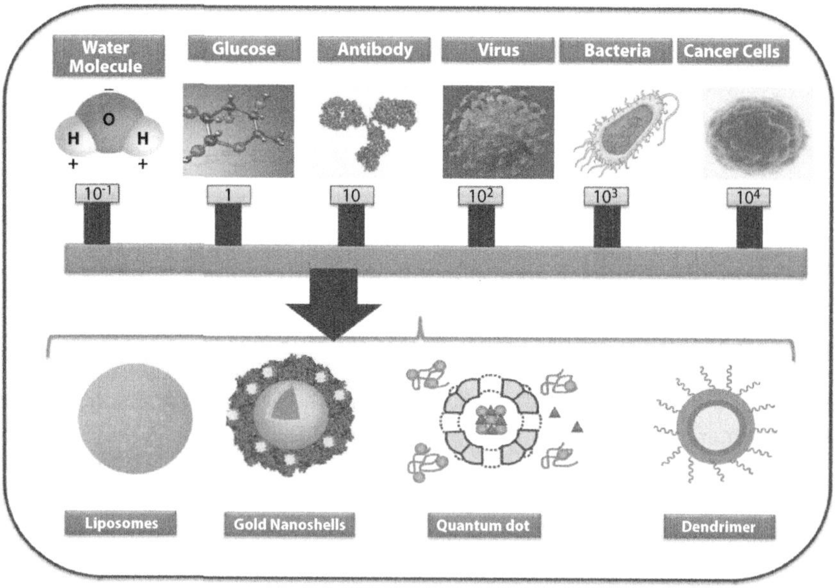

Figure 14.1 Nanoparticles of different sizes used in food science nanotechnologies.

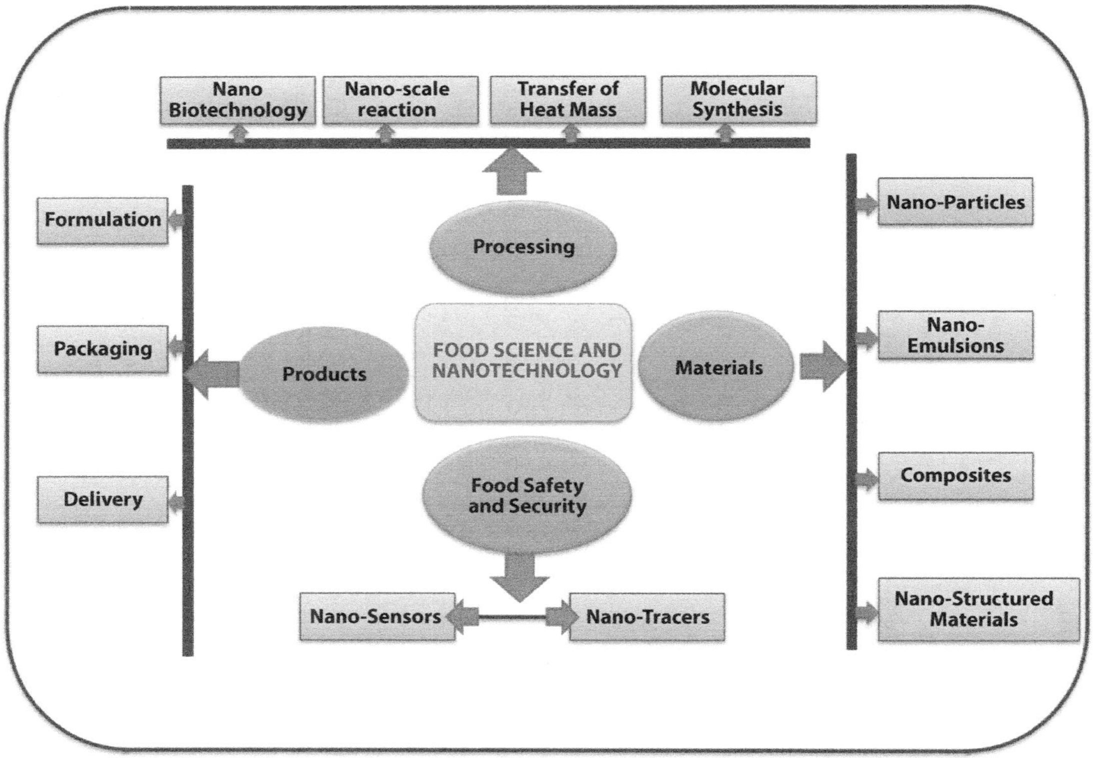

Figure 14.2 Nanotechnology in food science and technology.

used across a wide range of fields such as medicine, electronics, farming, and food processing [8]. Food waste is very expensive for the food industry. According to the Food and Agriculture Organization of the United Nations, more than 1.3 billion metric tons of consumable food is lost or lost in supply chains each year due to poor post-harvest processes, storage and transportation, as well as retail and consumer food waste [9]. In order to ease the food crisis resulting from population growth and environmental concerns, it is also necessary, apart from increasing food production, to reduce food waste. Food waste leads to microbial contamination and spoilage by reducing food quality and protection, reducing food supply, and increasing the likelihood of foodborne disease [10]. Nanotechnology is used in all aspects of the food industry, including food production, manufacturing, storage, and distribution (Figure 14.2). The use of nanosensors to detect any disease or containment in the food improves food safety. The enhanced food packaging from nanotechnology offers traditional plastic-based packaging and extends the shelf life of the food products to its functional elements such as antimicrobial properties. It also involves food contaminant identification, flavor development, and color creation [11]. In addition, nano-based delivery systems enhance food ingredients' nutritional value. Nanomaterials encourage plant growth in addition to their uses in the food industry. For example, TiO_2 has been shown to improve plant growth, gold nanoparticles have been shown to increase seed yield in *Arabidopsis*, and cellulose nanocrystals have been shown to improve seed germination due to their high water uptake ability [12]. Nanomaterials, such as carbon nanotubes, have antimicrobial properties. When carbon nanotubes aggregates came into direct contact with *E. coli*, they punctured the cell, causing cellular damage or death [13, 14]. In order to prepare high quality, pollution-free food, nano-biosensors are commonly used in the identification of carcinogenous pathogens. The future use and applications of nanotechnology in the food industry, including the production, preservation, storage, and protection of better shelf life and quality, are discussed in this study. Future applications of nanotechnology in nutraceuticals, the identification of food pathogens, and the potential adverse effects of nanotechnology on human and animal health are discussed in this current chapter.

14.2 Food Processing and Nanotechnology

In the processing, supply, safety, and packaging of foods, nanotechnology is used to create nanofoods, which are produced using nanotechnology.

The use of nanotechnology in food processing after harvest has great potential. He or she can mask an unwanted odor or flavor or improve the bioavailability of food. Particle size distribution, cluster potential, and surface charge are also improved [15]. Meat, fruits, vegetables, cheese, fast foods, bakery goods, and pastry products can benefit from edible nanoscailings, which act as a gas and moisture barrier. This is a great way to add flavor and color to processed foods, as well as providing enzymes and antioxidants, as well as extending the shelf life of processed foods. It is possible to coat-baked goods with antibacterial nano-coatings [16, 17]. In the preservation of the flavor and red wine, nanofilters have been used to strip color from beetroot juice and remove lactose from milk to replace with other sugars, rendering the milk ideal for patients with lactose-intolerance. Filters on a nanoscale aid in the removal from boiling of milk and water bacterial organisms. Nanomaterials may be used for the filtration of milk and beer to grow nanosieves [11, 17]. Nanotechnologies are used to prevent food-borne diseases in the processing of healthy foods that contain low levels of fats, sugar, and salt [18]. Bionanoencapsulated quercetin (biodeterminable poly-D, L-lactid) has been used to increase tomato shelf life and should be expanded to improve the shelf life of other plants and fruit [5]. Similarly, nanocapsules with tuna fish oil, nano-nourished juices, nanoshakes, and nanoceuticals containing slim shakes on the market are few nano-processing foodstuffs available on a commercial basis that is commonly sold in the USA, Australia, China, and Japan [19, 20].

14.3 Food Packaging for Nanotechnology

Fast 55%–65% of the USD 130 billion spent on food and drink packaging in the United States goes to the packaging industry [21]. Active and intelligent packaging systems have increased enormously with recent years in the use of animal-based food items that are vulnerable to contamination. The goal is to reduce spoilage and bypass contamination in meat and muscle products, to enrich tenderness through enzymatic action, to reduce weight loss, and to maintain a red cherry color in red meats [22]. Nanosensors can detect contaminants, pesticides, and microbial contamination in food products through flavor and color changes [23]. When used as carriers and as defense against microbial spoilage, nanoparticles in the food industry have antimicrobial activity. It serves as a barrier to microbes through the controlled release of antimicrobials from the substance inside [24].

As a result, plastics are enhanced with nanoparticles (100 nm or less). Nanocomposites are involved. Nanocomposites are thermoplastic polymers that contain 2%–8% of nanoscale materials, such as carbon particles, nanoclay, and polymer resins. Compared to their macroscale counterparts, nanocomposites have a higher surface-to-volume ratio [25]. The processing of reactive oxygen species is carried out for silver in silver zeolite and for food storage, and material decontamination and the disinfection of medical items ceramics covered with silver zeolite are used. The continuous antimicrobial role of silver nanocomposites is higher than silver zeolite [26]. The use of carbon nanotubes makes carbon dioxide (CO_2) or bad tastes easier to eliminate. In the manufacture of bottles and other food packaging materials, nanoclays in nanocomposites (bentonite) greatly improve gas barrier features and thus prevent oxygen and humidity due to diffusion or drinking and spoils of foodstuffs [27, 28].

It has been shown that the introduction of terrestrial nanoparticles into the biopolymer of ethyllin–vinyl alcohol and polylactic acid (PLA) has improved the oxygen barrier and increased food material shelf life [29]. Organic nanoclaves built into the polymer matrix have mechanical strength and prevent gases, volatility, and miniaturization [30]. Interestingly, the biodegradable biodegradation of the PLA bionanocomposite formed by the integration of nanofillers in the biodegradable polymer PLA is faster than its nanofilling counterpart. The mechanical, thermal, and barrier properties of the packaging material are significantly improved by the inclusion of polymer-clay nanocomposites [31]. There is a significant impact of nanotechnology on oxygen and moisture migration, respiratory rate, microbial growth, and volatile flavors and aromas [32]. In food packaging, applying silicate nanoparticles is a barrier to gases or humidity, reducing the spoilage and dryness of food [33]. As a barrier against oxygen, nanocoating of one to two icons is applied to food packs. Nanoemulsions are used to decontaminate food and food processing equipment. In addition to removing pesticide residues from fruit and vegetable oils and protein oils, glycerin, including nanomicellum-based products, also removes pesticide residues from soil. Flavors and nanoemulsions have no effect whatsoever on the appearance of a drink [34, 35]. Different food pathogens, such as gram-negative bacteria, control nanoemulsions significantly.

14.4 Intelligent Food Packaging Systems

Detecting pathogens or contamination, intelligent packaging systems respond to environmental stimuli by alerting consumers. In developing nanosensors for the detection of food pollutants, nanoparticles are used. Food analysis, taste or color identification, drinking water, and clinical diagnosis are just a few of the many applications for which nanosensors are used [36]. The use of nanosensors to trace physical, chemical, and biological changes on food processing helps for food packaging. Nanosensors and nanodevices that have been specially developed to detect food pathogens, chemical agents, and toxic substances in intelligent packaging [37]. During storage and transportation, sensors and indicators in smart packaging systems monitor the quality of the packaged food and provide information on its quality. Many practical nanomaterials can be used as nanosensors and active packaging materials with significant mechanical and barrier effects can be used as nutrient delivery systems [11, 38]. Measuring meat's integrity, freshness, and time-temperature was made possible by sensors [39]. Smart or smart packaging has been confirmed to maintain food quality during delivery. The specific sensor records the response of internal or external environmental stimuli to modifications. In the food packaging applications, the most widely employed metrics are integrity (package integrity determined), freshness (product quality), and time-temperature (change depending on time and temperature) [40].

These methods do however have some demerit; optical sensor approaches in real-time packing or in broad use are more successful than these methods [41, 42]. This is primarily due to an unpleasant odor that is difficult to detect with the human nose. Rotting foods can also lead to intoxication. In order to detect food poisoning, biosensors with high sensitivity are needed [43]. So, the electronic nose system relies on chemical sensors linked to a data processing system to collect and analyze data. In order to measure the chemical and physical properties of fruit smells, an electronic nasal signal was used [44]. As an example, it is possible to detect differences in strawberry scent, osmotic dehydration, and milk consistency in production by using the electronic nose. Food industry uses this software for very precise volatile determination and quality control processes. Nanosensors for food protection and quality control applications have also been implemented in the European project GOODFOOD (2004–2007) [45].

14.5 Antimicrobial Properties of Nanoparticles

Numerous nanoproducts for regulation of microbial growth are available on the market. When tested against *E. coli* O157: H7, *S. typhimurium*, *Vibrio parahaemolyticus*, and *Listeria monocytogenes*, NanoCid® L2000 was effective [46]. Many nanomaterial-related antimicrobial disinfectants have been listed in the nanotechnology consumer product inventory. Nanosilver is the primary antimicrobial agent for most of the antimicrobial products mentioned [47].

The antibacterial behavior of the nanoparticulated silver alone and of the silver NP integrated into the carboxymethyl cells film demonstrated the fact that AgNPs in the films of carboxymethyl cells are bactericidal rather than AgNPs alone [48]. One effective agent for suppressing the growth of biofilms and improved food quality and food protection is nano-engineered surfaces (antimicrobial lacquers). The UV-C ultraviolet-light-activated TiO_2 effectively controls problems of biocontamination in poultry farming, food processing, and food transport [49]. The transfer of pathogen mainly via air can be done at different phases of slaughter and processing and in poultry-meat contamination [50]. An antimicrobial effect was observed by incorporating microbicidal materials such as silver and other metallic materials into nanofibers [49]. Nanoenabler membranes, nanophotocatalysts, and nanoadsorbents are used in wastewater treatment [51].

14.6 Synergistic Antimicrobial Effects of Nanoparticles

In contrast to the nanoparticle, the synergistic effect of combined nanoparticles is a powerful, anti-microbial activity [52]. The *B. cereus* spores on aluminum and polyester surfaces killed the silver doped TiO2 NP. The silver-doped TiO2 NP destroys the bacteria and molds that the air-filters capture [53]. As a result of the stabilization of silver nanoparticles with SDS or PVP, the antimicrobial activity of silver nanoparticles against *E. coli* and *S. aureus* was increased. Silver NP coatings were applied to refrigerators and containers [54]. TiO_2NP UV active photoactivation of TiO_2 reportedly leads to biocidal activities against toxic microorganisms [55, 56].

14.7 Nutraceutical Delivery and Bioavailability Applications

Bioactive substances found in food provide immunity and protection from disease. Foods with higher bioactive molecule concentrations had a limited capacity. Poor bioavailability, lower solubility and stability of the gastrointestinal tract, and lower permeability and retention period are the primary reasons for this phenomenon [57]. Nanomaterials typically comprise a large area per unit mass and reduced particle size, improving biological activity, bioavailability, and solubility in embedded feeding stuffs [58]. Increased bioavailability and minimized color changes in the end products, nanosized iron, and zinc/nutraceutical supplies were used. Curcumin, carotenoids, linoleic acid conjugates, and coenzyme Q10 have low bioavailability [5]. Low bioavailability is a result of physicochemical and physiological parameters, including bioaccessibility, absorption, and transformation. In principle, the bioavailability, solubility, and stability of most bioactive molecules, such as antioxidants, vitamins, micronutrients, polyphenols, carotenoids, and food ingredients, can be improved through nanotechnology [59]. The majority of biologically active compounds used in the treatment of diseases are hydrophobic in nature and therefore less bioavailable than other compounds.

The nanoparticles in real food molecules include proteins, surfactants, and polysaccharides as well as lipids and cholesterol. Dietary nanoparticles are created by combining different components of the diet's ingredients, such as nanoemulsions and lipid droplets [60]. In addition, the encapsulated nature of these nanoparticles increases the effectiveness, use, and stability of bioactive food components [61]. Nutritional supplements, bioactive substances, functional foods, and herbal foods are manufactured using nanoformulation [62]. Diverse methods have been used to deliver nutraceuticals. Nanotubes, nanofibers, fullerenes, nanosheets, and nanoparticles can be used to produce liposomes, cubosomes, microemulsions, biopolymeric nanoparticles, nanosensors, monolayers, microgels, and nanoparticles [63]. The benefits and toxicity of nanocarrier systems in food products must be understood. In recent years, a major advance is being made in the design and manufacture of many food grade nanoparticles in order to establish an efficient micronutrient delivery system in food manufacturing [64]. The bioavailability of foods can be improved by colloidal feeds (excipients), although they themselves have no biological activity. As a result, the absorption of bioactive substances in the systemic circulation increases, resulting in increased bioactivity and a powerful benefit to health. Bioaccessibility, absorption, or transformation profiles of gastrointestinal (GTI) bioactive compounds were designed into the composition and structure of excipients foods [56, 65].

14.8 Food Technology and Nanoencapsulation

Nutraceutical molecules are encapsulated using nanoformulation to increase their bioavailability and biodistribution [66]. To allow for controlled release of the nucleus, substances are packed into tiny structures using nanostructures, nanoemulsions, or nanocomposites. Various nanoencapsulation (nanoparticles, liposomes, nanospheres, micelles, nanocochleates, and nanoemulsion) were used, depending on the need. They can be used as nutrition complements to cover up poor tastes and increase bio-availability and efficient dispersal without surfactants or emulsifiers of insoluble supplements [67, 68]. The non-toxic naturally non-caloric nanoparticles stevioside were used as nutraceutical antidiabetic agents [69]. Encapsulating catechin and epicatechin in BSA nanoparticles, tea stability and bioavailability have been improved [70]. Bioactive component nanoformulation, which ensures gradual release, preserves the antioxidant capability and increases the efficiency of bio-active molecules. Most natural plants are sensitive to different environments [71]. In addition to protecting the bioactive components, the entanglement inside the nanoparticles improves their function and stability (vitamins, antioxidants, phytochemicals, protein, lipid, and carbohydrates). These nanocapsules release active substances such as normal food when dissolved [72, 73]. Curcumin encapsulation in hydrophobic starch causes greater anticancer activity [74]. During nano-encapsulation, nanocoating can be used as a vehicle for functions, while nanocapsules can be developed to include the use, in animal feeding stuffs, of nanoadditives and antimicrobials Some of the compounds are bifunctional molecules, such as octenyl succinic anhydride-to-polylysine, that can be used as surfactants or emulsifiers and can be used to encapsulate bioactive compounds or drug products or antigemicrobials [75, 76]. Solubility and bioavailability are enhanced by nanoencapsulation, and unnecessary interactions with other food components are reduced as a consequence. Lipid nanoencapsulation systems such as liposomes and archaeosomes [77]. The charges of nitrates, enzymes, antimicrobial nutrients, and food additives can be used as nutrients [78]. Nanoencapsulation of probiotics has also been documented in previous studies. Food-supplemented probiotics are believed to be live mixtures of bacteria species. Yogurt, fruit juices, fermented milk, and pudding are examples of probiotic foods that can be consumed. Encapsulation of the products extends the product's shelf-life. The probiotic species formed with the nanoencapsulation technique which is introduced into specific areas of the GI tract in which the specific receptors are associated. The probiotics can be used to control immune reactions as novo vaccines [79, 80].

14.9 Mediated Delivery That is Specific to an Environmental Context

Diagnostic sensors and controlled release systems require nanoparticles that can change their properties in response to specific environmental stimuli, such as pH, enzyme activity, and temperature. NPs that can alter their properties depending on the environment were conducted in several studies. For example, after contact with lipase, bioactive molecules that carry lipid nanoparticles were broken down, and under simulated GIT conditions, bioactive molecules were released [81, 82]. The nanoparticles are filled with a bioactive product. Some nanoparticles are riboflavin whey loaded with protein, curcumin zine loaded with nanoparticles, and resveratrol [83]. pH-sensitive nanoparticles were used in the pharmaceutical industry to deliver anti-carcinogenic tissue molecules.

14.10 Nanomaterials in Food and Toxicological Aspects

Nanotechnology is becoming increasingly important, and people are increasingly concerned with nanomaterials' toxicity and environmental consequences. Nanoparticles mediated toxicity promotes dynamic, kinetic and catalytic properties, functional, net particles, agglomeration, and functioning [84, 85]. There are no harmful nanoparticles on the packaging surface, but their translocation and absorption into food can have an impact on health. Its cytotoxicity and genotoxicity have been discussed previously [86]. As a result of skin penetration, ingested substances or inhaled medications or implanted medical devices, the animal system is accessed. Cells interact with biological macromolecules. Because NPs are non-dissolvable and non-degradable, their toxicokinetic properties are affected [87]. Due to a lack of customer knowledge, government guidelines, policies, and detection methods for nanotechnology risk assessment, a better understanding of nanomaterial-based toxicity characterization and regulatory processes is required. Metallic nanoparticles become more toxic as they become smaller [88]. Highly responsive materials are nanoparticles which can easily cross membrane and capillary barriers and lead to different kinetic and dynamic toxic features. There are some nanoparticles that attach to proteins or enzymes to promote the production of ROS in addition to oxidative stress. Mitochondrial and apoptosis degeneration ROS accumulate in mitochondria [89]. Most animals have been subject to NP-imposed toxicity studies, which show that NP can cause high toxicity in different bodies (liver, kidneys, and immunosystems. There have been few *in vivo* studies using silver NP in mammals (e.g., mice and rats) [90].

14.11 Conclusion and Future Perspectives

In the food sector, nanotechnology plays an important role by improving quality food production, by improving the taste and texture and providing advanced processing, packaging, and long-term storage. The nanomaterials and nanosensors provide consumers with knowledge on the food status and nutrient status through the detection of pathogens. The majority of the nanotechnology-based delivery systems is hydrophobic in nature and thus provides an enhanced bioavailability and a targeted supply of nanoactive compounds. The nanotechnology-based foods provide both government and industry with considerable challenges to ensure consumer trust and acceptance of nanofoods on the market. Recent reports have been extensively published about active use of nanocolloidal particles in various food industries, such as consistency, protection, nutrition, processing, and packaging. For the development of foods with improved quality and sustainability, which are safer and healthier, colloidal particle properties and conductivity are essential. As of now, only a few countries have standard rules for nanotechnology use in food products. Insufficient theoretical exploration of nanosystems causes problems in drawing conclusions on their effectiveness. In food packaging, the use of nanoparticles as ingredients is less risky than that of nanoparticles. Nanomaterials are always at risk of entering the food chains through air, water, and ground when producing and using DNA, membrane disruption and death of the cell. There have been very few studies *in vivo* on human and animal health effects of nanofoods to date. The marketing of nanofoods should have adequate labeling and legislation, which will help to increase consumer acceptance. When properly controlled and regulated, the use of these nanotechnologies can play a significant role in improving food processing and product quality that promotes human health and well-being.

References

1. Nasrollahzadeh, M. *et al.*, An introduction to nanotechnology. in: Interface Science and Technology, Elsevier, vol. 28, p. 1–27, 2019.
2. Santos, C.S. *et al.*, *Industrial applications of nanoparticles–a prospective overview*, vol. 2, pp. 456–465, 2015.
3. Jain, K.K., Nanotechnologies, in: *The Handbook of Nanomedicine*, pp. 7–61, Springer, Humana Press, 2008.

4. Edwards, S.A., *The nanotech pioneers: where are they taking us?*, John Wiley & Sons, 2008.
5. Nile, S.H. *et al.*, Nanotechnologies in food science: applications, recent trends, and future perspectives. *Nano-Micro Letters*, 12, 1, 1–34, 2020.
6. Ajiboye, T.O. *et al.*, Graphitic carbon nitride-based catalysts and their applications: a review. Nano-Structures & Nano-Objects, 24, 100577, 2020.
7. Nair, R. *et al.*, Nanoparticulate material delivery to plants. *Plant Sci.*, 179, 3, 154–163, 2010.
8. Duncan, T.V., The communication challenges presented by nanofoods. *Nat. Nanotechnol.*, 6, 11, 683–688, 2011.
9. Tricco, A.C. *et al.*, *Essential Medicines List (EML) 2019.*
10. Stoops, J. *et al.*, Microbial community assessment of mealworm larvae (Tenebrio molitor) and grasshoppers (Locusta migratoria migratorioides) sold for human consumption. *Food Microbiol.* 53, 122–127, 2016.
11. Sekhon, B.S., Food nanotechnology–an overview. *Nanotechnol. Sci. Appl.*, 3, 1, 2010.
12. Bajpai, V.K. *et al.*, Prospects of using nanotechnology for food preservation, safety, and security. *J. Food Drug Anal.*, 26, 4, 1201–1214, 2018.
13. Sekhon, B.S., Nanotechnology in agri-food production: an overview. *Nanotechnol. Sci. Appl.*, 7, 31, 2014.
14. Bhattacharya, T. *et al.*, Nutraceuticals and Bio-Inspired Materials from Microalgae and their Future PerspectivesCurr. Top. Med. Chem., 21, 12, 1037–1051, 2021.
15. Mousavi, S.R. and Rezaei, M., Nanotechnology in agriculture and food production. 1, 10, 414–419, 2011.
16. Han, J.H., Edible films and coatings: a review. *Innov. Food Pack.*, 213–255, 2014.
17. Tiruvenkadam, N. *et al.*, Nano Coating Start-ups Opportunities in India-A Review. *Bull. Sci. Res.*, 1, 2, 12–20, 2019.
18. Medina-Reyes, E.I. *et al.*, Food additives containing nanoparticles induce gastrotoxicity, hepatotoxicity and alterations in animal behavior: the unknown role of oxidative stress. 111814, 2020.
19. Sharma, S. *et al.*, Nanostructured materials for food applications: spectroscopy, microscopy and physical properties. *Bioeng.*, 6, 1, 26, 2019.
20. Poças, M.F. *et al.*, Smart packaging technologies for fruits and vegetables, in: Smart packaging technologies for fast moving consumer goods, pp. 151–166, 2008.
21. Annous, B.A., Fratamico, P.M., Smith, J.L., Scientific status summary. *J. Food Sci.*, 74, 1, R24–37, 2009.
22. Jamshidian, M. *et al.*, Poly-lactic acid: production, applications, nanocomposites, and release studies. *Compr. Rev. Food Sci. Food Saf.*, 9, 5, 552–571, 2010.

23. Ravichandran, R.J., Nanotechnology applications in food and food processing: innovative green approaches, opportunities and uncertainties for global market. *Int. J. Green Nanotechnol. Phys. Chem.*, 1, 2, P72–P96, 2010.
24. Sorrentino, A. *et al.*, Potential perspectives of bio-nanocomposites for food packaging applications. *Trends Food Sci. Technol.*, 18, 2, 84–95, 2007.
25. Honarvar, Z., Hadian, Z., Mashayekh, M.J., Nanocomposites in food packaging applications and their risk assessment for health. *Electron. Physician*, 8, 6, 2531, 2016.
26. Llorens, A. *et al.*, Metallic-based micro and nanocomposites in food contact materials and active food packaging. *Trends Food Sci. Technol.*, 24, 1, 19–29, 2012.
27. Purkayastha, M.D. and Manhar, A.K., Nanotechnological applications in food packaging, sensors and bioactive delivery systems, in: *Nanoscience in Food and Agriculture*, vol. 2, pp. 59–128, 2016.
28. Ahmed, S. *et al.*, A review on biogenic synthesis of ZnO nanoparticles using plant extracts and microbes: a prospect towards green chemistry. *J. Photochem. Photobiol. B, Biol.*,166, 272–284, 2017.
29. Bang, G. and Kim, S.W., Biodegradable poly (lactic acid)-based hybrid coating materials for food packaging films with gas barrier properties. *Ind. J. Eng. Chem.*, 18, 3, 1063–1068, 2012.
30. Barick, A. and Tripathy, D.J., Thermal and dynamic mechanical characterization of thermoplastic polyurethane/organoclay nanocomposites prepared by melt compounding. *Mater. Sci. Eng. A*, 527, 3, 812–823, 2010.
31. Díez-Pascual, A.M. and Diez-Vicente, A.L., ZnO-reinforced poly (3-hydroxybutyrate-co-3-hydroxyvalerate) bionanocomposites with antimicrobial function for food packaging. *Appl. Mater. Interfaces*, 6, 12, 9822–9834, 2014.
32. Emanuel, N. and Sandhu, H.K., Food Packaging Development: Recent Perspective. *JoTCSTA*, 6, 3, 13–29, 2020.
33. Mihindukulasuriya, S. and Lim, L.-T., Nanotechnology development in food packaging: A review. *Trends Food Sci. Technol.*, 40, 2, 149–167, 2014.
34. Singh, N., An overview of the prospective application of nanoemulsions in foodstuffs and food packaging. *ASIO J. Microbiol. Food Sci. Biotechnol. Innova.*,1, 20–25, 2015.
35. Nandy, B.C. *et al.*, Recent Advances and Potential Applications of Nanoemulsions in Food Stuffs: Industrial Perspectives, Academic Press. p. 315, 2019.
36. Kulshreshtha, N.M., Shrivastava, D., Bisen, P.S., Contaminant sensors: nanotechnology-based contaminant sensors, in: *Nanobiosensors*, pp. 573–628, Elsevier, Academic Press, 2017.
37. Berekaa, M.M., Nanotechnology in food industry; advances in food processing, packaging and food safety. *Int. J. Curr. Microbiol. Appl. Sci.*, 4, 5, 345–357, 2015.
38. Sozer, N. and Kokini, J.L., Nanotechnology and its applications in the food sector. *Trends Biotechnol.*, 27, 2, 82–89, 2009.

39. Bibi, F. *et al.*, A review: RFID technology having sensing aptitudes for food industry and their contribution to tracking and monitoring of food products. *Trends Food Sci. Technol.*, 62, 91–103, 2017.
40. Biji, K. *et al.*, Smart packaging systems for food applications: a review. *J. Food Sci. Technol.*, 52, 10, 6125–6135, 2015.
41. Kerry, J., O'grady, M., Hogan, S., Past, current and potential utilisation of active and intelligent packaging systems for meat and muscle-based products: A review. *Meat Sci.*, 74, 1, 113–130, 2006.
42. Morris, M.A. *et al.*, Development of active, nanoparticle, antimicrobial technologies for muscle-based packaging applications. *Meat Sci.*, 132, 163–178, 2017.
43. Ditta, A., How helpful is nanotechnology in agricultur? *Adv. Nat. Sci. Nanosci. Nanotechnol.*, 3, 3, 3–5, 2012.
44. Sanaeifar, A. *et al.*, Application of MOS based electronic nose for the prediction of banana quality properties. *Measurement*, 82, 105–114, 2016.
45. Dasgupta, N. *et al.*, Nanotechnology in agro-food: from field to plate. *Food Research International*, 69, 381–400, 2015.
46. Zarei, M., Jamnejad, A., Khajehali, E., Antibacterial effect of silver nanoparticles against four foodborne pathogens. *Jundishapur J. Microbiol.*, 7, 1, 2014.
47. Vance, M.E. *et al.*, Nanotechnology in the real world: Redeveloping the nanomaterial consumer products inventory. *Beilstein J. Nanotechnol.*, 6, 1, 1769–1780, 2015.
48. Siqueira, M.C. *et al.*, Evaluation of antimicrobial activity of silver nanoparticles for carboxymethylcellulose film applications in food packaging. *J. Nanosci. Nanotechnol.*, 14, 7, 5512–5517, 2014.
49. Khan, S.T., Al-Khedhairy, A.A., Musarrat, J., ZnO and TiO 2 nanoparticles as novel antimicrobial agents for oral hygiene: A review. *J. Nanopart. Res.*, 17, 6, 1–16, 2015.
50. Rouger, A., Tresse, O., Zagorec, M., Bacterial contaminants of poultry meat: sources, species, and dynamics. *Microorganisims*, 5, 3, 50, 2017.
51. Rodrigues, S.M. *et al.*, Nanotechnology for sustainable food production: promising opportunities and scientific challenges. *Environ. Sci. Nano*, 4, 4, 767–781, 2017.
52. Usha, C., Sakthieaswari, P., Lakshmi, P.K., Medicine at Nanoscale: A New Paradigm for the Emerging Trend of Multidrug Resistance in Bacteria, in: *Handbook of Research on Nano-Strategies for Combatting Antimicrobial Resistance and Cancer*, pp. 133–158, IGI Global, 2021.
53. Vohra, A. *et al.*, Enhanced photocatalytic inactivation of bacterial spores on surfaces in air. *J. Ind. Microbiol. Biotechnol.*, 32, 8, 364, 2005.
54. Cho, K.-H. *et al.*, The study of antimicrobial activity and preservative effects of nanosilver ingredient. *Electrochimica Acta*, 51, 5, 956–960, 2005.
55. Eid, A.M. *et al.*, Endophytic Streptomyces laurentii mediated green synthesis of Ag-NPs with antibacterial and anticancer properties for developing functional textile fabric properties. *Antibiotics*, 9, 10, 641, 2020.

56. Roy, S. and Rhim, J.-W., Preparation of bioactive functional poly (lactic acid)/curcumin composite film for food packaging application. *Int. J. Biol. Macromol.*, 162, 1780–1789, 2020.
57. Vasconcelos, T., Sarmento, B., Costa, P., Solid dispersions as strategy to improve oral bioavailability of poor water soluble drugs. *Drug Discovery Today*, 12, 23–24, 1068–1075, 2007.
58. Acosta, E., Bioavailability of nanoparticles in nutrient and nutraceutical delivery. *Curr. Opin. Colloid Interface. Sci.*, 14, 1, 3–15, 2009.
59. Liu, Q. *et al.*, Food-grade nanoemulsions: preparation, stability and application in encapsulation of bioactive compounds. *Molecules*, 24, 23, 4242, 2019.
60. McClements, D.J. and Rao, J., Food-grade nanoemulsions: formulation, fabrication, properties, performance, biological fate, and potential toxicity. *Crit. Rev. Food Sci. Nutr.*, 51, 4, 285–330, 2011.
61. McClements, D.J., Encapsulation, protection, and delivery of bioactive proteins and peptides using nanoparticle and microparticle systems: A review. *Adv. Colloid Interface Sci.*, 253, 1–22, 2018.
62. Jampilek, J., Kos, J., Kralova, K., Potential of nanomaterial applications in dietary supplements and foods for special medical purposes. *Nanomaterials*, 9, 2, 296, 2019.
63. Manickam, V. *et al.*, Genotoxicity of Nanomaterials in Food, Environmental Chemistry Letters. in: *Nanoscience in Food and Agriculture 4*, pp. 141–180, Springer, 2017.
64. Jafari, S.M. and McClements, D.J., Nanotechnology approaches for increasing nutrient bioavailability, in: *Advances in Food and Nutrition Research*, vol. 81, pp. 1–30, 2017.
65. Artiga Artigas, M., *Design of nanostructured delivery systems to enhance the functionality of food bioactive compounds*, Doctoral dissertation, Universitat de Lleida, 2018.
66. Huang, Q., Yu, H., Ru, Q., Bioavailability and delivery of nutraceuticals using nanotechnology. *J. Food Sci.*, 75, 1, R50–R57, 2010.
67. Walsh, J. *et al.*, Playing hide and seek with poorly tasting paediatric medicines: do not forget the excipients. *Adv. Drug Deliv. Rev.*, 73, 14–33, 2014.
68. Augustin, M. and Sanguansri, L., Challenges in developing delivery systems for food additives, nutraceuticals and dietary supplements, in: *Encapsulation technologies and delivery systems for food ingredients and nutraceuticals*, pp. 19–48, Elsevier, Woodhead Publishing, 2012.
69. Gupta, E. *et al.*, Stevioside and rebaudioside A–predominant ent-kaurene diterpene glycosides of therapeutic potential: a review. *Czech J. Food Sci.*, 34, 4, 281–299, 2016.
70. Cai, Z.-Y. *et al.*, Bioavailability of tea catechins and its improvement. *Bioavailability of tea catechins and its improvement. Molecules*, 23, 9, 2346, 2018.

71. Wang, S. *et al.*, Application of nanotechnology in improving bioavailability and bioactivity of diet-derived phytochemicals. *J. Nutr. Biochem.*, 25, 4, 363–376, 2014.
72. Pinheiro, A.C. *et al.*, Chitosan/fucoidan multilayer nanocapsules as a vehicle for controlled release of bioactive compounds. 115, 1–9, 2015.
73. Mora-Huertas, C.E., Fessi, H., Elaissari, A., Polymer-based nanocapsules for drug delivery. *Int. J. Pharm.*, 385, 1–2, 113–142, 2010.
74. Yu, H. and Huang, Q., Enhanced *in vitro* anti-cancer activity of curcumin encapsulated in hydrophobically modified starch. *Food Chem.*, 119, 2, 669–674, 2010.
75. Magalhães, W.L.E. *et al.*, Nanotechnology applied to improve functionality in food, in: *Nutrient Delivery*, pp. 177–219, Academic Press, Elsevier, 2017.
76. Yu, H. *et al.*, Synthesis and characterization of novel antimicrobial emulsifiers from ε-polylysine. *J. Agric. Food Chem.*, 58, 2, 1290–1295, 2010.
77. Omerović, N. *et al.*, Antimicrobial nanoparticles and biodegradable polymer composites for active food packaging applications, 2021.
78. Reza Mozafari, M. *et al.*, Nanoliposomes and their applications in food nanotechnology. *J. Liposome Res.*, 18, 4, 309–327, 2008.
79. Momin, J.K. *et al.*, Potential of nanotechnology in functional foods. *Emir. J Food Agric.*, 25, 1, 2013.
80. Pathak, K. and Akhtar, N., 9 Nanoprobiotics: Progress and Issues, in: *Nanonutraceuticals*, 2018.
81. Chen, X. *et al.*, Coencapsulation of (–)-Epigallocatechin-3-gallate and quercetin in particle-stabilized W/O/W emulsion gels: Controlled release and bioaccessibility. *J. Agric. Food Chem.*, 66, 14, 3691–3699, 2018.
82. Bai, L. *et al.*, Oil-in-water Pickering emulsions via microfluidization with cellulose nanocrystals: 2. *In vitro* lipid digestion. *Food Hydrocoll.*, 96, 709–716, 2019.
83. McClements, D.J., Nanotechnology Approaches for Improving the Healthiness and Sustainability of the Modern Food Supply. *ACS Omega*, 5, 46, 29623–29630, 2020.
84. Jain, A. *et al.*, Nanomaterials in food and agriculture: an overview on their safety concerns and regulatory issues. *Crit. Rev. Food Sci. Nutr.*, 58, 2, 297–317, 2018.
85. Chen, J. *et al.*, Improving nutraceutical bioavailability using mixed colloidal delivery systems: lipid nanoparticles increase tangeretin bioaccessibility and absorption from tangeretin-loaded zein nanoparticles. *RSC Adv.*, 5, 90, 73892–73900, 2015.
86. Caballero-Díaz, E. and Cases, M.V., Analytical methodologies for nanotoxicity assessment. *Trends Analyt. Chem.*, 84, 160–171, 2016.
87. Chawla, R. *et al.*, Antimicrobial edible films in food packaging: Current scenario and recent nanotechnological advancements-a review. *Carbohyd. Polym. Technol. App.*, 2, 100024, 2021.

88. Schrand, A.M. *et al.*, Metal-based nanoparticles and their toxicity assessment. *Nanomed. Nanobiotechnol.*, 2, 5, 544–568, 2010.
89. Bhat, A.H. *et al.*, Oxidative stress, mitochondrial dysfunction and neurodegenerative diseases; a mechanistic insight. *Biomed. Pharmacother.*, 74, 101–110, 2015.
90. Panyala, N.R., Peña-Méndez, E.M., Havel, J., Silver or silver nanoparticles: a hazardous threat to the environment and human health? *J. Appl. Biomed.*, 6, 3, 2008.

15

Edible Film on Food With Smart Incorporation of Health-Friendly Supplements

Animesh Naskar[1], Ivi Chakraborty[2]*, Sebak Ranjan Roy[1] and Tanima Bhattacharya[3,4]

[1]*Department of Food Science & Technology, Maulana Abul Kalam Azad University of Technology, Haringhata, District Nadia, West Bengal, India*
[2]*Department Post Harvest Technology, Faculty of Horticulture, BCKV, P.O Mohanpur, District-Nadia, West Bengal, India*
[3]*College of Chemistry and Chemical Engineering, HUBEI University, Wuhan, China*
[4]*Department of Science, Engineering, and Novel Global Community Educational Foundation, Hebersham, NSW, Australia*

Abstract

Packaging is not only an integral part of any food rather it serves multivariate purposes especially in case of processed food. Despite their importance, in most cases, packaging materials are considered to be useless, waste items causing severe health and environmental hazard as petroleum-derived polymers having non-renewable and/or non-biodegradable nature are widely used for this purpose. A viable alternative to it is biopolymer-based edible packaging developed from agar, sodium alginate, glycerol, inulin, etc., including polysaccharides and polypeptides, one of the most promising approaches to these issues. These edible films not only improve physico-mechanical properties but also often incorporate unique functional, sensory, and nutritional characteristics and extend the storability of food. This type of edible bio-film material is a promising future food packaging option because it is biodegradable, compostable, made from renewable resources, produces zero waste, is completely single-use, and is less expensive. This chapter summarizes the production techniques, efficacy, and appropriateness of diverse types of film-forming resources like plant and animal residues, gums and flours, as well as their essential properties in comparison to their conventional, synthetic

**Corresponding author*: ivcpht@gmail.com

Annu, Tanima Bhattacharya and Shakeel Ahmed (eds.) Nanotechnology in Intelligent Food Packaging, (361–382) © 2022 Scrivener Publishing LLC

counterparts, and their application on various food items, as well as the benefits and limitations of edible films acceptable to the food industry.

Keywords: Edible film, food supplements, food coating, film composite, food science, food technology, health supplements

15.1 Introduction

Plastic use in our daily lives became unavoidable in the second half of the 20th century due to numerous advantages such as cost, lightweight, water resistance, durability, and mechanical performance such as heat stability, strong tensile strength and flexibility, excellent hurdle to oxygen (O_2), carbon dioxide (CO_2), and aroma compounds [1]. However, the usage of substandard plastics for packaging food and non-food even medical substances leaves harmful effects on human health and ecosystem including aqua lives, resulting in overall environmental problems. As plastic is non-biodegradable in nature, it is critical to employ alternate materials. Generally, the conventional packages are made up either of a single layer of petroleum-based plastic materials such as polyvinyl chloride, polyethylene, polyethylene terephthalate, polypropylene, polystyrene, or polyamide or obtained from multi layers of usually plastics (polypropylene, polyethylene, etc.), a metallic layer mostly aluminium, paper, etc., in addition to materials utilized for gluing and printing. However, most of the plastics are based on fossil resources causing environmental hazards in long run. According to research, approximately 8,300 × 10^6 metric tonnes of virgin plastics were produced in 2017, and approximately 6,300 × 10^6 metric tonnes of plastic waste was generated in 2015, with around 9% recycled, 12% incinerated (causing massive air pollution), and 79% accumulating in the environment [2, 3].

One of the viable solutions to this health and environmental hazards involves the use of biopolymer packaging that is both biodegradable as well as edible [4]. The use of environment-friendly edible films or coatings that efficiently protect diverse food products is an exceptionally viable option when it comes to a matter of food safety and utilization of renewable resources.

Food packaging systems serve a variety of purposes, including separating food from its surroundings, decreasing exposure to spoiling causes (such as microbes, O_2, moisture, and off-flavors), and avoiding undesirable compounds from being lost, thereby prolonging food shelf life, as well as containment, information, and marketing.

Edible coatings, which include edible films, sheets, coatings, and pouches, are a thin or slime layer of safe and edible substance. It is designed to serve as a food-safe coating, and as a result, it becomes an integral part of the food and is consumed along with those food products [5]. These layers are frequently applied to the food surface in liquid form, by submerging the product in the structural matrix' film-forming solution. In nature, edible films can be characterized by their self-sustaining structures, but edible coatings cling to the surface of the food [6]. Edible packaging materials are derived from diverse edible components, for instance, polysaccharides, proteins, lipids, and/or resins, that is inherently renewable and biodegradable and converts into compost and other biologically recyclable substances [7].

15.2 Advantages and Limitations of Edible Films

Although edible packaging materials cannot completely substitute conventional packaging systems, the competence of primary edible packaging can be increased through the utilization of non-edible secondary packaging to provide additional protection from the environment and to prevent macro and microbial attacks as well as foreign particles. The materials using for edible packaging can be accounted as appealing alternatives for a variety of reasons, including the propensity to defend the food by using their barrier and perfunctory capabilities to limit the release of active substances and mass transfer among the diverse food components, in addition to improving sensory features [5].

Furthermore, edible films and coatings can operate as moisture, oxygen, carbon dioxide, or vapor barriers, preventing lipid oxidation and color deterioration and improving product quality. Besides, the usage of edible films comprising essential oils and plant extracts with antioxidant and antibacterial characteristics is gaining popularity [1, 8–12].

15.3 Consumer Acceptance

Consumer acceptance challenges, such as sensory qualities, cultural and religious constraints, as well as safety and marketing around the utilization of film-coated food, especially when it comes to the composition of the film, have a considerable impact on the potential uses of edible materials. Although as food components, to remain undetected during ingestion, the material employed in edible packaging must be required to have neutral sensory qualities or be suited for coating food [5].

Consumers are becoming more conscious of what they eat and carefully check on the labels for the content of the product, expiry date, etc.

As per FDA, the list of ingredients and allergens on the labels are mandatory, since, for the susceptible sets of customers having allergies or other health issues and with specific cultural/religious limitations, several edible film-forming components may cause complications. In fresh produce and food-service applications, the obligation for allergen labeling is limited to the allergens (top eight) and is poorly implemented. Another safety concern is about possible alterations of microbiota in the product. For such cases, food products that are packaged/coated and have unique edible packaging uses must also be addressed [5].

Price, customer reluctance to utilise novel substances, and particular awareness of any additional instructions necessary for opening, cooking, consuming, or their disposal all play a role in the commercialization of edible packaging.

15.4 Functions and Property of Film Forming Substances

15.4.1 Stability and Shelf-Life

Despite the fact that knowledge is necessary for the flourishing sale of such food items, there is a lack of information on the firmness and durability of edible films made up of living/edible components.

In terms of legality and stability criteria, food packaging films made from any edible material could be deemed dehydrated foods. The majority of these substrates are nutrient-rich enough to function as microbial growth media. Unless they are wrapped aseptically in outer packing material, they should be kept stable largely by removing water. Stability should be maintained primarily by water removal unless they are aseptically wrapped within an external packing material. The water activity of peach puree is 0.95, whereas the water activity of edible films using the same puree is 0.43, which is below the microbial growth threshold [13]. Similarly, films have made from grape pomace extract obtained from wine processing and mango puree having low levels of water activity (below 0.6) [14, 15]. Considering these low levels of water activity, the films should be stable in microbiological concerns if properly maintained under dry conditions. Fruits like acerola, mango, and red mombin purees applied to thermoplastic starch (TPS) films had elevated water activity ranging from 0.66 to 0.70 [16]. Sugar crystallization can also cause fruit and vegetable-based edible films to perform poorly physically. However, edible films from mango puree did not show any such negative attributes for up to 3 months at 30 °C [15].

15.4.2 Mechanical Properties

Insufficient flexibility is one of the important factors that may promote premature breaking during the operation, handling, processing or production, and storage. Therefore, good mechanical qualities are one of the most important criteria for the use of edible films in food packaging [1–15].

15.4.3 Barriers to Environment

Controlled mass transfer considered between the food and its external environment is required for edible coatings and films. Permeability is a crucial factor to consider when choosing or creating edible packaging materials, and measurements should be taken under the conditions that a packaged product may encounter.

15.4.4 Barrier to Moisture

In order to prevent the moisture transfer from refined food items to the environment, edible packaging may be accounted for in this regard. Alteration of packaged food's water activity (a_w) may lead to microbial growth problems, unattractive modification of texture, and deteriorating chemicals and inhibiting various enzymatic interactions. Films constructed from various edible packaging materials have lower water vapor permeability (WVP) values than hydrocolloid-based films.

15.4.5 Oxygen Barrier Property

Food deterioration is caused by the oxidation of food constituents such as lipids and myoglobin discoloration in the pieces of fresh meat, as well as enzymatic browning in fruits (fresh-cut). Edible packaging preserves the quality of O_2-sensitive foods by the attributes of limited oxygen permeability (OP) and extends their shelf life while reducing the use of costly non-recyclable O_2-barrier plastics.

Furthermore, advanced edible films' development having the feature of definite gas permeability within specific storage situations can generate a customized environment, suppressing horticultural product respiration rates and/or ethylene may produce during the storage and circulation [17].

15.4.6 Barrier to Aroma

VOC (volatile organic compound) barriers are vital in generating quality volatile flavor and preventing the penetration of exterior off-flavors into final packaged food items during their storage and allocation is important. Hydrophilicity in connection with the polysaccharide and polypeptide-based edible films makes them superior hurdles to non-polar aroma components [18, 19].

15.5 Film Forming Process

In general, edible packaging formulations contain at least one natural polymer, having the efficacy of generating the structural matrix in an incessant and cohesive manner. Cohesion forces between film-forming polymer molecules, such as non-covalent and covalent bonds, determine the qualities of edible films and desired film coatings, as well as their use in food products. Structural chemistry of biopolymers, processes of film-formation, physico-chemical parameters involved in film-formation, and plasticizer and additive concentrations all influence cohesive strength. Edible coatings can be created by melting solid fats, waxes, or resins and then solidifying them, followed by solubilization in an organic solvent. Finally, solvent evaporation may be accounted for or forming a water emulsion followed by water evaporation as required. The formation of edible films based on hydrocolloid, on the other hand, can be accomplished using two primary technologies: wet and dry procedures.

Solution casting is a wet technique that uses coacervation method to precipitate hydrocolloid. Cooling after thermal coagulation or gelation can also cause it (like in the case of gelatine or agar). Coacervation can be simple or complex that is dependent on the quantity of biopolymers present in the solvent. A lipid substance, as well as maybe a surfactant, is mixed in the emulsion-based film formation system as desired [7]. Water and organic solvent (ethanol) are the only food-grade solvents especially involved in edible food packaging. The more advanced form of the wet method is known as dry thermoplastic extrusion which is predominantly dependent on the thermal features of the biopolymer in making the film.

To make a premade stand-alone film, spread the degassed casting solution on an appropriate surface of base material in the form of thin layer followed by its drying to release. Another process comprises subsequent stages such as spraying, dipping, draining, enrobing, brushing, coating, and removing solvent or cooling (lipid-based coatings) are used to generate a direct coating on food goods or medicines [7–17]. With the aim of making self-supporting films, solution casting has been developed by various drying methods like air drying and infrared application.

To gain the intended functions, formulation of the film-coating and physico-chemical parameters are required to be optimized to achieve film matrix' adequate cohesiveness, as well as precise attachment and exposure on the food surface. Coating thickness is determined by the techniques applied and the casting solutions' viscosity. The adhesion of coating to the food surface is predominantly dependent on chemical affinity among the coatings and their opinionated surfaces [17]. Hydrophilic film-forming fluids, for example, adhere poorly to hydrophobic product surfaces. Surface-active ingredients, like emulsifiers, can be placed on the food or mixed into the film-forming fluid to improve adherence in these circumstances. Solution casting is a common method for evaluating bio-based materials' film-forming ability. Standard procedures are used to measure the qualities of cast films in order to forecast how coatings made using the same formula will perform on food.

The dry technique is appealing because it is less expensive and has a better throughput than the wet process for producing standalone films. By combining heat, pressure, and shear in the extruder's barrel, plasticized biopolymers are heated over their glass transition temperature to generate a homogeneous melt under low water content conditions. Apart from film formation, this soft, stretchy melt can be moulded into a variety of useful shapes by applying pressure along with heat. The level of conformational alterations, agglomeration, and chemical cross-linking in thermal extrusion is influenced by the supply rate of dry biopolymer, glycerol concentration, temperature, and screw speeds.

15.6 Film/Coat Formulation and Various Components Used in Coating

Usually, edible films can be characterized by their self-sustaining structures, but edible coatings cling to the surface of the food [6]. A variety of biopolymers have been studied in order to construct appropriate structures of films or coatings.

Proteins, polysaccharides, and lipids should all be present when it comes to the formulation of edible films and coatings because they can form a cohesive three-dimensional matrix. Various active substances such as antimicrobial agents, antioxidants, coloring compounds, flavors, and nutraceuticals are also included in the system of film formulation to get better firmness, quality, and safety of packaged foods. In order to improve the quality, stability, and safety of packaged foods, various active compounds such as antimicrobials, antioxidants, color agents, flavors, and nutraceuticals are also incorporated into film-forming solutions. Those ingredients may also give the coat antibacterial, antifungal, or antioxidant properties [20, 21]. Fruits and vegetables are having a variety of health-promoting substances, so films made from these plant resources may be an unconventional way of nutritional intake including pigments, dietary fiber, and polyphenols with antioxidant potential [14, 22].

A biopolymer extracted from any biomass has physical and chemical properties that are used to make edible films influence the final qualities of developed films [23].

Polysaccharides have the potential to be valuable raw materials for biopolymers that are environment-friendly [24]. The polysaccharide-based edible coating acts as a permeable membrane that selectively blocks the exchange of O_2 and CO_2 gases, lowering the rate of fruit respiration [25]. Agar (algal biopolymer), sodium alginate, glycerol, starch-based films, etc., are used for making edible packaging [26]. Of course, other polysaccharides are also may be used for this purpose. Agar is utilized as a substitute for traditional plastic materials because of its low thermal and mechanical qualities, thermoplastic nature, biodegradability, biocompatibility, and moderate water resistance. Nanotechnology is very useful in the creation of biopolymer films to improve both the characteristics and the cost [1, 6, 27]. Aloe vera gel consists of polysaccharides that contain many functional components such as Acemannan which can inhibit post-harvest damage to fresh food products, has antidiabetic, anticancer, and antimicrobial activity, and increases the proliferation of injured cells [28–30]. Nano starch made from kidney beans has better water solubility, oil absorption efficacy, competency of water absorption, and swelling power than native starch, according to Sharma *et al.* (2021) [31]. The inclusion of nano starch during the formulation of film increased the composite film characteristics. The addition of nano starch to the film formulation lowered the water vapor transmission rate and enhanced the composite film's burst strength.

The most prevalent set of biopolymers employed in the creation of edible materials is hydrocolloids, which include both polysaccharides and proteins. Plants, animals, and microbes are all good sources of them. Cellulose derivatives, amylose, chitosan, alginates, carrageenans, pullulan, and pectins are the most popular polysaccharides utilized in the manufacturing of edible films and coatings. At the same time, the most common proteins soybean proteins, wheat gluten, corn zein, sunflower proteins, gelatin, whey, casein, and keratin are utilized for the same [32]. Those materials, on the other hand, are hydrophilic in nature. As a result, a variety of triglycerides (fats and oils) are mixed into a hydrocolloid matrix to improve their ability to keep water vapor out. Waxes, triacylglycerol, free fatty acids, monoglycerides in acetylated form, and vegetable oils are the most common [21].

The film-forming ability and stability of some novel substances were reviewed by Galus *et al.* (2020) [21]. A few of them are mentioned here: whole grain flours (e.g., quinoa, chia, and amaranthus) and pulses like lentils, chickpeas, grass peas, and mung beans. Biopolymers derived from food waste are known for their film-forming ability, which is one of the major focusing areas of recent researches. Fruit and vegetable wastes, which are often processed into flour, are thought to be interesting substrates for use as film-forming components. Furthermore, when compared to processing fruits or vegetables, the cost of processing residues in terms of energy is low because there are no novel technologies used, and the variations in the processing are linked to well-established procedures.

The most commonly utilized polysaccharides for film-forming applications are cellulose and starch, although a variety of structurally complex biopolymers are generated by microbes like bacteria and fungi, particularly polysaccharides (xanthan, pullulan, etc.) are receiving growing interest. Furthermore, the type of lipid resources (oils and fats, waxes, etc.) can be included in polysaccharide and protein lattices to produce the composite films and coatings with improved moisture barrier properties [33].

15.6.1 Protein

Proteins, like polysaccharides, are among the most significant macromolecules found in nature including animal proteins, plant-derived proteins, collagen, silk fibroin, and milk protein. Due to the exposure of polar moieties, it provides a large number of possible functional capacities as a biodegradable membrane by forming many intermolecular bonds [34, 35]. Proteins are natural polymers that can form amorphous three-dimensional structures that are primarily sustained by non-covalent interactions. The heat sensitivity, structural diversity, and hydrophilic properties of proteins all influence the functional qualities of the final materials. Different vegetable and animal proteins, such as maize zein, soy protein, gelatin, casein, etc., are extensively employed as biodegradable polymers [36].

SPI-based nano-composite films were developed by grafting with nano-silica via silane agent (KH560) and cellulose nanofibers [37], and SPI and modified montmorillonite using melt extrusion [38, 39].

Cost, functional qualities, flexibility, tension tolerance, optical characteristics like opacity hurdle effect against the gas flow, availability in nature, structural resistance to moist environment, and sensory adequacy are all factors that influence the employment of biopolymers, particularly gelatin, involved in packaging of extremely perishable food products. The newest developments in gelatin-based films and coatings are also discussed in this article.

15.6.2 Gelatin

Gelatin is a naturally occurring water soluble protein with no discernible odor and a random polypeptide chain structure in an aqueous medium. This is made by the partial hydrolysis of collagen found mostly in the skins, bones, connective tissues, and tendons of vertebrate and invertebrate animals. Collagen's structure is made up of hard bar-like molecules grouped in threads and interconnected by covalent linkages.

In the 1930s, to make gelatin, pig skin was employed as basic raw material. For large-scale production in the food industry, it is still the most significant material. It is still the most important material for large-scale food industrial production. The expensive utilization of this gelatin in pharmaceutical industry is usually extracted from cattle bones. Nevertheless, fish gelatin production has expanded in recent years in an effort to diversify away from other gelatin resources, highlighting over 1.5% of whole gelatin production.

By-products generated from the fishing industries have recently been viewed as possible sources of gelatin instead of disposable waste. Conversely, one of the main disadvantages of fish gelatin is that it has less stable rheological qualities than gelatin derived from mammalian sources [40]. Besides, because fish and poultry gelatin production is currently limited, the resulting goods are more expensive than those made from mammalian gelatins.

The instability of the collagen triple-helix produces soluble gelatine. The qualities of gelatine are determined by two key factors: the initial collagen's features and the extraction procedure. However, the amount of collagen converted to gelatine is determined by the pre-treatment with physico-chemical environment including hot water extraction, pH of the solution, temperature, and extraction time.

15.6.3 Lipids

Many edible lipid compounds have been used to give gloss to food and to protect it from moisture transfer. Lipid and resin complexes, unlike other macromolecules, are not biopolymers. They do not have a huge molecular structure since they do not comprise a lot of repetitive units linked together by covalent bonds. As a result, they are delicate and rarely form self-supporting and cohesive film structures. Not only are the lipids, the resins included in edible film-forming ingredients to offer composite films with a moisture barrier due to their low polarity [41]. However, there are drawbacks to using lipids in this regard, including their textural characteristics, oily surface, waxy flavor, and probable rancidity. Examples of lipid materials used to coat film are waxes, glycerol esters, resins, etc.

15.6.4 Composite Materials

Composite materials in edible packaging can be developed by combining bio-components for specific uses, with the goal of utilising complementing functional characteristics or overcoming defects [42]. At low relative humidity, edible films made of biopolymers (polar) are generally effective gas barriers with reasonably acceptable mechanical properties. Proteins and polysaccharides, on the other hand, produce water-sensitive films having poor performance to prevent moisture penetration and significantly deteriorated gas hurdle and mechanical characteristics when exposed to high humidity. Hydrophobic lipids, on the other hand, are good in preventing moisture migration. To date, the majority of composite films that have been examined have combined lipid molecules, a structural matrix based on hydrocolloids. In this formulation, lipid components restrict water transport, while the hydrocolloid components give strength and structural integrity by acting as selective gas barriers [43].

In general, composite films are prepared in a hydrocolloid atmosphere as a stable lipid emulsion or as a bilayer of hydrocolloid-lipid. Over the hydrocolloid-based film, the bilayer composite films consist of separate secondary layer with lipid components. To make bi-layer films, desired lipid solution or a molten lipid is cast or laminated onto a dry, premade polysaccharide or protein film, comparable to bi-layered composite plastic films. The emulsion approach, on the other hand, requires dissolving the lipid in the film-forming solution before casting the film. Phase separation causes the bi-layer film to form later in the drying process since the film-forming solution has degraded into an unstable emulsion [44].

Other than these major film-forming elements, a number of additives are incorporated to obtain better the overall quality of the developed film in relation to their formulations, stability, storability, etc., e.g., plasticizers, emulsifier, antimicrobials, organic acids, and their salts, chitosan, plants extracts, bacteriocins, enzymes, antioxidants, and bio-preservatives.

15.6.5 Waste Utilization to Form the Film

Andrade *et al.* (2016) investigated a few edible films made from flour of various residues from different cereals, pulses and legumes [Figure 15.1], including those from different fruits and vegetables like orange, watermelon, lettuce, carrot, spinach, and cucumber [21, 45]. Without the use of plasticizers, they were able to produce uniform, flexible films with promising properties. Furthermore, the addition of potato skin residual flour improved mechanical resistance. The researcher examined and developed films using ripe banana corn starch and peel flour. One more noteworthy example is the usage of citrus fruit by-products [46]. Mariniello *et al.* (2010) and We *et al.* (2019) studied a film-forming material based on grapefruit albedo and found that the generated films had good functional qualities [47, 48].

Valorization of wastes (like Pomace after oil extraction and leaf extracts) has become an important aspect in the olive oil industry in order to increase profit and to reduce environmental pollution. The recovery of leaf phenolic compounds, which are of great importance to the food industry due to their antioxidant and antibacterial properties, is one viable strategy. The use of an emulsification-internal gelation process to encapsulate Olive Leaf Extract (OLE) was successful in progressively releasing polyphenols during salad dressing storage, boosting product protection against oxidation [49].

Brito *et al.* (2019) developed edible films from solid residues left over after making an isotonic drink, which included various fruits and vegetables [50]. The films were uniformly yellowish and pliable and had good water solubility. The pectin-enriched composition had a lot of scientific and functional potential for making biodegradable films. In addition, when pectin is added in small amount in the films, it provides improved color, barrier properties, and lesser hygroscopicity. Usually, plant residues are made up of a variety of components. Diverse residue fractions may have versatile applications, such as components for film formulation, dietary fiber, depending on their particle size and composition.

15.6.6 Root Plants

As edible films and coatings, tuber and root starches have exposed their potential. Starches or flour obtained from uncommon root plants, on the other hand, are being studied in film formulation practice. In this circumstance, starches from canna, ulluco, and yam were used to test new film-forming materials [21].

Gutiérrez *et al.* (2018) extracted film-forming starch from the guinea arrowroot tuber. The authors also looked into composite films made from waste such as flour and extract of grape waste from the winemaking process [51]. Cross-linking reactions mediated by these natural fillers decline the films' hydrophilic nature.

However, when compared to films made from common polysaccharides or proteins, most flour films showing heterogeneous arrangements with lesser moisture and mechanical resistance. As a result, their implications may be influenced by the product's features and purpose.

15.6.7 Plant Gums

Gum arabic, gum karaya, guar gum from cluster bean, and tragacanth gum are just a few of the natural, plant-derived gum exudates that have been identified in recent times. Gum arabic is one of the most well-known and oldest natural gums, having been used for about 5,000 years. Gum arabic has a wide range of applications in the food, paint, and textile industries because of its emulsifying, stabilising, thickening, and binding properties [52].

Furthermore, gum arabic has the potential to be used as a defensive edible covering for extending the longevity of food items, by removing their watery and oily exposure. Gum arabic–coated tomatoes, apples, and strawberries had a longer shelf life with beneficial characteristics such as a considerable delay in changing of their color, loss of weight, inflexibility, titratable acidity, soluble solids concentration, the content of ascorbic acid, and percentage of decay [53–55].

Almond gum is a new type obtained from the almond tree's trunk and branches. Trees create exudate gums through gummosis in response to numerous stressors. Exudate gums are recognized to have intriguing physico-chemical, as well as biological properties, making them ideal substrates in use as food additives [56]. To improve the quality of edible coated strawberries, coatings have been made from *Cajanus cajan* seeds [57].

15.6.8 Wild Plants

Until far, the most often employed polysaccharides for film formation or coating have been starches and cellulose derivatives. Further researches have looked into various wild plants having complex polysaccharides as a potential source of new edible covering material. *Opuntia cactus*, a xerophyte that can be utilized to make a high-quality edible covering, is one of them [58].

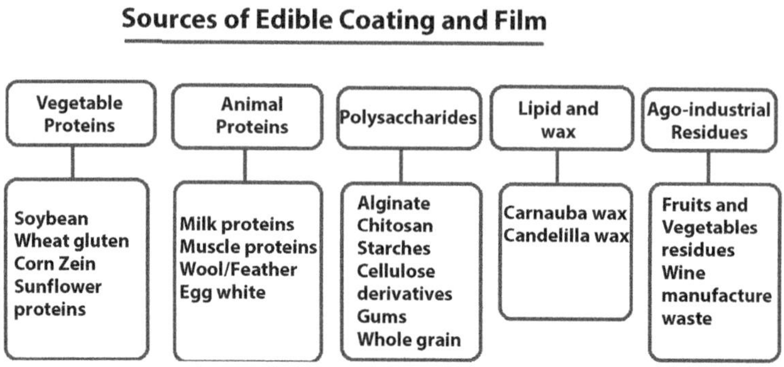

Figure 15.1 Sources of edible films and coating residues [21].

Polysaccharides derived from the Opuntia cactus were used as an edible coating on citrus fruits that could be preserved for up to 35 days at 5°C and 90% relative humidity. In terms of physico-chemical qualities, and general appearance, cactus-based edible covering extended the longevity of *Kinnow mandarin.*

15.7 Applications of Edible Film/Coating

15.7.1 Fresh Storage of Whole Produce

The application of edible coatings accompanied with various active substances (antimicrobials, antioxidants, anti-browning, fortifying agents, etc.) represents an alternative preservation technology by providing several beneficial effects like acting as a semi-permeable membrane for necessary modification and regulation of internal gas composition, limiting moisture loss, lowering the rate of respiration and weight loss while keeping the hardness of the fresh food and giving the coated products a glossy finish. One of the main causes of fruit and vegetable deterioration is promoting the coatings' functional performance beyond their barrier capabilities, hence improving stability, quality, and safety [26]. Fresh pineapples are rich in source of several other bioactive molecules and specifically the bromelain enzyme, an anti-inflammatory agent that acts as remedial measure against acute sinusitis, sore throat, arthritis, gout, skin, and digestive ailments. Many studies have been conducted on the use of edible coatings for pineapple storage. The effect of a 2% Semperfresh (sucrose-fatty acid ester) edible covering on pineapple fruit postharvest quality during cold storage was investigated and compared to a commercial coating material, Sta-fresh (10%) 7055 (paraffin-polyethylene wax). When compared to 1% Semperfresh-coated fruit and control, coating with Semperfresh and Sta-fresh 7055 reduced internal browning and preserved higher ascorbic acid content. Coating treatments increased the storage life of pineapple fruit by up to 5 weeks, compared to only 28 days for the control fruit [59]. Tomato and peach postharvest quality has been improved using sodium alginate as an edible covering [60, 61].

15.7.2 Extending Shelf Life of Fresh Cuts

Raphaela Gabri Bitencourt *et al.* (2014) created an edible coating infused with mint essential oil with the goal of determining its efficacy in suppressing *in vitro* microbial formation and improving quality and durability of fresh-cut pineapple [62]. An alginate-based antimicrobial edible coating was developed and used to keep fresh-cut pineapple fresh for longer [63]. Oms-Oliu (2008) used polysaccharide-based edible film and observed the extended longevity of fresh-cut melon [64].

15.7.3 Enhancing the Storability of Processed Food Items with Edible Film Coating

An edible film or coating is a kind of lining material with a thickness of less than 0.3 mm that is formed from a combination of biopolymers and different additives. Usually, the edible coating is formed directly on the food, while the edible film is previously made and then adhered to any product. Díaz-Montes and Castro-Muñoz (2021) reported the application of such biopolymers for preserving the food quality. Depending on the biomaterials used and the types of biologically active compounds incorporated in the film formulation, specific properties, such as sensorial, antimicrobial, antioxidant, physical, physicochemical and nutritional characteristics in coated products supposed to improve (Kumar *et al.*, 2021). Recently, the physiological and functional properties of edible films (including polysaccharides, chitosan, animal and plant proteins and lipids) for commercial application to meat and meat products have been studied by Song *et al.* (2021). More emphasis on developing the right combination of blended materials is needed because the effectiveness of edible films for meat products or any other kind of foods depends on biopolymer types and bioactive compounds to enhance their functional properties. Generally, the edible films and coatings are biodegradable and eco-friendly packaging that has the potential to replace the primary packaging for food material. However, as the edible films and coatings materials may have a sufficient number of nano-particles which enter the human body along with the food and because of their tiny size, they can enter the human organ and damage them, e.g. 20 nm size silver nanoparticles can enter the human lungs. Not only this, even they can initiate genotoxicity by changing the DNA material. Therefore, the size, amount and composition of nanoparticles/active agents used to develop the edible film packaging are crucial (Chawla *et al.* 2021) factors of consideration.

15.8 Use of Nanoparticles as Biopolymer and Nanolaminates

Food-grade biopolymers made up of proteins or polysaccharides can be developed using nanoparticles. This may be put to encapsulate and release useful substances in response to different environmental cues. Polylactic acid is a frequent component of several biodegradable biopolymeric nanoparticles. Polylactic acid is commonly used to encapsulate medicines, vaccines, and kinds of proteins; however, it has drawbacks such as rapid bloodstream release and liver and kidney retention. Nanolaminates are significantly thin food-grade films (each layer of 1–100 nm) made up of two or more nanometer-sized layers. These are very useful in the manufacture of edible films and have a variety of essential food-industry uses, such as protecting the quality of a variety of food products. These films defend the food items on exposure to moisture, gases, and lipids, as well as serving as transporters of colors, tastes, antioxidants, minerals, and antimicrobials, and improving the textural aspects of foods [65, 66].

15.9 Conclusion

In conclusion, edible film production is primarily limited to the laboratory scale and is thought to be expensive in comparison with plastic films. To enhance the practicality of commercialized edible packaging, more cost-effective research and larger-scale production is required. Commercialization feasibility is determined by the complexity of the manufacturing process and the quantity of the investment. Furthermore, food makers insist that items have a lengthy shelf life. As edible packaging materials are essentially biodegradable, their protective functions last for a shorter period of time than traditional packaging. As a result, extensive research is required to achieve the appropriate stability, durability, and safety of edible packaging during storage/use circumstances, as well as attractiveness, cost-effectiveness, and ease of production/development even in large quantities.

References

1. Otoni, C.G., Avena-Bustillos, R.J., M. C., H., Azeredo, M.V., Moura, M.R., Mattoso., L.H.C., McHugh, T.H., Recent Advances on Edible Films Based on Fruits and Vegetables -A Review. *Compr. Rev. Food Sci. F.*, 16, 1151–1168, 2017.
2. Geyer, R., Jambeck, J.R., Law, K.L., Production, use, and fate of all plastics ever made. *Sci. Adv.*, 3, 7–11, 2017.
3. Giosafatto, V., Al-Asmar, A., D'Angelo, A., Roviello, V., Esposito, M., Mariniello, L., Preparation and characterization of bioplastics from grass pea flour cast in the presence of microbial transglutaminaze. *Coatings*, 8, 435, 2–12, 2018.
4. Puscaselu, R., Gutt, G., Amariei, S., Biopolymer-based films enriched with *Stevia rebaudiana* used for the development of edible and soluble packaging. *Coatings*, 9, 360, 2019a.
5. Janjarasskul, T. and Krochta, J.M., Edible Packaging Materials. *Annu. Rev. Food Sci. Technol.*, 1, 415–48, 2010.
6. Hassan, B., Chatha, S.A.S., Hussain, A.I., Zia, K.M., Akhtar, N., Recent advances on polysaccharides, lipids and protein based edible films and coatings: A review. *Int. J. Biol. Macromol.*, 109, 1095–1107, 2018.
7. Krochta, J.M., Protein as raw materials for films and coatings: definitions, current status, and opportunities, in: *Protein-Based Films and Coatings*, vol. 1, A. Gennadios (Ed.), pp. 1–40, CRC Press, New York, 2002.
8. Gennadios, A., Hanna, M.A., Kurth, L.B., Application of edible coatings on meats, poultry and sea foods: a review. *LWT-Food Sci. Technol.*, 30, 4, 337–350, 1997.
9. Ojagh, S.M., Rezaei, M., Razavi, S.H., Hosseini, S.M.H., Effect of chitosan coatings enriched with cinnamon oil on the quality of refrigerated rainbow trout. *Food Chem.*, 20, 193–198, 2010.
10. Corrales, M., Fernandez, A., Han, H.J., Antimicrobial Packaging Systems, in: *Innovations in Food Packaging*, pp. 133–170, 2014.
11. Aitboulahsen, M., Zantar, S., Laglaoui, A., Chairi, H., Abdelhay, A.A., Bakkali, M., Zerrouk, M.H., Gelatin-Based Edible Coating Combined with Mentha pulegium Essential Oil as Bioactive Packaging for Strawberries. *Hindawi J. Food Qual.*, 1–7, 2018.
12. Uçak, I., Physicochemical and antimicrobial effects of gelatin-based edible films incorporated with garlic peel extract on the rainbow trout fillets. *Prog. Nutr.*, 21, 1, 232–240, 2019.
13. McHugh, T.H., Huxsoll, C.C., Krochta, J.M., Permeability properties of fruit puree edible films. *J. Food Sci.*, 61, 1, 88–91, 1996.
14. Deng, Q. and Zhao, Y., Physicochemical, nutritional, and antimicrobial properties of wine grape (cv. Merlot) pomace extract-based films. *J. Food Sci.*, 76, 3, E309–17, 2011.

15. Sothornvit, R. and Rodsamran, P., Effect of a mango film on quality of whole and minimally processed mangoes. *Postharvest Biol. Technol.*, 47, 3, 407–15, 2008.
16. Dantas, E.A., Costa, S.S., Cruz, L.S., Bramont, W.B., Costa, A.S., Padilha, F.F., Druzian, J.I., Machado, B.A.S., Caracterizac¸˜ao e avaliac¸˜ao das propriedades antioxidantes de filmes biodegrad´aveis incorporados com polpas de frutas tropicais. *Cienc. Rural*, 45, 1, 142–148, 2015.
17. Baldwin, E.A., Surface treatments and edible coatings in food preservation, in: *Handbook of Food Preservation*, vol. 21, M.S. Rahman (Ed.), pp. 478–508, CRC Press, Boca Raton, FL, 2007.
18. Fabra, M.J., Hambleton, A., Talens, P., Debeaufort, F., Chiralt, A., Voilley, A., Influence of interactions on water and aroma permeabilities of ι-carrageenan-oleic acid-beeswax films used for flavor encapsulation. *Carbohydr. Polym.*, 76, 2, 325–332, 2009.
19. Hambleton, A., Debeaufort, F., Bonnotte, A., Voilley, A., Influence of alginate emulsion-based films structure on its barrier properties and on the protection of microencapsulated aroma compound. *Food Hydrocoll.*, 23, 8, 2116–2124, 2009.
20. Salgado, P.R., Ortiz, C.M., Musso, Y.S., Di Giorgio, L., Mauri, A.N., Edible films and coatings containing bioactives. *Curr. Opin. Food Sci.*, 5, 86–92, 2015.
21. Galus, S., Kibar, E.A.A., Gniewosz, M., Krasniewska, K., Novel Materials in the Preparation of Edible Films and Coatings-A Review. *Coatings*, 10, 674, 2–14, 2020.
22. Espitia, P.J.P., Avena-Bustillos, R.J., Du, W.X., Chiou, B.S., Williams, T.G., Wood, D., McHugh, T.H., Soares, N.F.F., Physical and antibacterial properties of acai edible films formulated with thyme essential oil and apple skin polyphenols. *J. Food Sci.*, 79, 5, M903–910, 2014.
23. Plackett, D., *Biopolymers-New Materials for Sustainable Films and Coatings*, John Wiley & Sons, Ltd., Hoboken, NJ, USA, 2011.
24. Puscaselu, R., Gutt, G., Amariei, S., Rethinking the Future of Food Packaging: Biobased Edible Films for Powdered Food and Drinks. *Molecules*, 24, 2–14, 3136, 2019b.
25. Winarti, C., Application of Edible Coating Based on Sago Starch with Addition of Vitamin C to Paprika: Consumer Preferences and Microbiological Quality, 21, 1, 68–76, 2011.
26. Sapper, M. and Chiralt, A., Starch-based coatings for preservation of fruits and vegetables. *Coatings*, 8, 152, 1–19, 2018.
27. Sagnelli, D., Kirkensgaard, J., Giosafatto, C.V., Ogrodowicz, N., Kruczała, K., Mikkelsenf, S., Maigret, J.E., Lourdin, D., Mortensen, K., Blennow, A., All-natural bio-plastics using starch-betaglucan composites. *Carbohydr. Polym.*, 172, 237–24, 2017.

28. Suriati, L., Candra, I.P., Supardika, I.K., Aloe-Gel Coating for Delaying Physicochemical Change of Fresh-Cut Mango. *SEAS (Sustainable Environ. Agric. Science)*, 05, 01, 58–65, 2021.
29. Suriati, L., Utama, I.M.S., Harsojuwono, B.A., Gunam, I.B.W., Incorporating Additives for Stability of Aloe Gel Potentially as an Edible Coating. *AIMS Agric. Food*, 5, 3, 327–336, 2020.
30. Vandana, A.K., Suresha, G.J., Archana, T.J., Swamy, G.S.K., Influence of environment friendly bio-preservative Aloe vera gel coating on quality parameters of jamun fruits. *Acta Horticulturae, 1241. Proc. III International Symposium on Underutilized Plant Species*, International Society for Horticulturae Science, 2019.
31. Sharma, I., Sinhmar, A., Thory, R., Sandhu, K.S., Kaur, M., Nain, V., Pathera, A.K., Chavan, P., Synthesis and characterization of nano starch-based composite films from kidney bean (*Phaseolus vulgaris*). *J. Food Sci. Technol.*, 58, 2178–2185, 2021.
32. Falguera, V., Quintero, J.P., Jiménez, A., Muñoz, A., Ibarz, A., Edible films and coatings: Structures, active functions and trends in their use. *Trends Food Sci. Technol.*, 22, 291–303, 2011.
33. Pérez-Gago, M.B. and Rhim, J.W., Edible coating and film materials: Lipid bilayers and lipid emulsions, in: *Innovations in Food Packaging*, 2nd ed, pp. 325–350, Elsevier, Amsterdam, The Netherlands, 2013.
34. Hua, Y.F., Cui, S.W., Wang, Q., Mine, Y., Poysa, V., Heat induced gelling properties of soy protein isolates prepared from different defatted soybean flours. *Food Res. Int.*, 38, 377–385, 2005.
35. Ojijo, V. and Ray, S.S., Processing strategies in bionanocomposites. *Prog. Polym. Sci.*, 38, 1543–1589, 2013.
36. Ramos, M., Valdés, A., Beltrán, A., Carmen, M.G., Gelatin-Based Films and Coatings for Food Packaging Applications. *Coatings*, 6, 41, 2016.
37. Qin, Z., Mo, L., Liao, M., He, H., Sun, J., Preparation and Characterization of Soy Protein Isolate-Based Nanocomposite Films with Cellulose Nanofibers and Nano-Silica via Silane Grafting. *Polymers*, 11, 1835, 2–15, 2019.
38. Kumar, P., Sandeep, K.P., Alavi, S., Truong, V.D., Gorga, R.E., Effect of Type and Content of Modified Montmorillonite on the Structure and Properties of Bio-Nanocomposite Films Based on Soy Protein Isolate and Montmorillonite. *J. Food Sci.*, 75, 5, 46–56, 2010a.
39. Kumar, P., Sandeep, K.P., Alavi, S., Truong, V.D., Gorga, R.E., Preparation and characterization of bio-nanocomposite films based on soy protein isolate and montmorillonite using melt extrusion. *J. Food Eng.*, 100, 480–489, 2010b.
40. Alfaro, A.T., Balbinot, E., Weber, C.I., Tonial, I.B., Machado-Lunkes, A., Fish gelatin: Characteristics, functional properties, applications and future potentials. *Food Eng. Rev.*, 7, 33–44, 2014.
41. Greener, I.K. and Fennema, O., Barrier properties and surface characteristics of edible, bilayer films. *J. Food Sci.*, 54, 6, 1393–1399, 1989.

42. Krochta, J.M. and De Mulder-Johnston, C., Edible and biodegradable polymer films: challenges and opportunities. *Food Technol.*, 51, 2, 61–67, 1997.
43. Gontard, N., Duchez, C., Cuq, J.L., Guilbert, S., Edible composite films of wheat gluten and lipids: Water vapour permeability and other physical properties. *Int. J. Food Sci. Technol.*, 29, 1, 39–50, 1994.
44. Perez-Gago, M.B. and Krochta, J.M., Emulsion and bi-layer edible films, in: *Innovations in Food Packaging*, vol. 22, J.H. Han (Ed.), pp. 384–402, Elsevier Academic, New York, 2005.
45. Andrade, R.M.S., Ferreira, S.L., Gonçalves, É.B.A., Development and characterization of edible films based on fruit and vegetable residues. *J. Food Sci.*, 81, 412–418, 2016.
46. Arquelau, P.B.F., Silva, V.D.M., Garcia, M.A.V.T., de Araújo, R.L.B., Fante, C.A., Characterization of edible coatings based on ripe "Prata" banana peel flour. *Food Hydrocoll.*, 89, 570–578, 2019.
47. Mariniello, L., Giosafatto, C.V., Di, P.P., Sorrentino, A., Porta, R.J., Swelling, mechanical, and barrier properties of albedo-based films prepared in the presence of phaseolin cross-linked or not by transglutaminase. *Biomacromolecules*, 11, 2394–2398, 2010.
48. We, H., Lei, Y., Zhu, R., Zhao, M., Lu, J., Xiao, D., Jiao, C., Zhang, Z., Shen, G., Li, S., Preparation and characterization of bioactive edible packaging films based on pomelo peel flours incorporating tea polyphenol. *Food Hydrocoll.*, 90, 41–49, 2019.
49. Jolayemi, O.S., Stranges, N., Flamminii, F., Casiraghi, E., Alamprese, C., Influence of Free and Encapsulated Olive Leaf Phenolic Extract on the Storage Stability of Single and Double Emulsion Salad Dressings. *Food Bioprocess. Technol.*, 14, 93–105, 2021.
50. Brito, T.B., Carrajola, J.F., Gonçalves, E.C.B.A., Martelli-Tosi, M., Ferreira, M.S.L., Fruit and vegetable residues flours with different granulometry range as raw material for pectin-enriched biodegradable film preparation. *Food Res. Int.*, 121, 412–421, 2019.
51. Gutiérrez, T.J., Herniou-Julien, C., Álvarez, K., Alvarez, V.A., Structural properties and in vitro digestibility of edible and pH-sensitive films made from guinea arrowroot starch and wastes from wine manufacture. *Carbohydr. Polym.*, 184, 135–143, 2018.
52. Patel, S. and Goyal, A., Applications of Natural Polymer Gum Arabic: A Review. *Int. J. Food Prop.*, 18, 986–998, 2015.
53. El-Anany, A.M., Hassan, G.F.A., Ali, F.M.R., Effects of edible coatings on the shelf-life and quality of Anna apple (*Malus domestica* Borkh) during cold storage. *J. Food Technol.*, 7, 5–11, 2009.
54. Tahir, H.E., Xiaobo, Z., Jiyong, S., Mahunu, G.K., Zhai, X., Mariod, A.A., Quality and postharvest-shelf life of cold-stored strawberry fruit as affected by gum Arabic (*Acacia senegal*) edible coating. *J. Food Biochem.*, 42, 12527, 2018.

55. Ali, A., Maqbool, M., Ramachandran, S., Alderson, P.G., Gum Arabic as a novel edible coating for enhancing shelf-life and improving post-harvest quality of tomato (*Solanum lycopersicum* L.) fruit. *Postharvest Biol. Technol.*, 58, 42–47, 2010.
56. Bashir, M. and Haripriya, S., Assessment of physical and structural characteristics of almond gum. *Int. J. Biol. Macromol.*, 93, 476–482, 2016.
57. Robles-Flores, G., Abud-Arcgilla, M., Ventura-Canseco, L.M.C., Meza-Gordillo, R., Grajales-Lagunes, A., Ruiz-Cabrera, M.A., Gutiérrez-Miceli, F.A., Development and evaluation of a film and edible coating obtained from the Cajanus cajan seed applied to fresh strawberry fruit. *Food Bioprocress. Tech.*, 11, 2172–2181, 2018.
58. Riaz, S., Sultan, M.T., Sibt-e-Abass, M., Irman, M., Ahmad, R.S., Hussain, M.B., Shariati, M.A., Kosenko, I.S., Kleymenova, N.L., Egorova, G.N., Extraction of polysaccharides from opuntia cactus for its potential application in edible coating to improve the shelf life of citrus (Kinnow mandarin) fruit. *J. Microbiol. Biotechnol. Food Sci.*, 8, 745–750, 2018.
59. Nimitkeatkai, H., Srilaong, V., Kanlayanarat, S., Effect of edible coating on pineapple fruit quality during cold storage. *Acta Hortic.*, 712, 712, 643–647, 2006.
60. Zapata, P.J., Guillén, F., Martínez-Romero, D., Castillo, S., Valero, D., Serrano, M., Use of alginate or zein as edible coatings to delay postharvest ripening process and to maintain tomato (*Solanum lycopersicon* Mill) quality. *J. Sci. Food Agric.*, 88, 1287–1293, 2008.
61. Maftoon azad, N., Ramaswamy, H.S., Marcotte, M., Shelflife extension of peaches through sodium alginate and methyl cellulose edible coatings. *Int. J. Food Sci. Technol.*, 43, 951–957, 2008.
62. Minh, N.P., Nhi, T.T.Y., Hue, D.N., Ha, D.T.T., Chien, V.M., Quality and shelf life of processed pineapple by different edible coatings. *J. Pharm. Sci. Res.*, 11, 4, 1441–1446, 2019.
63. Mantilla, N.V., Development of an Alginate-based Antimicrobial Edible Coating to Extend the Shelf-life of Fresh-cut Pineapple. A Thesis has been submitted to Texas A & M University for the degree of Master of Science, 2012.
64. Oms-Oliu, G., Using polysaccharide-based edible coatings to enhance quality and antioxidant properties of freshcut melon. *LWT Food Sci. Tech.*, 41, 10, 1862–1870, 2008.
65. Díaz-Montes, E.; Castro-Muñoz, R., Edible Films and Coatings as Food-Quality Preservers: An Overview. *Foods*, 10, 249, 2021. https://dx.doi.org/10.3390/foods10020249
66. Kumar, S.; Mukherjee, A.; Dutta, J., Chitosan based nanocomposite films and coatings: Emerging antimicrobial food packaging alternatives. *Trends Food Sci. Technol.*, 97, 196–209, 2020. https://doi.org/10.1016/j.tifs.2020.01.002.

67. Song, D.H.; Hoa, V.B.; Kim, H.W.; Khang, S.M.; Cho, S.H.; Ham, J.S.; Seol, K.H., Edible Films on Meat and Meat Products. *Coatings*, 11, 1344, 2021. https://doi.org/10.3390/
68. Chawla, R., Sivakumar, S. and Kaur H., Antimicrobial edible films in food packaging: Current scenario and recent nano-technological advancements - a review. *Carbohydrate Polymer Technologies and Applications*, 2, 2021, 100024. https://doi.org/10.1016/.
69. Ameta, S.K., Rai, A.K., Ameta, S.C., Hiran, D., Ameta, R., Biogenic Nano-Particles and their Use in Agro-ecosystems, in: *Use of Nanomaterials in Food Science*, pp. 457–488, Springer Link, Springer Nature, Singapore Pte Ltd, 2019, https://link.springer.com.
70. Krishnakumar, T., Nanotechnology in Food Processing and Pacakging, in: Nanotech in Food World, 2020.

16

Future of Food Packaging: Intelligent Packaging

Jincy Abraham

J.D. Birla Institute, Kolkata, West Bengal, India

Abstract
The new dimensions of food packaging are influenced by consumer preferences for food safety, innovations, and convenience. In an era of technological advances, food packaging has evolved into smart packaging. Both active and intelligent food packaging as an important component of smart packaging provides the customers with an optimum food product experience. Active food packaging aims at improving the shelf-life of the product by addition of certain components into the package. While intelligent packaging helps the food product to navigate the product cycle until it reaches the consumer. Intelligent food packaging includes tools that help during the product cycle, for example, indicators, RFID tag, barcode labels, and biosensors. A few of the different types of indicators used in intelligent food packaging include time-temperature indicators, leak indicators, freshness indicators, gas indicators, and pH indicators. Advantages of intelligent food packaging include monitoring the internal environment of the package, monitoring the logistics of the product, protection from theft, improving pathogen detection, increased product value, reduction in food wastage while extending the shelf life and providing convenience to the consumer while expanding market. The application of nanotechnology along with further research can ensure sustainable and commercially viable consumer accepted intelligent food packaging.

Keywords: Food packaging, biosensors, time-temperature indicators, radio frequency identification devices, nanotechnology, food packaging, barcodes

Email: jincys.mail@gmail.com

Annu, Tanima Bhattacharya and Shakeel Ahmed (eds.) Nanotechnology in Intelligent Food Packaging, (383–418) © 2022 Scrivener Publishing LLC

16.1 Introduction

Age old packaging systems are no longer sufficient in the current times due to increased complexity in the products developed, higher consumer expectations, economic policies and environmental sensitivity. Innovations in food packaging provide enhanced functionality to meet a variety of needs of the customers. Some key features of these innovations include low preservative processed foods, products complying with regulatory requirements, tracking systems for monitoring the food cycle, and also providing protection against lawsuits [1]. Brody and others [36] defined intelligent packaging as a system that sensed and communicated. Clarke [37] defined intelligent packaging as one that included logic capability, and Rijk [45] (2002) defined intelligent packaging as one that monitored the condition of packaged foods to provide information about the food quality during transport and storage. Intelligent packaging (Figure 16.1) gives importance to consumer's health which is influenced by the product package interaction as part of intelligent packaging [24, 25].

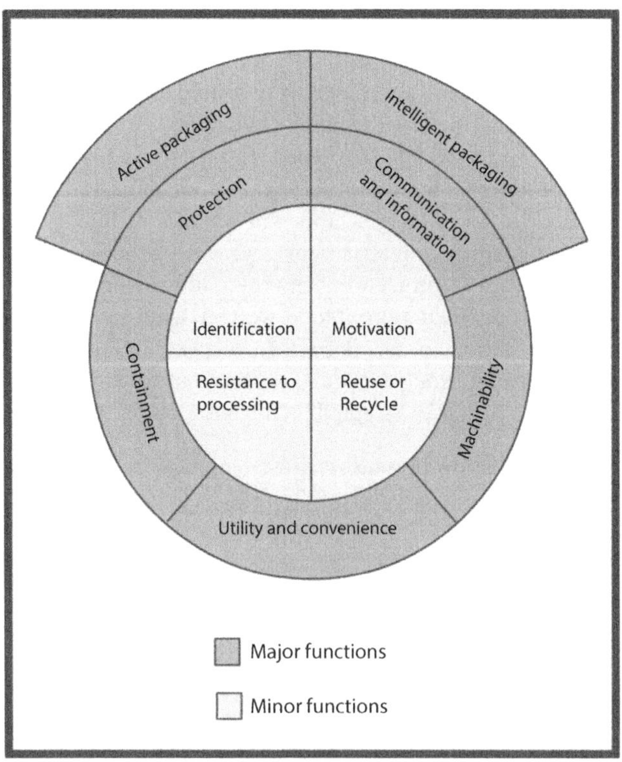

Figure 16.1 Functions of packaging [25].

As per Yam (2005), intelligent packaging is characterized as a packaging framework that is equipped for functions like sensing, detecting, recording, tracking, communicating, and applying scientific logic to work with dynamic to expand time span of usability, enhance safety, work on quality, give data, and caution about potential dangers. The uniqueness of intelligent packaging is in its ability to convey that the package and the food move constantly together all through the inventory network cycle, and the package is the food's best buddy and is in the best situation to communicate the state of the food. It is imperative to underscore that intelligent packaging is a framework that includes the package, the food product, the outer climate, and different contemplations [2].

In addition to the basic communication function of conventional packaging, intelligent packaging also provides additional information to the buyers as it is able to monitor and record minor changes taking place both inside and outside of the package. The function of intelligent packaging associated with assessing the quality of the food inside the package is linked with the use of indicators and other intelligent tools like sensors [24]. According to Kerry *et al.*, intelligent packaging is used "to monitor the conditions of packaged foods like meat to provide information on the quality of the packaged good during transport and storage". Intelligent packaging systems (Figure 16.2) use communication functions to facilitate decision-making aimed at preserving food quality, extension of shelf life, and improving overall food safety. These packaging systems consist of hardware components like

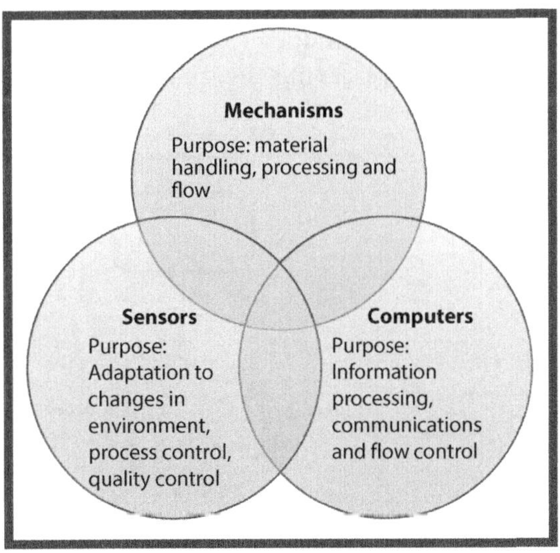

Figure 16.2 Intelligent system components [28].

time-temperature indicators (TTIs), freshness and/or ripening indicators, radio frequency identification (RFID) systems, and gas detectors [4].

Intelligent packaging is associated with the advancements made in the tools developed. It provides signals for monitoring and analyzing the freshness of product which is usually food. It provides information regarding the properties of the food inside the container and the container environment, thus helping the retailer, customer, and manufacturer by providing useful information. Common indicators address issues like ripeness controllers and indicators, signaling gas spillage, bio probes, time-temperature monitors, toxin indicators, and radio frequency indicators. The idea of intelligent packaging in real sense is to assess viability and strength the active packaging systems. It offers detailed knowledge in logistics and maintains quality of food by identifying critical control points by the use of printed labels attached to the packaging material. Smart packaging systems involve self-cooling and self-heating systems used for regulating temperatures [24].

Advances in nanotechnology and nanomaterials empower the improvement of new better active and intelligent packaging systems. These packaging systems provide great benefits to the food industry to improve freshness, shelf-life of food, while also allowing monitoring to control the storage conditions from the place of manufacture till the place of consumption by the consumer [5].

16.2 Tools of Intelligent Packaging

The various devices and techniques (Figure 16.3) used in intelligent packaging contribute to preserving the quality of different food with

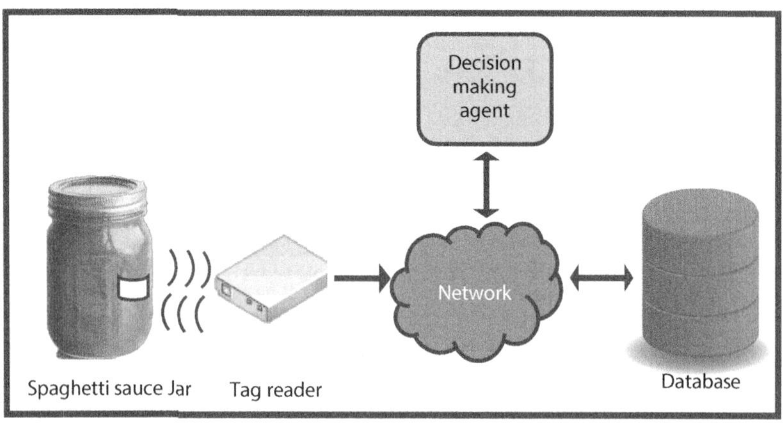

Figure 16.3 Intelligent jar of spaghetti sauce [46].

Table 16.1 Food quality and safety indicators used in intelligent packaging [26].

Indirect indicators	Direct indicators
Time-temperature indicators and critical temperature indicators	Toxins and biogenic amines
Leak indicators	Volatile compounds, organics, ethanol, nitrogen, and sulfuric compounds
Storage conditions: temperature, time, and oxygen	Pathogenic bacteria
	Microbial growth and spoilage

regard to monitoring microbial activity and providing consumer convenience. Intelligent packaging helps with logistics in the food industry while curbing issues like food frauds and promoting the product marketability [6]. Intelligent packaging systems support food safety management systems like Hazard Analysis and Critical Control Points' (HACCP) and Quality Analysis and Critical Control Points' (QACCP) systems in maintaining food safety. It helps to detect processes that affect the quality attributes of the product and helps to efficiently improve the food quality [4, 22].

Some of the indicators of food quality are mentioned in Table 16.1.

16.3 Indicators

Substances that demonstrate the absence, presence, or concentration of another substance or the degree of reaction between multiple substances by means of a characteristic change are known as indicators [4].

16.3.1 Quality or Freshness Indicators

The change in the quality of the product during its shelf-life is indicated by freshness indicators. A permanent color change in the product helps the consumers to easily interpret. It gives quality information about product resulting from chemical changes or microbial activity inside the product. Visual information regarding the microbial quality of the product is available due to reactions between the microbial growth metabolites and the integrated indicators within the package [3, 10, 12].

Yoshida *et al.* [51] developed a colorimetric chitosan bio-based pH indicator with a potential to be used as indicators of metabolites formed from microbial growth like L-lactic acid, n-butyrate, acetic acid, and D-lactate. In case of meats, production of carbon dioxide during storage is an indication of quality decline. Researchers at Sejong University developed carbon dioxide indicators consisting of aqueous solutions of whey protein isolate or chitosan. Changes in transparency by the pH-dependent whey indicated the presence of carbon dioxide [14]. Majority of the quality indicators are based on detection of volatile metabolites produced during the aging of foods, for example, amines, diacetyl, hydrogen sulfide, and ammonia. Some other types of quality indicators respond to organic acids or volatile biogenic amines formed due to microbial degradation of protein rich foods like cadaverine from lysine, histamine from histidine, and putrescine from arginine. One of the drawbacks of freshness indicators are color changes that indicate contamination can occur in products devoid of any visible sensory or quality deterioration. Hence, the presence of some target metabolite may not necessarily be an indication of low quality food [4].

Scientists in New Zealand developed RipeSense™ sensor which enables consumers to choose fruits that best appeals to their palate. Volatile gases like complex mixture of esters, alcohols, ketones, aldehydes, and lactones that contribute to the fruit aroma on ripening are detected by the indicator. For unripe fruit that is crisp in the texture, the indicator is red colored. As the fruit ripens, the color changes from orange to yellow till the time the fruit is very ripe and juicy in texture. As the fruit softens, it produces more aroma, and the indicator changes color accordingly. Pears and avocados are fruits whose ripeness is difficult to assess as they do not show marked change on their skin color on ripening; hence, the use of freshness indicators in these fruits has been of great help to the consumers.

Freshness indicators monitor the food quality during the stages of both storage and transportation until it reaches the customer. Loss of freshness can be due to unfavorable conditions during the product cycle. Hence, the indicators provide information on microbial growth, presence of microbial metabolites or chemical changes occurring in the product. Some of the quality depicting metabolite includes organic acids, glucose, biogenic amines, ethanol, carbon dioxide, volatile nitrogen compounds, adenosine triphosphate (ATP) degradation products, and sulfuric compounds. The freshness indicators must be placed inside the packaging to be able to be in contact with the compounds [25].

16.3.2 Time-Temperature Indicators

According to the regulation of European Commission (EC) 450/2009 [38], specific rules are mentioned for active and intelligent materials that come in contact with food. According to the European Commission 450/2009, time-temperature indicators are responsible to evaluate the minimum amount of time a product has spent above the threshold temperature and also provide information on whether a threshold temperature has been exceeded or not. This provides time-temperature history of the product. Visual indications of temperature history are obtained from the labels during the different stages of storage and distribution and also provide information on temperature fluctuations in case of frozen products [4]. An important factor in determining the shelf life of a food product is temperature. Fluctuations in temperature can be responsible for deterioration in the product quality. Time-temperature indicators used in the cold chain or food supply chain provide useful information about the safety of the food [16]. The kinetics of deteriorations either physical or chemical, as well as microbial activity in food is influenced by temperature. TTIs are usually small adhesive labels stuck to the shipping containers or retail packages. These labels are also used as "freshness indicators" for estimating the shelf life of perishable foods. Customers can see visually distinct changes like an increase in color intensity or diffusion of a dye, and these changes are temperature dependent [2]. The principle function of this indicator is based on the detection of temperature and time dependent on chemical, enzymatic, mechanical, electrochemical, or microbial changes in a food. For example, physical or chemical responses are based on acid base reactions or polymerization toward time and temperature. While, in contrast, biological responses are based on biological changes like in case of microorganisms, spores, or enzymes in relation to time and temperature. The measured values are usually indicated as a visible response, like mechanical deformations or color changes [17, 18].

Time-temperature indicators are simple and inexpensive gadgets stuck to the package and can be distinguished as follows:

1. Critical temperature indicators indicate whether the product was heated above or cooled below the permitted temperature.
2. Partial history indicators indicate if a product has been exposed to temperature fluctuations, leading to changes in product quality.
3. Full history indicators record the complete temperature profile along the food life cycle [11].

TTIs have simple functionality and are user-friendly devices. Fresh-Check from Lifeline technologies is an example of one such device. The functioning of the device is based on a polymerization reaction that results in a color change in the indication range. If the color of the active center matches the outer ring, then the product needs to be consumed immediately, while a clear center indicates freshness. If the product is not fresh, then the TTI has a dark center [15]. TTIs are based on irreversible chemical, physical, and microbial or enzymatic changes and their response is expressed in the form of either color changes or mechanical deformations and provides information about the cumulative time-temperature history of the product [19].

Commercially available TTIs are enzymatic, diffusion, and polymer-based systems. Some examples of diffusion-based time-temperature indicators are 3M Monitor Mark® and Freshness Check® of 3M Company, USA. An example of commercial enzymatic TTI is the VITSAB® in which a color change is induced by a drop in pH that results from the controlled enzymatic hydrolysis of a lipid substrate. A commercially available TTI dependent on polymerization reaction is the Lifelines Freshness Monitor® Fresh-Check (lifelines Technology Inc., USA). Indicators in the form of barcodes printed with fading inks that disappear due to temperature abuse include the FreshCode (Varcode Ltd.) and Tempix (Tempix AB) [4, 13].

American Society for Testing and Materials (ASTM) F1416 Standard Guide for Selection of TTIs provide information regarding selection criteria for the commercially available TTIs for usage on commodities like food and pharmaceuticals [39]. ASTM standards points out that it is the duty of the product manufacturer to determine the shelf life of the product at the optimum temperatures and then refer with the TTI manufacturer to opt for the optimum indicator that closely resembles the quality of the product.

Lifelines Technologies (Morris Plains, New Jersey) had demonstrated an idea of use of a laser optical wand to scan a TTI-barcode label to immediately receive product information along with temperature history. Due to expensive scanners and lack of technology to support the information high way in the supply chain, this idea was not accepted by the industry at that time. In the modern times, availability of wireless technologies along with powerful and affordable scanner provides a favourable opportunity for companies to adopt advanced TTI systems for monitoring and controlling the food quality. Bioett (Lund, Sweden) has developed a TTI/barcode system in which information can be read by a hand-held scanner displayed on a computer monitor and then downloaded into a database for further analysis. KSW Microtec (Dresden, Germany) has developed a battery-powered TTI/RFID tag in which thin-film batteries are printed on a flexible substrate. Infratab (Oxnard, California) also developed a battery powered TTI/RFID tag which uses a microchip to detect and integrate temperature over time to determine the shelf life of the product [2].

16.3.2.1 Chemical TTIs

TTIs are mostly based on a chemical reaction rate that is influenced by temperature leading to color changes. The color intensity thus represents a time-temperature integral measurement.

The OnVu™ TTI(BASF Future Business GmbH, Germany) consists of a temperature-sensitive, photochromic ink made of benzylpyridines that are activated by ultraviolet light and makes the color change to dark blue color. The self-adhesive label then turns progressively lighter as time passes and the ambient temperature rises. The product reaches the end of its shelf life when the activated color attains the same shade as of the reference. Temperature sensitivity can be controlled by checking the type of photochromic chemical used along with the intensity of the ultraviolet radiation applied during the activation process.

The Fresh-Check® TTI from Temptime consists of a small circle of polymer surrounded by a reference ring and is based on solid-state polymerization reactions. The indicator contains diacetylene monomers which deepen in color due to cumulative temperature exposure. If the polymer center is darker than the surrounding reference ring, then consumers are advised not to consume the food. The indicators are shipped from the manufacturer already activated and have no means for in-field activation and thus respond to temperature of the storage. They are stored at −24°C to minimize indicator response before usage.

16.3.2.2 Physical TTIs

Physical TTIs are based on diffusion of colored chemical substances as a function of time and temperature. An example of one such indicator is the MonitorMark™ TTI from 3M which is a partial-history indicator consisting of a pad with a blue dye within a carrier substance. On removing the slip-tab, it brings the pad and the wick into contact, and the dye remains within the pad until the carrier substance undergoes a phase change due to temperature exposure. The response indicated is measured by reading the distance the dye has migrated from the windows in the indicator.

16.3.2.3 Physicochemical TTIs

Devices which exploit the chromatic properties of pH indicators by coupling the pH indicator transition with a diffusive mechanism are physicochemical TTIs. The TT Sensor™ TTI from Avery Dennison depends on the diffusion of an organic acid between two polymer layers with a base layer of a rectangular label with a circular area (the active zone) containing the fluorescent pH indicator. The non-active part of the label contains a reference pink color. When the activation layer sticks to the lower active zone, the acid diffuses showing color changes from the initial yellow to the reference pink.

16.3.2.4 Biochemical and Biological TTIs

Many TTIs have been developed which are based on biological systems or biochemical reactions coupled with pH indicators. These indicators function by an acid-producing bacteria causing color change by the transition of a pH indicator. Usually, it consists of a culture medium along with a pH indicator within a sealed transparent cover attached to the food package. The culture medium used is generally a strain of cold-resistant, non-pathogenic, acid-producing bacteria, for example, *Lactobacillus*. One of the benefits of such a monitoring system is that it allows the indicator device to accurately mirror the deterioration of the food, hence providing precise results. Some more of the examples of some indicators are as follows.

The CheckPoint® TTI from VITSAB (Visual Indicator Tag Systems AB) is a full-history indicator which is based on the hydrolysis of a lipid substrate (e.g., glycerol trihexanoate) in fluid suspension by a proprietary lipase enzyme. The pH indicator dye gradually changes color as the hydrolysis reaction proceeds. The color changes range from initial deep green to bright yellow to orange and finally red. Color changes are compared visually with a reference color scale and can also be used in automatic optical reading systems.

Color change due to change in pH resulting from the growth of selected strains of acid-producing, non-pathogenic bacteria is how (eO)® TTI from CRYOLOG functions and consists of a nutritive medium-based gel and an indicator dye. This type of indicator is usually seen in frozen foods wherein the lactic acid bacteria are inoculated in the label according to the specifications of the food that need to be traced. The label is frozen until it is activated by bring-it-back to room temperature. As the bacteria in (eO) grow depending on the time-temperature profile which results in the decreases of the pH of the medium and thus the indicator changes color from green to red. Some of the examples of strains of patented bacteria that are capable of growing at low storage temperatures (0°C–8°C) include *Leuconostoc mesenteroides*, *Carnobacterium piscicola*, and *Lactobacillus fuchuensis* [19, 20].

16.3.3 Integrity and Gas Indicators

The main function of a leak indicator is to ensure package integrity during both production and distribution cycle [4].

Changes taking place in gas concentration within the package maybe due to enzymatic activity in the food like during respiration of fruits and vegetables or due to deliberate absorption or generation of gas by added absorbers or generators or due to loss of package integrity because of faults in the package. An ideal oxygen indicator must be cost effective, non-toxic, non-water-soluble components, food grade, and have simple analytical instrumentation that can be easily interpreted by an untrained personal. It should also have a long shelf life under ambient aerobic conditions and must be active only when the package has been sealed and is majorly oxygen free. Ideally, it should exhibit an irreversible response toward oxygen to avoid false indications. These indicators are best applied as an ink, either directly to the package or as a label that can be printed on both paper and plastic. Principles, on which most of the commercially available types of gas indicators are based, are either luminescence or colorimetric changes [19, 20].

Modified atmospheric packaging of foods with low initial oxygen shows changes in color of the redox dyes with changes in oxygen concentration. The problem with this system is that the device is highly sensitive and any residual oxygen in the package is also susceptible to the indicator. Contrary to this, the indicator may not be very precise as the oxygen may also be consumed by the natural microbes present within the food. Oxygen indicator tablets were made by Ageless Eye® (Mitsubishi Gas Chemical Company) that indicates the absence or presence of oxygen by changing the color. It indicates oxygen level of 0.5% or more by changing color of the tablet to blue and turns to pink during lack of oxygen (<0.01%). The color transition takes 3 hours and sometimes more but the presence of oxygen is indicated within 5 minutes or less. Reversible and non-reversible oxygen indicator labels were launched by EMCO Packaging (UK) for the visual indication of pack integrity [4].

Gas composition in the package headspace can change due to gas generation by spoilage microorganisms, respiration of fresh produce, or gas transmission through the packaging material. Indicators in the form of a package label on packaging films can monitor changes in the gas composition, thus providing useful information about the product quality and safety. Oxygen can be responsible for color changes, oxidative rancidity, or microbial spoilage in food and hence oxygen indicators are the most common gas indicator in food packaging. Other than oxygen, water vapor, carbon dioxide, ethanol, hydrogen sulfide, and other gas indicators are also available. Researchers Hong and Park have used carbon dioxide indicator to measure the degree of fermentation in kimchi products during storage and distribution. This indicator consists of a chemical dye in a polymeric film and a carbon dioxide absorbent. Integration of gas indicators with barcode labels or RFID tags enables gas indicator signals to be transmitted both visually and electronically. Innovations in smart ink and printing technology allow gas indicators to be read automatically using optical systems from a distance [2].

Removal of ethylene from within the package helps in prolonging the shelf life and preserving the quality of fresh foods and this is achieved by use of ethylene adsorbers and absorbers within the food package. One of the most commonly used ethylene absorber is potassium permanganate that works by oxidizing ethylene to ethanol and acetate. Another way of removing ethylene from the package is by adsorption with zeolite or activated carbon. These are generally supplied either as sachets or added within the package [24].

16.3.4 Radio Frequency Identification Device

RFID involves use of radio frequencies to decipher information from a distance with lesser obstruction or disorientation on a small device known as a tag or transponder which is attached to the product package that needs to be tracked [21] as shown in Figure 16.4. The device consists of a transistor circuit with a microchip, a substrate or encapsulation material, and an antenna. Most RFID are employed in secondary or tertiary packaging and are responsible for an increase in convenience and efficiency in supply chain management and traceability. Smart labels are RFID tags that simply identify the package. A RFID chip provides information regarding the aggregate temperature history of the product and the shelf life. Hence, this information can be used for intelligent inventory management by first selling products with least shelf life remaining and hence preventing in product wastage and loss [19, 20].

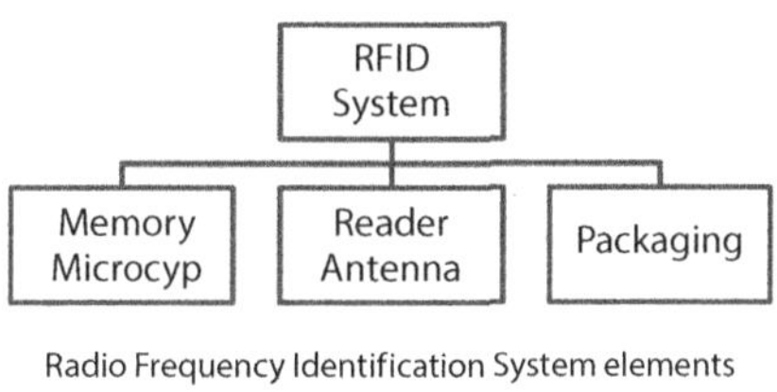

Figure 16.4 Radio frequency identification system elements [26].

16.3.4.1 Radio Frequency Identification System Elements

An automation identification technology that uses wireless sensors for identification of items and to collect data about the product without human intervention using tags and readers is the basis of RFID. It is an advanced form of data carrier for product identification and traceability automatically. The RFID device tags have an identification number (ID) within which the reader uses to assess information regarding the ID number from a database and act upon it accordingly. One of the various parts of a typical RFID device system (Figure 16.5) is a reader that emits radiowaves to capture data from a RFID device tag within which is a minuscule microchip connected with an antenna. The data collected is then transferred to a host computer for analysis and decision making [4].

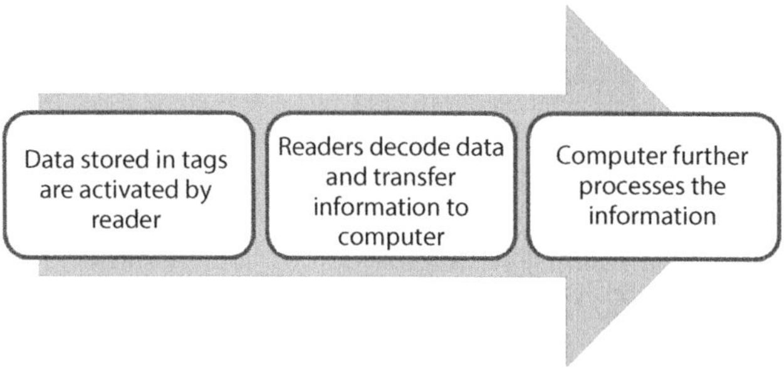

Figure 16.5 The working principles of an RFID system [26].

RFID in intelligent packaging are radiowave-based system that can wirelessly track items and consists of readers, radar, computer hardware, software, networking, and database. Use of wireless technique for data transfer between reader and radar makes this a superior technology over others [24].

The RFID tags are classified into passive tags and active tags. Passive tags have no battery and rely on the power supply from the reader [9]. The coiled antenna within the tag forms a magnetic field when radiowave from the reader encounters with passive RFID tag. The less expensive passive tag sends the information encoded in the tags memory by drawing energy from it and has a reading range of upto 15 feet. While the more expensive active tags come with its own internal battery for powering the microchip's circuitry and broadcasting signals to the reader with a reading range of 100 feet or more. A third type, the semi-passive RFID tags either power the electronics that enable the tag to modulate the electro-magnetic waves emitted by the reader antenna or use battery to maintain memory in the tag. Factors on which the reading range depends on include the power of the reader, the frequency of operation, and the possible interference from metal objects [2, 13].

Because of its ability to identify, categorize, and manage the flow of goods RFID finds application in traceability control and supply chain management processes. It provides fast automated processes at the supply chain level like exception management and information sharing (Figure 16.6). RFID with tags available in the market are non-integrated, mountable, and no flexible sensor and can be used to monitor the relative humidity, temperature, pressure, light exposure, and the pH of products. The tags can detect interruptions in the cold chain which can have a negative impact on the food quality [4, 22].

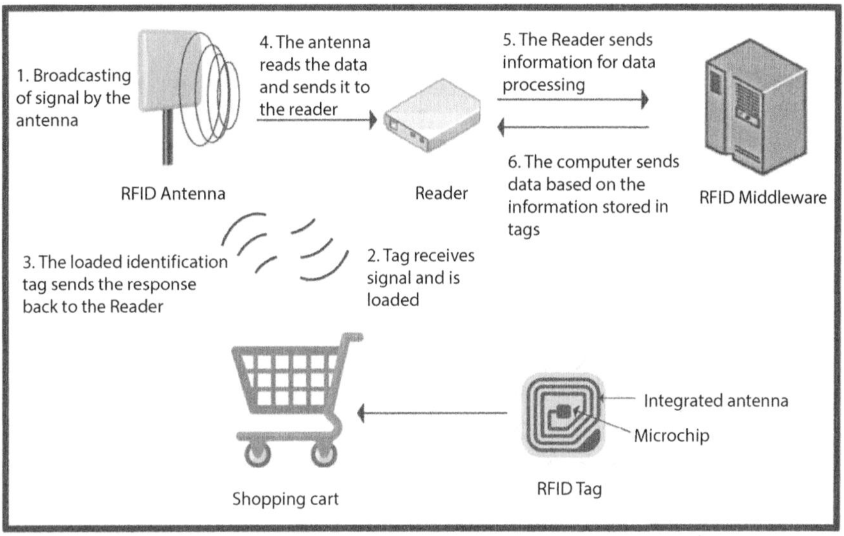

Figure 16.6 Schematic representation of the RFID System [27].

Some of the unique characteristics of RFID device tags are availability of large data storage capacity (upto 1 MB for high-end RFID tags) that can be used to store information on temperature, relative humidity, cooking instructions, and nutritional information. Other features include that the tag need not be oriented toward the reader for data transfer as the radiowaves travel through a wide array of non-metallic materials and provide useful real-time information updates about the product quality as it passes through the supply chain. These tags have both advantages and disadvantages and continue to be used either alone or in combination with other tools of intelligent packaging like the barcode and TTI or a biosensor. Various applications like use in supply chain management, asset tracking, feed pattern of live stocks, and security control have led to the use of these devices by major retailers, companies, government agencies, and researchers Table 16.2. Major retailers like Metro Group and Wal-Mart stores have issued mandates to their leading suppliers to use RFID tags on their distribution packages [2].

Table 16.2 Radio frequency of a RFID system [30].

Frequency	Read range	Data speed	Applications
Low Frequency	0–5 cm	Low	Emergency Card Access Control
High Frequency	10 cm–1 m	Low to Moderate	ID Cards Payment
Ultra-High Frequency	3 m–7 m	Moderate to High	Warehouse Management Tracking
Microwave	10 m–15 m (passive)	High	Electronic Toll Collection
	20 m–40 m (active)		Container Tracking

Some examples of the commercially available devices are as follows:

- A credit card sized, battery-operated RFID temperature monitoring tag that captures and delivers the temperature history of the product was manufactured by Sealed Air Corporation, Saddle Brook, New Jersey, and is known as TurboTag® T-700.
- Credit card sized flat PakSense™ Ultra Wireless Labels from Boise, Idaho, can digitally record the temperature and time of a product during its life cycle.
- Semiactive RFID tags from Infratab Inc. in Oxnard, California, called Freshtime™ can observe the temperature and integrate it with time to determine the shelf life of the product. The battery operated tags have an optional visual display which shows color changes depending on the state of the product. The color changes ranges from green, yellow, and red with green indicating freshness, while red indicates the product to be unsafe. The temperature ranges for functioning of these tags are −25°C to 70°C.

- Identec Solutions AG in Lustenau, Austria created Thei-Q310 RFID tags that have sensors for temperature, humidity, light intrusion, and shock. The thresholds can be specified by the user for the breach conditions and is also available with external sensors and an LED for visual recognition. Some features of this tag include storage capacity of upto 13,000 temperature readings in its memory and a battery life of more than 6 years [19, 20].

16.3.5 Barcodes

The most popular yet least expensive form of data carriers is barcodes Table 16.3. In the 1970s, the UPC (Universal Product Code) barcode was introduced which is a linear symbology with a pattern of bars and spaces to represent 12 digits of data and has become ubiquitous in the grocery store for facilitating stock reordering, inventory control, and checkout. Its small storage capacity allows only a limited amount of information like manufacturer identification number and the item number. Many standards have been developed into commonly accepted languages known as "symbologies," to enable barcodes to communicate with scanners and printers [9].

A new family of barcode symbologies called the Reduced Space Symbology (RSS) by UniformCode Council 2004 was introduced to encode more data in a given smaller space. In case of loose produce items like oranges or apples with limited space, the RSS-14 Stacked Omnidirectional barcode encodes full fourteen-digit Global Trade Item Number (GTIN) with a narrow symbol. The RSS Expanded Barcode can encode upto 74 alphanumeric characters and finds application in variable measure products like seafood and meat that are sold by weight and requires a larger data capacity to encode additional informations like batch/lot number, package weight, and packed date. Two-dimensional barcodes like the PDF 417 (Portable Data File) carries upto 1.1 kilobytes data within a

Table 16.3 Comparison between RFID and barcodes [30].

Parameters	Barcode	RFID
Read Range	Directly With No Obstacle	Longer
Price	Economical	Expensive
Read/Write Tag Data	No	Yes

space of a UPC barcode and allows encoding of additional information like cooking instructions, online address of food manufacturer, graphics, and nutritional information. Advantages of a portable data include immediate availability without the need to access an external database. The Uniform Code Council introduced a new symbology called Composite Symbology by combining a 2D barcode like the PDF 417 with a linear barcode like UPC and provides more versatility. Applications of innovative wireless handheld barcode scanners include product identification and traceability in the food industry (Figure 16.7) [35].

Figure 16.7 Some common types of barcodes used in the intelligent food packaging sector [35].

16.3.6 Sensors

A sensor is a device used for detecting, locating, or quantifying energy or matter by giving a signal for the detection or measurement of either a chemical or physical property to which the device responds and can provide continuous output of signals. The two main functional parts of a sensor are a transducer and a receptor [4].

16.3.6.1 Biosensors

Biosensors detect, record and transmit information related to biological reactions and consist of a bioreceptor and transducer that recognizes the target analyte and converts the biochemical signals into quantifiable electronic response, respectively. Bioreceptors can either be an organic or biological material like hormones, enzymes, antigens, nucleic acids, and microbes, while the transducers may be acoustic, optical, or even electrochemical [2].

Owing to its unique optical properties and rapid response to external stimuli, polydiacetylenes (PDAs) have been popular for use as sensors. They are a class of linear conjugated polymers having alternating double and triple bonds that form thin films which show color changes from deep blue to red in response to different stimuli like biological molecules, pH and temperature. Researchers have developed surfactant-functionalized polydiacetylenes thin films to detect *Staphylococcus aureus* and *E. coli* in apple juice. Polydiacetylenes thin films synthesis is simple and have great potential as biosensors for detection of toxins and food pathogens in the food industry [19, 20].

Biosensors can be incorporated in food packaging films as part of intelligent packaging. Flex Alert Company Ltd. with Vancouver-based partners has developed a commercially available biosensor capable of detecting pathogens like *E. coli*, *Salmonella*, and *Aflatoxins* in food products like fresh fruit, seeds, dried nuts, coffee beans, and even wine barrels. Active biosensors send real-time information during the product cycle about the presence of toxins by using wireless communication and alerts manufacturer, consumer, and distributors [26]. *Salmonella* and *E. coli* can be detected by using nanoporous silicon-based biosensor, while *Bacillus cereus* and *E. coli* can be detected by prototype nanobiosensor. Nanocomponents help detect multiple pathogens faster and more precisely and in modern times have been integrated with ultra-thin polymer substrates for RFID chips that are capable of detecting moisture and temperature of the product along with presence of pathogens [24]. Scientists prepared a biosensor by integrating curcumin (Cur) with a taragum (TG)/polyvinyl alcohol (PVA) film which showed promising results as the color response was visible within one to three minutes in the NH_3 environment. This film was used to evaluate the quality of shrimps and the results were positive [6]. Hence, sensors can be used in the food industry to reduce food wastage and deal with risk of diseases caused by changes in the foods.

CH-based hydrogels are hydrophilic biomaterial with a variety of application prospect in tissue engineering and delivery systems. These can respond to external stimuli and convert environmental signals to a measurable signal output through its swelling behavior. These have great potential in various applications in the intelligent food packaging system due to its favorable antioxidant activity, good biodegradability property, and antibacterial property [31].

Bacteria produce alkaline metabolites like ammonia, biogenic amines, dimethylamine, and trimethylamine during food spoilage that accumulates within the package. Negative changes in food quality correlated with the growth of pathogenic microorganisms show changes in pH. Bacterial nanocellulose (BNC) detects the pH changes due to the presence of a pH sensitive dye. The dye can be either natural like extracts from plants anthocyanins, curcumin, or synthetic ones like bromocresol green/purple, cresol and methyl red, xylenol, chlorophenol, and bromothymol blue. These are typically used as colorimetric indicators that are immobilized on biopolymer matrix. The pH detectors monitor the quality of packaged food and provide qualitative information via visual changes in color. Moradi *et al.* [40] developed an indicator which was based on BNC and black carrot anthocyanins and the label was fabricated by dipping and coating with BNC film into a dye solution. The colorimetric indicator detected pH changes due to growth of microbes responsible for deterioration in fish fillets. As the pH increased from acidic to alkaline due to food spoilage the color changed from red to grey. Researchers like Mohammad Alinejhad *et al.* [41] tested colorimetric indicator as a sensor for monitoring the freshness of packed frozen shrimps. They prepared the label by incorporating natural dye extracts from *Echiumamoenum* flowers into the BNC matrix. The response of the composite to the pH range 2 to 12 demonstrated color variations from violet to yellow. The microbial spoilage of proteinaceous foods lead to high levels of volatile nitrogen-based compounds causing an increase in pH which leads to color change of the dye which is immobilized in the film made of BNC. Another pH-sensitive smart label indicator was created by Dirpan *et al.* [42] by immersion of BNC films into a solution of bromophenol blue. pH changes from dark blue to green was seen in case of immobilized synthetic pigment which was used in packed mangoes as a freshness sensor. BNC biosensors in intelligent food packaging systems find application in food safety. The technology allows the detection of a biological analyte followed by converting biochemical signals into an electrical response via transducers. Ghasemi *et al.* [43] developed nanobiosensors based on conductometric principles that measured the activity of pathogens in food. The biosensor was developed by immersing BNC membrane in a solution of TiO_2–Ag-pyrrole followed by adding oxidizing compound like $FeCl_3$ to initiate the polymerization of polypyrrole. The main function of BNC biosensor was to create OH^- ions due to interactions between BNC/polypyrrole/TiO_2–Ag film surface with the pathogen present in food which was then recorded as a change in the electrical conductivity of the composite [32].

16.3.6.2 Gas Sensor

Gas sensors are able to detect the presence of volatile compounds or gases like oxygen, carbon dioxide, volatile amines, and other gases, which indicate the loss of freshness of the food. These sensors can be used to detect the presence of various gaseous analyte in the package. Different types of gas sensors are carbon dioxide sensors, oxygen sensors, ethanol sensor, water vapor sensor, metal oxide semiconductor field effect transistors, organic conducting polymers, and piezoelectric crystal sensors. Types of wireless sensors include active and passive sensors. The various features of an active wireless sensor include an onboard power supply like a battery, an energy harvester, or a super-capacitor and are capable of transmitting signals upto several hundreds of meters. While passive wireless sensors lack an onboard power supply and rely on their power supply from the electromagnetic field generated by the sensor reader. Read/write or read-only memory is present in both active and passive wireless sensors [4].

Kuswandi *et al.* [12] developed a system of intelligent packaging in which BNC was combined with methyl red and served as a gas detector based on the principles of colorimetry changes which were observed due to volatile amines emission because of food spoilage. Sukhavattanakul and Manuspiya developed a hybrid membrane based on BNC with silver nanoparticles (AgNPs) and alginate-molybdenum trioxide nanoparticles (MoO_3NPs) formed by physical adsorption. The developed film was used to detect hydrogen sulfide (H_2S), which is released in the food due to microbial spoilage or lipid oxidation. The H_2S sensors mechanism is based on the color reactions that occur due to conversion of Ag to Ag_2S to form atomic hydrogen changes in the Mo oxidation state and reduction of MoO_3NPs to a colored sub-oxide by atomic hydrogen [35].

Vilela *et al.* [44] developed humidity sensor composite material control moisture levels in food products during transportation and storage. The material was developed via free radical polymerization of sulfobetaine methacrylate (SBMA) along with the BNC fibrous network in the presence of polyethylene glycol diacrylate which was used as a cross-linking agent. The composites fabricated had protogenic moieties like sulfonic acid groups in their structure which were responsible for current conductivity on being exposed to humidity. Monitoring of changes in ionic conductivity was possible due to this system as a function of liquid content [32].

Optical oxygen sensors follow the principle of absorbance changes caused by direct contact with the analyte. Optochemical sensors sense gas analytes like carbon dioxide, hydrogen sulfide, and volatile amines that help to detect the food quality. The optochemical sensing system includes fluorescence-based system with a pH-sensitive indicator, an absorption-based colorimetric sensing, and energy transfer approach using phase fluorimetric detections. In case of non-vegetarian products like fish, meat, and poultry, pH-sensitive dyes are used in sensors for detecting basic volatile amines. Indicators consisting of methyl red/cellulose membrane or curcumin/bacterial cellulose membrane show color changes due to the presence of volatile amines in spoilt fishes [4]. Lokke *et al.* [47] studied the use of wireless sensor networks to establish the respiration rate of oxygen in fresh florets of broccoli at various temperatures of 5°C, 10°C, and 20°C and also at modified gas compositions of decreased O_2 and increased CO_2 levels. Wireless Sensor Networks offer the potential to monitor the postharvest respiration rate of oxygen of horticultural produces during its shelf life, leading to better quality consumer products. The polyanaline (PANI)–based RFID sensors are formulated with camphor sulfonic acid (CSA) for sensing NH_3 and H_2O vapors with a detection limit of 500 parts per trillion (ppt) for monitoring of fish freshness [19, 20, 47].

16.3.6.3 Chemical Sensor

The chemical sensor is a chemical selective coating capable of detecting the activity, occurrence, concentration, and composition of a particular chemical or volatile substance through surface adsorption. Transducers observe and convert particular chemicals into signals. Depending on the external power requirement for measurement, transducers can be either active or passive type. Carbon nanomaterials like nanoparticles, graphite, grapheme, nanotubes, and nanofibers are applied in chemical sensors due to their exceptional electrical and mechanical properties along with high precise surface area. Nano-based sensors can be used for detection of chemical contaminants, pathogens, food spoilage, product tampering, and traceability. Optical transducers do not require electrical power and can decipher from a distance by use of visible, ultraviolet, or infrared light. Optical transducers made of silicon are composed of optical circuits integrated with silicon semiconductor material [4].

16.3.6.4 Printed Electronics

An emerging technology using electrically functioning inks on flexible substrates having unique properties like being portable, rollable, foldable, bendable, and light weight is known as printed electronics. Printed transducer with a receptor on top is generally present on the flexible printed chemical sensors. Molecular imprinting is a technique in which the analyte molecules are incorporated into a pre-polymeric mixture and allowed to form bonds. After the polymer has formed, the analyte molecules are removed leaving behind a cavity in its shape, thus enabling the identity of the targeted molecule as the shape of the cavity is specific to the molecule of concern [4].

16.3.6.5 Electronic Nose

Systems used to imitate the mammalian olfactory system within an instrument designed to obtain reproducible measurements, allowing detection of aroma mixtures present in the food are known as electronic nose. The nose system consists of a range of either chemical or physical biosensors with partial specificity and statistical methods enabling the cognition of simple or complex flavors. These systems are useful in the quality assessment of fresh yellow fin tuna and vacuum packed beef. The aroma released by fruits and vegetables can be an indicator of the food quality and can be detected by electronic nose. Gomez *et al.* [50] using PEN 2 E nose studied the volatile compounds formed during different stages of tomato and were able to differentiate the stages of ripeness. In a study, the quality of modified atmosphere packaged broiler chicken cuts was evaluated, and the results were compared with the results obtained from microbiological, sensory, and headspace gas composition analysis. The results from the electronic nose were more precise [4]. Salinas *et al.* developed an optoelectronic nose made up of seven sensing materials prepared by the amalgamation of pH indicators and chromogenic reagents along with different chemical recognition properties for monitoring the quality of pork sausages. The results indicated that it was able to monitor the spoilage process of the food [7]. Use of electronic nose ensures an effective monitoring of food quality by a rapid detection device installed on packages [7]. Dua *et al.* [48] developed a flexible, lightweight chemi-resistor made of thin film fabricated using ink jet printing on flexible plastic surfaces to detect chemically aggressive vapors at parts per billion levels at room temperature [30, 48].

16.3.7 Thermochromic Inks

Thermochromic inks are temperature-sensitive and printed on packages to indicate the optimum temperature for product consumption. The inks change color depending on the composition at specific temperatures. Sometimes, for marketing purpose, certain colors are chosen that reveal hidden messages like "drink now" or "too hot" on the product.

This technology first became popular with wine labeling and ever since has been adopted for promotions and special occasions.

The Food Sentinel System™ from SIRA Technologies in Pasadena, California, utilizes thermochromic ink to visually identify products affected by temperature fluctuations by having a colored extra wide bar in the lower bar code which makes it unreadable at checkout.

Thermochromic liquid crystals (TLC) printing inks have specific temperature-dependent optical properties due to which the color changes gradually due to specific structural changes in the microencapsulated active thermochromic material within the activation region only. In the activation region the color changes like a rainbow and is known as the "the color play". The range of color changes for each temperature within the activation region of the used TLC ink is obtained by spectrometric measurements [34].

16.3.8 Microwave Doneness Indicators

Devices that detect and visually indicate the state of doneness of foods heated within a microwave oven is known as Microwave Doneness Indicators (MDIs) (Figure 16.8). One of the criteria for optimum utility of the doneness indicator is that the entire food must be heated uniformly. If the food gets heated non-uniformly, then the regions around the edges form the hottest regions, causing the doneness indicator to trigger before

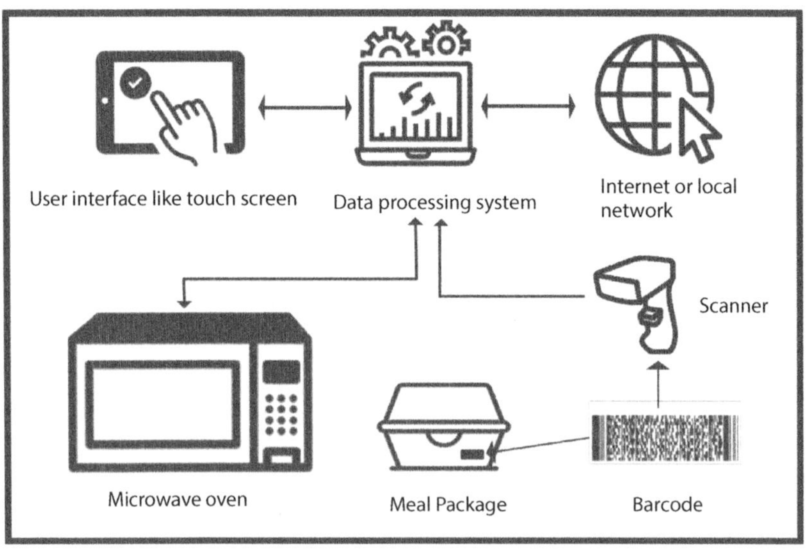

Figure 16.8 Intelligent microwave oven system [2].

the cooler regions have achieved the optimum temperature. The cover of the container generally contains the doneness indicator to demonstrate a visual indication of doneness. As the food heats within the microwave oven, along with the food the headspace is also heated and simultaneously the heat is then transferred to the cover. The indicator system depends on the relationship between the temperature of the food and that of the lid.

Many doneness indicators can provide false indications as these devices maybe heated by the microwaves. A solution to this problem is the shield doneness indicator. Because field distributions in microwave ovens are complicated, the relationship between the field experienced by the food and the field experienced by the sensor on the lid will generally be different and will vary from oven to oven. As a result, indicator systems sensitive to the field would work satisfactorily in some ovens but give false indications in others. Because the temperature of the cover generally reaches its equilibrium value before a target temperature has been achieved throughout the product, the sensor must incorporate a time-dependent mechanism. In essence, the detector is activated at a particular temperature and a progressive change in color occurs over the desired time period. An example of this is the plastic containers of syrup for pancakes labeled with a thermochromic ink to indicate the optimum temperature of the syrup following microwave heating.

If temperature sensor is placed in contact with a metallic surface in a microwave oven, then it experiences greatly reduced field intensity because the electric fields parallel to the metal surface are very small at or near the surface of the metal. As a result, such a sensor responds primarily to temperature as it has little or no opportunity to absorb microwave energy. The major drawback of doneness indicators is the difficulty in observing a color change without opening the microwave oven. Hence, as a solution, an innovative heating sensor that emitted an audible signal was developed in Japan. It consisted of a whistling device on a portion of the cover that generated sound when gas passed quickly through the device and worked best in case of heating high moisture foods. Package with a tuned notch in the cover that whistles once the product is done was developed by a Japanese company. Another example can be use of a microchip that can provide an audible or visual signal that acts as an indicator [11].

16.4 Tools for Protection Against Theft, Counterfeiting, and Tampering

In order to curb the incidences of counterfeiting and theft, special inks and dyes, holograms, electronic tags, and laser labels have been adopted by the food industry but their usage is minimal due to low unit value of packaged food commodities.

Tampering has been a concern for food manufacturers for many years, and there are occasional well-publicized cases where major branded products have been maliciously contaminated with a poisonous substance and the company held to ransom. However, there is no tamper-proof or tamper-evident package that will stop a determined person from contaminating a product. Devices such as the button on metal closures attached to glass containers, which have been retorted and the tamper-evident band located on the skirt of screw closures attached to glass and plastic bottles have been used for many years but do not qualify as intelligent packaging.

Intelligent packaging has a function to protect the product and hence can demonstrate when the package has been tampered with (Figure 16.9). Some examples of these technologies include transparent labels on package on being tampered show permanent color change with words like "stop" or "opened" becomes visible. Another example includes labels that release a dye on being ruptured [10, 19, 20].

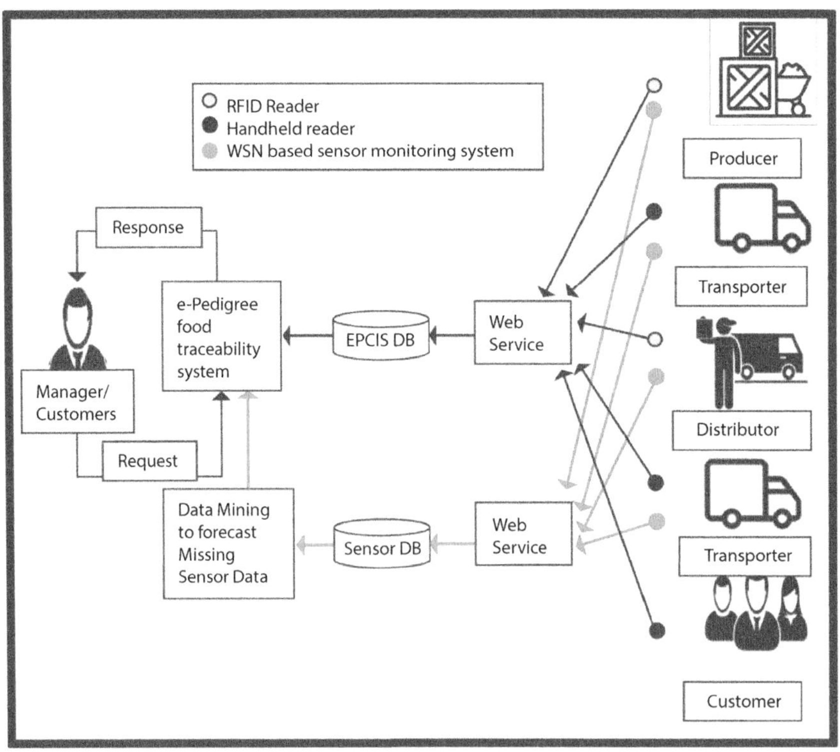

Figure 16.9 Traceability system in food supply chain [49].

16.5 Nanotechnology in Intelligent Packaging

A promising new innovative technology like the nanotechnology in the food packaging industry contribute to food quality and safety by providing mechanical and barrier properties, detecting pathogens. The nanolayer of aluminum that coats the interior of snack food packages is an example. The different techniques by which the nanomaterials are produced are self-assembly, crystallization, solvent extraction/evaporation, layer-by-layer deposition, biomass reactions, and microbial synthesis.

The most common nanomaterials used in food packaging are nanocomposites. These function in food packaging materials by improving the strength, antimicrobial properties, barrier properties, and stability to temperature fluctuations. In the year 1990s, the use of nanocomposites for food packaging began with the use of montmorillonite clay in a wide variety of polymers like nylon, polyvinyl chloride, polyethylene, and starch. The nanocomponents have one dimension less than 1-nm wide, while the quantity of nanoclays varies from 1% to 5% by weight. The nanocomposite high surface area is responsible for contributing unique properties to the packages. Durethan is a transparent nanocomposite coating and the plastic film produced by Bayer and contains clay nanoparticles dispersed within the plastic. Mixtures of silicate nanoparticles and polyamide film have the property to prevent volatiles like carbon dioxide, oxygen, and moisture from contacting fresh foods. These particles prevent the process of diffusion by acting as an impermeable barrier and have other properties like being lightweight, more heat-resistant, strong, reduced transportation costs, and enhancing the shelf life of foods.

Nanotechnology has dealt with the problem of offlavor and oxidation due to packaging of beer in plastic bottles. A subsidiary of Amcol International Corporation, Nanocor, designed nanocomposites in plastic beer bottles that provide shelf life of upto 6 months. Barrier nylons are produced by blending oxygen scavengers and nanocomposites in case of multilayer, co-injection blow-molded polyethylene terephthalate (PET) bottles. Plastic bottles with nanocrystals prevent the loss of carbon dioxide and incorporate oxygen into the bottles and promote the shelf life of beer and soft drinks by eighteen months. Cylindrical carbon nanotubes with antimicrobial properties can improve the mechanical strength of food packaging materials. These carbon nanotubes caused cellular damage of *Escherichia coli* by rupturing their cells and destroying them. The carbon nanotubes act as building blocks in food packaging and are used in the form of nano-wheels. By incorporating self-aggregated inorganic alumina platelets, the barrier and mechanical properties of plastics were improved.

Nanotechnology in food packaging has various applications (Figure 16.10) like detection of allergens, chemicals, pathogens, toxins, and microbes. Nanovesicles in nanotechnology help with detection of microbes like *E. coli* 0157:H7, *Salmonella* spp., and *Listeria monocytogenes.* While Liposome nanovesicles are used for detection of peanut protein allergens. Association of antibodies to *Staphylococcus enterotoxin* Bontopoly (dimethylsiloxane) chips led to formation of biosensors with a detection limit of 0.5 ng/ml. A NanoBioluminescence detection spray with an engineered luminescent protein is capable of binding to the surface of microbes like *Salmonella* and *E. coli* for detection in foods. A nanoporous silicon-based biosensor can detect *Salmonella* and *E. coli*, while other prototype nanobiosensors can detect *Bacillus cereus* and *E. coli* and can also detect multiple pathogens rapidly and precisely.

Nanocomponents along with RFID chips and biosensors are capable of sensing the moisture or temperature of a product along with detecting food-borne pathogens. Deoxyribonucleic acid (DNA) biochips have the

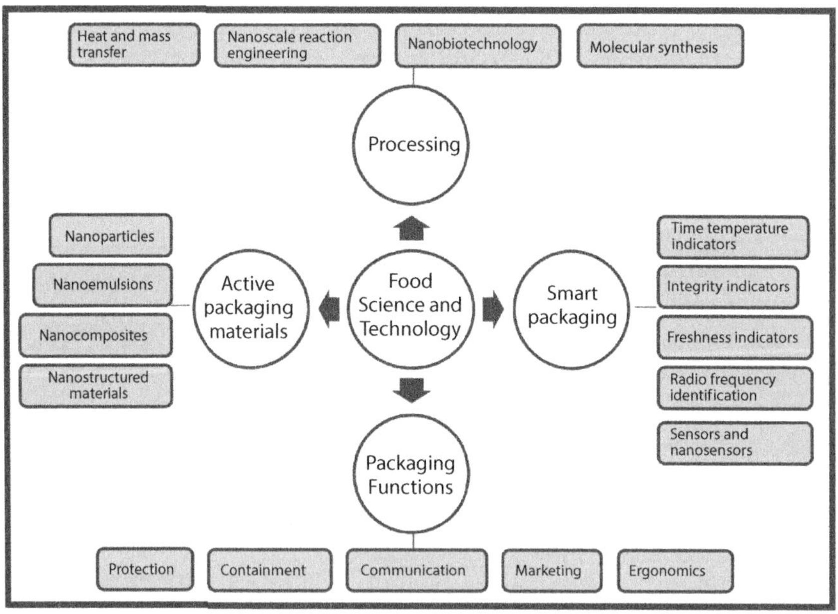

Figure 16.10 Application matrix of nanotechnology in food science and technology [27].

property to repair itself in case of damage and can also detect pathogens. DNA-coated carbon tubes with nano-size diameters are used to provide nanosensors that are capable to detect odor and tastes change in a product. One strand of the DNA acts like the sensor while the carbon nanotube serves as the transmitter in these.

Electronic tongue nanosensors can be used to stimulate color changes in food packages to provide information to the consumers regarding food spoilage and can also detect substances in parts per trillion. Polymeropal films are color-changing films consisting of photonic crystals used to produce color changes in food packaging materials. Synthetic DNA barcodes detect and monitor pathogens in food by producing fluoresce under ultraviolet light [24].

Cellulose nanofibers are a biodegradable and bio-renewable nanomaterial that decreases costs as well as environmental impact. Nanodiamonds are examples of other nanocomposites with antibacterial and anti-inflammatory properties that can be used as both a food additive and biosensor and protect the food from microbial toxins spoilage. Nanodiamond particles have various functions in food packaging like improving durability and flexibility, resisting humidity and temperature changes, and improving anaerobic conditions and antimicrobial properties [33].

16.6 Safety and Regulatory Issues

The main aim of food legislation is maintaining food safety along with the changing industrial and consumer demands. In order to deal with the global concerns of food safety, the packaging industry has adopted various innovative techniques to deal with the regulatory and legal requirements. One of the problems faced during the food supply chain is the distance of the source to the point of consumption which can result in changes in the product quality. As a solution to this problem many logistic companies use strip chart recorder inside certain marked boxes per batch to observe the temperature along with the traditional labels for traceability information. RFID tags with temperature sensors integrated with physical and chemical sensors are used as temperature managed traceability systems.

To prevent misunderstanding and misuse by the consumers proper labeling must be done, for example, to prevent sachets present in the package from being consumed. The migration process of intelligent substances must comply with food legislation and the toxicity of the migrating substances must be considered. Adapting mass transfer modeling tools and migration tests to monitor the phenomena of migration within the package needs careful consideration. Exposure of the consumers to the levels of toxicity of the migrating substances is also of significance.

The European Union (EU) regulations determine the suitability of intelligent packaging systems which come in direct contact with foods. As a result, the legislations which apply to the conventional packaging materials are also applied to intelligent packaging systems under the European Union [24]. The European Food Safety Authority safety assessment focuses on major three risks associated with the dietary exposure of chemical which include migration of active or intelligent substances, migration of their degradation and/or reaction products and their toxicological properties.

European Union Regulation 1935/2004/EC under the European Union legislation allows use of active packaging in Europe with agents that could migrate into foods but must comply with the tests. The regulation defines active materials as "articles and materials that are intended to extend the shelf life or to maintain or improve the condition of packaged food." It sets the framework for the European Food Safety Authority (EFSA) evaluation process and also contains general provisions on the safety of active and intelligent food packaging [23]. A new regulation, the European Union Regulation 450/2009/EC provides details of specific rules for the safe use of active and intelligent materials and articles. Countries like Australia, Japan, and the United States have more flexible regulations that permit technological innovations in the food packaging sector when compared to the European Union which passed demanding packaging directives to be complied with [8].

Environmental policies make the manufacturer accountable for the waste and costs of recovering and recycling, in the process contributing to addition expenses in logistics while ensuring sustainability. Use of returnable transport units comply with waste regulations while providing operational and ecological benefits. Intelligent logistics with automated collaboration and transfer of information help solve problems within the grocery supply chain [24].

Combining several single intelligent packaging devices into a single one ensures maximum benefits from this technology. This guarantees food safety but has limitations with regard to cost and legislations [6].

16.7 Future Trends

Nanotechnology in food packaging is the future to guarantee safe food packages for both the consumers and the environment. To address the challenges of designing packaging materials based on the requirements of the food product and availability of packaging materials, reverse engineering approach is employed. Hence, to solve challenges in food packaging customized stimulated or controlled release of active agents for specific target indicators in the package is done [24].

16.8 The Industrial Internet of Things

A concept aimed at providing a global interconnected network infrastructure for connecting objects between the virtual and physical world with controlling and tracking devices equipped with sensors and indicators is known as the Industrial Internet of Things (IIoT). Packages attached with RFID tags can be tracked along its product cycle as part of this technology. E-commerce delivery services also form an apt example. Not only electronics but equipments, household items, manufacturing machines, as well as food packages to even an entire manufacturing unit can be a part of the IIoT [29].

16.9 Real-Time Capabilities

Real-time monitoring and controlling the conditions of the product affect the food safety and consumer health and promote waste reduction. A reliable IIoT infrastructure is associated with Information and Communication Technologies embedded in both product package and logistics and refers to the virtual-physical production along with delivery networks at both horizontal and vertical levels. The horizontal levels are found within a company while the vertical levels form across multiple companies. Digitization of the product lifecycle and logistics can be achieved with use of artificial intelligence, which plays a significant role in improving machine learning, data mining, and in the decision support systems, and are able to adjust processes in real time on the basis of the available data.

16.10 Cybersecurity

Cybersecurity and data privacy issues is a major concern with regard to the emerging technologies and needs to be immediately addressed to avail the optimum benefits from intelligent technology. The current internet security is characterized by attacks that are constantly evolving, voluminous, persistent, extremely fast, and highly complex and impose significant challenges on preventive security services. Hence, methodologies that enable automatic detection and response to cyber-attacks must be employed synergistically with prevention to achieve effective in-depth defense strategies and robust cybersecurity systems [1].

16.11 Conclusion

Answering consumer demands and the need for food safety led to innovative formulation of smart, active, bioactive, and intelligent food packaging techniques. The developed innovations led to improvement of food safety and quality, feasibility, and bioactivity of the functional components. Novel and innovative packaging techniques applications help reduce consumer complaints and have a positive impact on health. Conventional packaging has been globally replaced with innovative food packaging techniques and is here to stay [24].

References

1. Schaefer, D. and Cheung, M.W., Smart packaging: Opportunities and challenges. *Proc. CIRP*, 72, 1022–1027, 2018.
2. Yam, K.L., Takhistov, P.T., Miltz, J., Intelligent packaging: Concepts and applications. *J. Food Sci.*, 70, 1, 4–7, 2005.
3. Kerry, J.P., O'Grady, M.N., Hogan, S.A., Past, current and potential utilisation of active and intelligent packaging systems for meat and muscle-based products: A review. *Meat Sci.*, 74, 1, 113–130, 2006.
4. Biji, K.B., Ravishankar, C.N., Mohan, C.O., Gopal, T.K.S., Smart packaging systems for food applications: A review. *J. Food Sci. Technol.*, 52, 6130–6132, 2015.
5. Pereira de Abreu, D.A., Cruz, J.M., Losada, P.P., Active and intelligent packaging for the food industry. *Food Rev. Int.*, 28, 2, 174–178, 2012.
6. Sohail, M., Sun, D.-W., Zhu, Z., Recent developments in intelligent packaging for enhancing food quality and safety. *Crit. Rev. Food Sci. Nutr.*, 58, 15, 2–10, 2018.

7. Realini, C.E. and Marcos, B., Active and intelligent packaging systems for a modern society. *Meat Sci.*, 98, 3, 14–19, 2014.
8. Restuccia, D., Spizzirri, U.G., Parisi, O.I., Cirillo, G., Curcio, M., Lemma, F., Puoci, F., Vinci, G., Picci, N., New EU regulation aspects and global market of active and intelligent packaging for food industry applications. *Food Control*, 21, 1425–1435, 2010.
9. Ghaani, M., Cozzolino, C.A., Castelli, G., Farris, S., An overview of the intelligent packaging technologies in the food sector. *Trends Food Sci. Technol.*, 51, 1–11, 2016.
10. Heising, J.K., Dekker, M., Bartels, P., Van Boekel, M., Monitoring the quality of perishable foods: Opportunities for intelligent packaging. *Crit. Rev. Food Sci. Nutr.*, 54, 5, 645–649, 2014.
11. Han, J.H., Ho, C.H.L., Rodrigues, E.T., *Intelligent packaging, Innovations in Food Packaging*, Elsevier Academic Press, Oxford, 2005.
12. Kuswandi, B., Jayus, J., Abdullah, A., Heng, L.Y., Ahmad, M., A novel colorimetric food package label for fish spoilage based onpolyaniline film. *Food Control*, 25, 1, 184–189, 2012.
13. Hogan, S.A. and Kerry, J.P., Smart packaging of meat and poultry products, in: *Smart packaging technologies for fast moving consumer goods*, J. Kerry, and P. Butler, (Eds.), pp. 33–59, JohnWiley& Sons Ltd, WestSussex, England, 2008.
14. Jung, J., Puligundla, P., Ko, S., Proof-of-concept study of chitosan based carbon dioxide indicator for food packaging applications. *Food Chem.*, 135, 4, 2170–2174, 2012.
15. Müller, P. and Schmid, M., Intelligent packaging in the food sector: A brief overview. *Foods*, 8, 1, 16, 2019.
16. Fang, Z., Zhao, Y., Warner, R.D., Johnson, S.K., Active and intelligent packaging in meat industry. *Trends Food Sci. Technol.*, 61, 60–71, 2017.
17. Dobrucka, R. and Cierpiszewski, R., Active and intelligent packaging food research and development- A review. *Polish J. Food Nutr. Sci.*, 64, 1, 7–15, 2014.
18. Maschietti, M., Time temperature indicators for perishable products. *Recent Pat. Eng.*, 4, 2, 129, 2010.
19. Ahvenainen, R., *Novel food packaging techniques*, Woodhead Publishing Limited, Cambridge, 2003.
20. Robertson, G.L., *Food Packaging Principle and Practise*, CRC Press Taylor & Francis Group, Boca Raton, 2013.
21. Kumar, B., Anumba., C.J., Domdouzis, K., Radio-Frequency Identification (RFID) applications: A brief introduction. *Adv. Eng. Inform.*, 21, 4, 350–351, 2007.
22. Vanderroost, M., Ragaert, P., Devlieghere, F., De Meulenaer, B., Intelligent food packaging: The next generation. *Trends Food Sci. Technol.*, 39, 1, 47–62, 2014.

23. Dainelli, D., Gontard, N., Spyropoulos, D., Zondervan-van den Beuken, E., Tobback, P., Active and intelligent food packaging: Legal aspects and safety concerns. *Trends. Food. Sci.Technol.*, 19, 103–112, 2008.
24. Majid, I., Nayik, G.A., Dar, M.S., Nanda, V., Novel food packaging technologies: Innovations and future prospective. *J. Saudi Soc Agric. Sci.*, 17, 454–462, 2018.
25. Suppakul, P., Intelligent Packaging, in: *Handbook of Frozen Food Processing and Packaging*, pp. 835–853, 2014.
26. Nemes, S.A., Szabo, K., Vodnar, D.C., Applicability of agro-industrial by-products in intelligent food packaging. *Coatings*, 10, 550, 1–21, 2020.
27. Fuertes, G., Soto, I., Carrasco, R., Vargas, M., Sabattin, J., Lagos, C., Intelligent packaging systems: Sensors and nanosensors to monitor food quality and safety. *J. Sens.*, 2016, 1–8, 2016.
28. Hall, L.E., Intelligent Packaging and material handling. *Intelligent Robots and Computer Vision IX: Algorithms and Techniques, 1381*, 1990.
29. Bouzembrak, Y., Kluche, M., Gava, A., Marvin, H.J.P., Internet of Things in food safety: Literature review and a bibliometric analysis. *Trends Food Sci. Technol.*, 94, 54–64, 2019.
30. Wang, L., Wu, Z., Cao, C., Technologies and fabrication of intelligent packaging for perishable products. *Appl. Sci.*, 9, 4858, 1–17, 2019.
31. Yang, J., Shen, M., Luo, Y., Wu, T., Chen, X., Wang, Y., Xie, J., Advanced applications of chitosan-based hydrogels: From biosensors to intelligent food packaging system. *Trends Food Sci. Technol.*, 110, 822–832, 2021.
32. Ludwicka, K., Kaczmarek, M., Białkowska, A., Bacterial nanocellulose—A biobased polymer for active and intelligent food packaging applications: Recent advances and developments. *Polymers, 12*, 10, 2209, 2020.
33. Neethirajan, S., Ragavan, V., Weng, X., Chand, R., Biosensors for sustainable food engineering: Challenges and perspectives. *Biosensors, 8*, 1, 23, 2018.
34. Jakovljević, S.M., Lozo, B., Gunde, M.K., Packaging added value solutions by thermochromic liquid crystal-based printed labels. *Conference: 32nd International Conference on Digital Printing Technologies (NIP)*, 2016.
35. Ghaani, M., Cozzolino, C.A., Castelli, G., Farris, S., An overview of the intelligent packaging technologies in the food sector. *Trends Food Sci. Technol.*, 51, 1–11, 2016.
36. Brody, A.L., Strupinsky, E.R., Kline, L.R., Active packaging for food applications, Lancaster: Technomic Publishing, 236, 2001.
37. Clarke, R., Radio frequency identification: smart or intelligent packaging?, *J. Packaging Sci. Technol. Jpn.*, 10, 5, 233-247,2001.
38. Commission Regulation (EC) No 450/2009, on active and intelligent materials and articles intended to come into contact with food, *Official Journal of the European Union*, 2009.
39. ASTM F1416-96 Standard Guide for Selection of Time-Temperature Indicators, STANDARD by ASTM International, 2019.

40. Moradi, M., Tajik, H., Almasi, H., Forough, M., Ezati, P., A novel pH-sensing indicator based on bacterial cellulose nanofibers and black carrot anthocyanins for monitoring fish freshness. *Carbohydr. Polym.*, 222, 2019.
41. Mohammad Alinejhad, S., Almasi, H., Moradi, M., Immobilization of Echium amoenum anthocyanins into bacterial cellulose film: A novel colorimetric pH indicator for freshness/spoilage monitoring of shrimp. *Food Control*, 113, 2020.
42. Dirpan, A., Latief, R., Syarifuddin, A., Rahman, A.N.F., Putra, R.P., Hidayat, S.H., The use of colour indicator as a smart packaging system for evaluating mangoes Arummanis (Mangifera indica L. var. Arummanisa) freshness. *IOP Conf. Ser. Earth Env. Sci.*, 157, 2018.
43. Ghasemi, S., Bari, M.R., Pirsa, S., Amiri, S., Use of bacterial cellulose film modified by polypyrrole/TiO2-Ag nanocomposite for detecting and measuring the growth of pathogenic bacteria, *Carbohydr. Polym.*, 232, 2020.
44. Vilela, C., Moreirinha, C., Domingues, E.M., Figueiredo, F.M.L., Almeida, A., Freire, C.S.R., Antimicrobial and conductive nanocellulose-based films for active and intelligent food packaging. *Nanomaterials*, 9, 2019.
45. Rijk R,. Proceedings of a Conference on active and intelligent packaging development, 19th International Conference: Plastics & Polymers in Contact with Foodstuffs, *Pira Intl.*, 2002.
46. Wong C.Y., McFarlane D., Zaharudin Ahmad A., Agarwal V., The Intelligent Product Driven Supply Chain, *IEEE International Conference on Systems, Man and Cybernetics*, 2003.
47. Løkke M.M., Seefeldt H.F., Edwards G., Green O. 2011. Novel wireless sensor system for monitoring oxygen, temperature and respiration rate of horticultural crops post harvest. *Sensors,* 11, 8456–8468.
48. Dua, V., Surwade, S.P., Ammu, S., Agnihotra, S.R., Jain, S., Roberts, K.E., Park, S., Ruo, R.S., Manohar, S.K, All-Organic Vapor Sensor Using Inkjet-Printed Reduced Graphene Oxide., *Angew. Chem. Int.*, 49, 2154–2157, 2010.
49. Ganjar Alfian, Jongtae Rhee, Hyejung Ahn, Jaeho Lee, Umar Farooq, Muhammad Fazal Ijaz, M. Alex Syaekhoni, Integration of RFID, wireless sensor networks, and data mining in an e-pedigree food traceability system, *J. Food Eng.,* 212, 65–75, 2017.
50. Gòmez AH, Hu G, Wang J, Pereira AG, Evaluation of tomato maturity by electronic nose, *Computers Electr. Agric.*, 54, 44–52, 2006
51. Yoshida M.P. Cristiana, Maciel Borges V. Vinicius, Mendonça Eleonora D Mariana, Franco Telma Teixeira, Chitosan biobased and intelligent films: Monitoring pH variations, *LWT-Food Sci. Technol.*, 55, 83–89, 2014.

Index

Printed and bound by CPI Group (UK) Ltd, Croydon, CR0 4YY
31/07/2022
03138991-0005